Achieving sustainable production of pig meat

Volume 3: Animal health and welfare

It is widely recognised that agriculture is a significant contributor to global warming and climate change. Agriculture needs to reduce its environmental impact and adapt to current climate change whilst still feeding a growing population, i.e. become more 'climate-smart'. Burleigh Dodds Science Publishing is playing its part in achieving this by bringing together key research on making the production of the world's most important crops and livestock products more sustainable.

Based on extensive research, our publications specifically target the challenge of climate-smart agriculture. In this way we are using 'smart publishing' to help achieve climate-smart agriculture.

Burleigh Dodds Science Publishing is an independent and innovative publisher delivering high quality customer-focused agricultural science content in both print and online formats for the academic and research communities. Our aim is to build a foundation of knowledge on which researchers can build to meet the challenge of climate-smart agriculture.

For more information about Burleigh Dodds Science Publishing simply call us on +44 (0) 1223 839365, email info@bdspublishing.com or alternatively please visit our website at www.bdspublishing.com.

Related titles:

Achieving sustainable production of pig meat Volume 1: Safety, quality and sustainability
Print (ISBN 978-1-78676-088-3); Online (ISBN 978-1-78676-091-3, 978-1-78676-090-6)

Achieving sustainable production of pig meat Volume 2: Animal breeding and nutrition
Print (ISBN 978-1-78676-092-0); Online (ISBN 978-1-78676-094-4, 978-1-78676-095-1)

Improving organic animal farming
Print (ISBN 978-1-78676-180-4); Online (ISBN 978-1-78676-182-8, 978-1-78676-183-5)

Chapters are available individually from our online bookshop: https://shop.bdspublishing.com

climate•SMART•publishing
IN AGRICULTURAL SCIENCE

BURLEIGH DODDS SERIES IN AGRICULTURAL SCIENCE

NUMBER 25

Achieving sustainable production of pig meat

Volume 3: Animal health and welfare

Edited by Professor Julian Wiseman
University of Nottingham, UK

burleigh dodds
SCIENCE PUBLISHING

Published by Burleigh Dodds Science Publishing Limited
82 High Street, Sawston, Cambridge CB22 3HJ, UK
www.bdspublishing.com

Burleigh Dodds Science Publishing, 1518 Walnut Street, Suite 900, Philadelphia, PA 19102-3406, USA

First published 2018 by Burleigh Dodds Science Publishing Limited

Library of Congress Control Number: 2017938496

British Library Cataloguing in Publication Data
A catalogue record for this book is available from the British Library

ISBN 978-1-78676-096-8 (print)
ISBN 978-1-78676-099-9 (online)
ISBN 978-1-78676-098-2 (online)
ISSN 2059-6936 (print)
ISSN 2059-6944 (online)

Typeset by Deanta Global Publishing Services, Chennai, India
Printed by Lightning Source

Contents

Series list

Title	Series number
Achieving sustainable cultivation of maize - Vol 1 From improved varieties to local applications Edited by: Dr Dave Watson, CGIAR Maize Research Program Manager, CIMMYT, Mexico	001
Achieving sustainable cultivation of maize - Vol 2 Cultivation techniques, pest and disease control Edited by: Dr Dave Watson, CGIAR Maize Research Program Manager, CIMMYT, Mexico	002
Achieving sustainable cultivation of rice - Vol 1 Breeding for higher yield and quality Edited by: Prof. Takuji Sasaki, Tokyo University of Agriculture, Japan	003
Achieving sustainable cultivation of rice - Vol 2 Cultivation, pest and disease management Edited by: Prof. Takuji Sasaki, Tokyo University of Agriculture, Japan	004
Achieving sustainable cultivation of wheat - Vol 1 Breeding, quality traits, pests and diseases Edited by: Prof. Peter Langridge, The University of Adelaide, Australia	005
Achieving sustainable cultivation of wheat - Vol 2 Cultivation techniques Edited by: Prof. Peter Langridge, The University of Adelaide, Australia	006
Achieving sustainable cultivation of tomatoes Edited by: Dr Autar Mattoo, USDA-ARS, USA & Prof. Avtar Handa, Purdue University, USA	007
Achieving sustainable production of milk - Vol 1 Milk composition, genetics and breeding Edited by: Dr Nico van Belzen, International Dairy Federation (IDF), Belgium	008
Achieving sustainable production of milk - Vol 2 Safety, quality and sustainability Edited by: Dr Nico van Belzen, International Dairy Federation (IDF), Belgium	009
Achieving sustainable production of milk - Vol 3 Dairy herd management and welfare Edited by: Prof. John Webster, University of Bristol, UK	010
Ensuring safety and quality in the production of beef - Vol 1 Safety Edited by: Prof. Gary Acuff, Texas A&M University, USA & Prof.James Dickson, Iowa State University, USA	011
Ensuring safety and quality in the production of beef - Vol 2 Quality Edited by: Prof. Michael Dikeman, Kansas State University, USA	012
Achieving sustainable production of poultry meat - Vol 1 Safety, quality and sustainability Edited by: Prof. Steven C. Ricke, University of Arkansas, USA	013
Achieving sustainable production of poultry meat - Vol 2 Breeding and nutrition Edited by: Prof. Todd Applegate, University of Georgia, USA	014
Achieving sustainable production of poultry meat - Vol 3 Health and welfare Edited by: Prof. Todd Applegate, University of Georgia, USA	015

Acknowledgements

We wish to acknowledge the following for their help in reviewing particular chapters:

- Chapter 2: Professor Finn Skou Pedersen, Aarhus University, Denmark; Professor Montserrat Torremorell, University of Minnesota, USA

Introduction

Pig meat is the most widely-consumed meat in the world, accounting for 40% of the world's overall meat consumption. The leading producers are China, the EU, USA, Brazil, Russia, Vietnam, Canada, Japan, the Philippines and Mexico. Consumption is growing, both in developed markets such as the United States and particularly in developing countries in Asia. Previous growth in pig production has been driven primarily by developments in breeding and the shift to larger, more intensive systems. These systems face a range of challenges in increasing production sustainably to meet rising demand. Pig production remains vulnerable to zoonotic and other diseases affecting pigs as well as the problem of antibiotic resistance. There is growing pressure to improve feed efficiency in the face of competition for raw materials and rising feed costs. At the same time, there is an increasing focus on reducing the environmental impact of animal production. Finally, consumers are increasingly concerned about animal welfare in intensive systems.

These challenges are addressed by *Achieving sustainable production of pig meat*. The three volumes are:

- Volume 1 Safety, quality and sustainability
- Volume 2 Animal breeding and nutrition
- Volume 3 Animal health and welfare

This volume, Volume 3, reviews the latest research on diseases affecting pigs and their management, as well as providing a comprehensive review of pig welfare across the life-cycle, from gilts and sows to weaned piglets and finishing pigs. The volume also includes generic welfare issues such as the role of pasture-based systems, humane transport, lairage and slaughter techniques.

Part 1 Animal health

The first part of the volume reviews the main diseases affecting pigs as well as ways of managing diseases and boosting pig immune function. Chapter 1 provides an overview of common bacterial, viral, and parasitic pathogens of pigs. Recent events such as the 2009 pandemic influenza outbreak, the continuous spread of African swine fever virus in Eastern Europe, and the recent introduction of several new pathogens into the United States and their spread to Canada, Mexico, Central, and South America have shown the ability of pig diseases to move quickly across borders, and the importance of global cooperation in improve the health and welfare of pigs. The chapter summarizes some of the most common bacteria (both gram-negative and gram-positive), viruses, and parasites found in pig production, and discusses which are particularly important due to their significant effect on production, importance from an international trade perspective, or zoonotic concern.

Continuing the theme of diseases in pigs, Chapter 2 addresses changing patterns of disease affecting pigs, with a particular focus on Porcine Reproductive and Respiratory Syndrome (PRRS) and Porcine Epidemic Diarrhoea (PED). PRRS virus and PED virus are two of the major viruses that affect pigs worldwide. The chapter examines the transmission, clinical presentation, evolution and spread of these two agents (PRRSv and PEDv), given their importance for pig production worldwide. The chapter discusses in detail the causes, transmission and spread of these two viruses in the pig population both worldwide, and more specifically, the United Kingdom (UK). The chapter provides readers with an overview of the complexity of these two agents and how this influences their clinical presentation and evolution over time and space.

Moving from respiratory systems to the gut, Chapter 3 examines the influence of the gut microbiome on developing immune and metabolic systems in the young pig. The immune and metabolic systems of young piglets develop after birth, and the rate and type of development is strongly associated with the rate and type of colonisation of the intestine with bacteria. The chapter describes the enteric and mucosal immune systems of pigs and presents evidence from experimental studies of the links between the immune system and microbiota. The chapter describes the composition of the microbiome in pigs and highlights its importance in the development of pig immune systems.

The focus of the final chapter in Part 1, Chapter 4, is disease identification and management on the pig farm. After examining the process of disease identification, the chapter provides an overview of disease management and control in pigs. The chapter evaluates both external and internal biosecurity before focusing on the role of vaccination and antimicrobial medication.

Part 2 Welfare issues

Part 2 reviews what we know about pig behaviour and appropriate welfare standards. It also assesses the welfare of different groups of pigs, from gilts and sows to weaned piglets and finishing pigs, as well as transport, lairage and slaughter.

Chapter 5 sets the scene by reviewing current understanding of pig behaviour. The behavioural patterns of domesticated pigs are well conserved from their ancestors. This suggests that the underlying motivational systems are similar to those of wild boar and feral domestic populations. The chapter describes how commercial conditions, whilst providing some welfare benefits, can constrain behavioural expression. The chapter reviews the behavioural ecology of pigs, introduces the concept of behavioural needs and considers the developmental and additive genetic basis behind individual differences in behaviour.

Complementing the previous chapter's focus on pig behaviour, Chapter 6 provides an overview of the challenges of defining and ensuring animal welfare in pig production. Raising pigs for consumption involves exercising control over the quality and duration of their lives, and it is widely accepted that there is a duty to exercise this control humanely. The chapter begins by reviewing conceptual frameworks which can be used to consider and assess animal welfare, and discusses the development of public opinion on the subject. The chapter then discusses common welfare concerns such as confinement, floor space, group size, injuries and surgical husbandry procedures.

Moving on from welfare challenges in pig production Chapter 7 considers pasture systems for pigs. The inclusion of pastures in outdoor pig systems contributes to improving welfare and the sustainability of these farming systems, since it reduces the environmental impact of pig production, as well as reducing nutrient load and soil compaction and thereby improving nutrient distribution. The chapter examines the characteristics and consequences of pasture pig systems (including forage), and provides a detailed case study of the pasture pig system developed at the Center for Environmental Farming Systems (CEFS), in Goldsboro, North Carolina in the United States.

Chapter 8 considers the welfare of a specific group of pigs, namely gilts/pregnant sows. Mismatches between the evolutionary biology of the sow and current commercial production systems give rise to welfare challenges of stereotyped behaviour development in confined systems and aggression in group housing systems. The chapter describes the welfare issues associated with individual confinement systems and examines the nature and significance of stereotyped behaviour in gestating sows. The chapter addresses the issue of hunger in the pregnant sow, and pressure to adopt group housing systems for pregnant sows. The chapter has a particular focus on the importance of social organisation in sows and looks in detail at the relationship between aggression in stable groups and the method of feed provision available. Finally, the chapter examines sows in extensive systems.

Complementing the previous chapter's focus on pregnant sows, Chapter 9 goes on to examine the welfare of another group of pigs: weaned piglets. Concerns about and demands for improved animal welfare and animal handling systems from authorities, non-government organizations (NGOs), markets, and the public in general are increasing. Among the phases of pork production with opportunities to improve welfare, the weanling period has the greatest opportunity to positively impact most animals. The chapter focuses on current practices that can be detrimental to piglet well-being, alternatives and/or improvements to these, and advances in technology that could improve animal well-being as well as profitability and sustainability.

Continuing the theme of pig welfare at specific life stages of the pig, Chapter 10 examines the welfare of pigs during finishing. Finisher pigs, those of a post-weaning age kept for slaughter, make up the majority of the world pig population of approximately 1 billion animals. The intensive production system for post-weaned pigs is characterised by higher animal density, larger farms, use of concentrated foods and control of the production environment, particularly temperature, humidity and lighting. The chapter explores the relationship between nutrition management and the welfare of finishing pigs and the links between physical and social environment and finisher welfare. The chapter addresses the importance of environmental enrichment and ways to avoid tail biting behaviour. Finally, the chapter describes methods of practical welfare assessment of finisher pigs.

Chapter 11 looks at issues surrounding the transport and lairage of pigs. Defects in meat quality cause huge economic losses for the swine industry each year. Reduction in stress prior to slaughter can help alleviate these costs. The chapter describes the main events that may contribute to pre-slaughter stress and their impacts on animal welfare and pork quality, from on-farm loading facilities and handling to transportation, lairage and handling at the slaughter facility. The chapter evaluates the effects of different pre-slaughter events on animal welfare and pork quality, and describes what producers, truck drivers, and slaughter facilities need to do in order to mitigate these factors.

Concluding the volume, and examining the final stage of pig welfare, Chapter 12 focuses on humane slaughter techniques for pigs. The chapter addresses three stages of the slaughter process, namely movement of pigs, stunning, shackling and sticking. The chapter considers the challenges associated with group-based handling of pigs on the day of slaughter and examines the importance of best practice in surveillance and documentation of animal welfare levels on the day of slaughter. Finally, the chapter shows how improved value of meat products can be achieved by the adoption of the best slaughtering practices.

Part 1

Animal health

Chapter 1

Diseases affecting pigs: an overview of common bacterial, viral and parasitic pathogens of pigs

Alejandro Ramirez, Iowa State University, USA

1 Introduction

Diseases affecting pigs can be quite complex. It is well recognized that often these conditions are multifactorial, especially as in the case of respiratory diseases, hence the term 'porcine respiratory disease complex' (PRDC) has been accepted (Brockeier, 2002). The concept of the disease triad (Fig. 1) emphasizes this complexity and the interaction between not only different pathogens, but the host, pathogen and the environment.

To maintain a sustainable pork production system, we must move away from the idea of one agent–one disease and look at the whole picture from a holistic point of view. Recent world events concerning pigs such as the 2009 pandemic influenza outbreak, the continuous spread of African swine fever virus (ASFv) in Eastern Europe and the introduction of several new pathogens into the United States (porcine epidemic diarrhoea virus and porcine deltacoronavirus) and their spread to Canada (limited), Mexico, Central and South America have emphasized that we live in one world with minimal borders. Diseases of pigs spread rapidly across various countries.

http://dx.doi.org/10.19103/AS.2017.0013.14

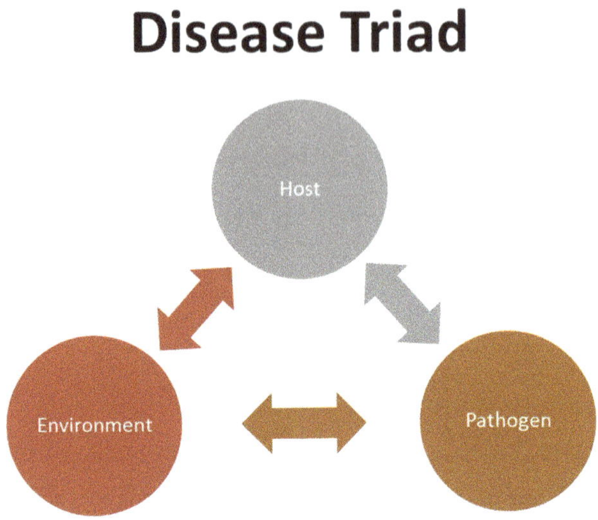

Figure 1 Depiction of the disease triad demonstrating the interactions between host, pathogen and the environment.

The concept of one world and one health requires all of us to work together to improve the health and welfare of pigs to achieve sustainable pork production.

The following section summarizes some of the most common bacteria, viruses and parasites found in pig production including those of greatest importance due to their effect on production, importance from an international trade perspective, as well as their zoonotic concern. It is not to be considered a comprehensive review by any means.

2 The most common bacterial pathogens in pig production: gram-negative bacteria

The advent of new diagnostic technologies such as polymerase chain reaction (PCR) testing have allowed pig veterinarians to diagnose a wide range of viral diseases, making bacterial infection appear to be 'old' pathogens. However, bacterial pathogens continue to significantly affect the health and well-being of pigs, and pig bacterial pathogens are of great importance today because of human concern regarding antimicrobial resistance. Worldwide, pig farmers and veterinarians are being pressured to use less antimicrobials stressing the importance of their responsible and judicious use and disease prevention instead.

2.1 *Actinobacillus pleuropneumoniae*

Actinobacillus pleuropneumoniae (APP) is a gram-negative coccobacillus and has global importance as the causative agent of pleuropneumonia in pig. Pleuropneumonia is characterized as a highly contagious disease with sudden onset, high mobility and mortality, only affecting pigs. Many different serotypes are recognized based on the RTX exotoxins secreted by the organism. Currently, four of these exotoxins are recognized (ApxI, ApxII, ApxIII, ApxIV) and their presence varies among serotypes (Chiers et al., 2010). These toxins cause severe internal pulmonary haemorrhage and cytotoxicity. Apart from these toxins, many other virulence factors have also been identified. The virulence of serotypes, as well (asymptomatic to high mortality) as their prevalence in different geographical regions, is highly variable (Clota et al., 1996; Mittal et al., 1996; Kucerova et al., 2005; Gottschalk, 2003). It is well recognized that isolates from most herds have more than one low-virulent serotype (Gottschalk, 2003).

One of the major challenges with APP is that pigs continue to carry the organism in their lungs and tonsils for several months (Desrosiers, 2004), thus creating an opportunity for repeated outbreaks of this clinical disease, as well as creating a challenging environment at all stages of pig production, from growing pigs until slaughter. Antimicrobials are used to treat the disease with several countries, especially the United States, demonstrating a pattern of resistance to beta-lactams (penicillin, amoxicillin). Antimicrobial treatment helps minimize mortalities during the early stages of an outbreak, but they do not eliminate carrier pigs and Sjölund et al. (2009) have suggested that the use of highly effective antimicrobials prevents good antibody response to the infection, thus leaving pigs susceptible to future re-infections.

Vaccination against APP is challenging due to the various different serotypes. They involve stimulation by several different Apx exotoxins as well as an outer membrane protein (Gottschalk, 2012).

The severity of APP infections is a significant burden on the sustainability and feasibility of an infected herd. This is especially true today with the current emphasis on antimicrobial stewardship, thus infected herds must undergo a de-population and re-population many times with APP-free pigs.

There are no zoonotic concerns regarding APP other than it can lead to extensive antimicrobial use in infected herds.

2.2 *Bordetella bronchiseptica*

Bordetella bronchiseptica is a gram-negative rod that is found throughout the world and infects many different mammalian species (Brockmeier et al., 2012). In pigs *Bordetella bronchiseptica* primarily causes pneumonia and atrophic rhinitis.

Several virulence genes which appear to require co-expression of the BvgAS genes are identified (Beier and Gross, 2008) and are also subject to phase variation. Early BvgAS genes are involved in bacterial attachment followed later (once a large number of bacteria have colonized the area) with gene expression for toxin production (Brockmeier et al., 2012). It is this toxin production (especially dermonecrotic toxin) that contributes to the progression of the disease, especially atrophic rhinitis (nasal turbinate and septal damage).

Vaccination of sows before farrowing can be used in conjunction with antimicrobials to prevent atrophic rhinitis in pigs. The aim of these interventions is to prevent or minimize early colonization by *Bordetella bronchiseptica*, allowing the pig's immune system to protect it as it matures as well as preventing the late stages of BvgAS gene expression, which produces the dermonecrotic toxin. Vaccination helps minimize disease but does not prevent infection. Antibiotics can help minimize disease transmission between pigs.

Bordetella bronchiseptica often works in conjunction with toxigenic *Pasteurella multocida* to cause the more severe disease known as progressive atrophic rhinitis (de Jong and Nielsen, 1990).

Although human infections can occur with *Bordetella bronchiseptica* they are rare, and pigs do not appear to be a concern for zoonosis.

2.3 *Brachyspira* spp.

The *Brachyspira hyodysenteriae* is a spirochetal bacterium that causes swine dysentery (SD). Herds infected with SD incur serious financial losses due to mucoid and bloody diarrhoea resulting in a significant number of deaths as well as poor growth performance.

SD affects the large intestine of grower and finisher pigs but rarely affects weaners (Hampson, 2012). With the advent of PCR technology and genetic sequencing, several new species of *Brachyspira* have been identified and shown to cause SD lesions in growing pigs (Burrough et al., 2012; Chander et al., 2012). So although technically SD is only associated with *Brachyspira hyodysenteriae*, today the phenotypic culture characteristics (especially hemolysis) of *Brachyspira* spp. appear to be a more sensitive indicator of potential to induce dysentery-like disease in pigs (Table 1) than molecular identification alone (Burrough et al., 2012).

There is a limited arsenal of antimicrobials which can be used to treat SD. Mice and rats can serve as an important reservoir for *Brachyspira hyodysenteriae* (and maybe other *Brachyspira* spp.), which make it difficult to completely eliminate an infection from a herd. The limited weapons available are expensive antimicrobials that can add a significant cost to SD control. Therefore, de-population and re-population with SD-free pigs may be necessary in conjunction with aggressive cleaning, disinfection of premises and extensive rodent control programmes. Currently there are no effective vaccines against *Brachyspira* spp.

There are no zoonotic concerns regarding *Brachyspira hyodysenteriae* other than there may need to be extensive antimicrobial use in infected herds.

Table 1 Clinical significance of some *Brachyspira* spp. in pigs (Burrough 2015, pers. comm.)

	Hemolysis	Clinical disease in pigs
Brachyspira hyodysenteriae	Strong beta-hemolysis	Swine dysentery
Brachyspira hampsonii	Strong beta-hemolysis	Dysentery-like disease
Brachyspira suanatina	Strong beta-hemolysis	Dysentery-like disease
Brachyspira pilosicoli	Weak beta-hemolysis	Spirochetal colitis (mild disease)
Brachyspira murdochii	Weak beta-hemolysis	Mild to non-pathogenic
Brachyspira intermedia	Weak beta-hemolysis	Mild to non-pathogenic
Brachyspira innocens	Weak beta-hemolysis	Non-pathogenic

2.4 *Brucella suis*

Brucella suis is a gram-negative coccobacillus of zoonotic importance in pigs, usually causing reproductive failure including long-term, non-fatal, granulomatous inflammation in several organs (including joints and reproductive organs such as testes). Its impact on production is mostly noted in acute outbreaks as frequently endemic brucellosis is mild enough that it can go undetected. Feral pigs are often found to be infected with *Brucella suis* throughout the United States (Leuenberger et al., 2007; Leiser et al., 2013).

The long duration of bacteraemia (5 weeks to 34 months) reported by Deyoe (1967 and 1972) suggests that immune response to infection is insufficient to eliminate bacteria from blood and intracellular niche, especially in macrophages (Olsen et al., 2012). Treatment or vaccination is not recommended because of the zoonotic potential of this organism, and the animal should be removed from the herd.

Although human brucellosis is extremely important, swine brucellosis is of lesser importance since milk from pigs is rarely consumed (Pappas et al., 2006). The zoonotic potential for infection due to prolonged bacteraemia and the relatively low infectious dose for humans is especially important due to work-related exposure and wild boar hunters.

2.5 *Escherichia coli*

Escherichia coli is a gram-negative aerobic rod that can be associated with many diseases in pigs including diarrhoea, septicaemia, mastitis and urinary tract infections. Often with pigs the different *E. coli* are identified by their fimbrial type which vary depending on the age of the pigs affected (Table 2). Vaccination can be effective if the correct fimbrial type is included in the vaccine. It is of particular interest to note that pigs under 20 days of age are significantly less susceptible to F18 *E. coli* due to the lack of receptors at this age.

Most *E. coli* are considered zoonotic prompting a food safety concern, although it is important to note that pigs are not considered a normal source for enterohaemorrhagic *E. coli* O157:H7 (Fairbrother and Nadeau, 2006). Table 3 lists some of the common enterotoxins associated with *E. coli* in pigs.

Table 2 Most common fimbrial types for *Escherichia coli* in pigs and age susceptibility

Neonatal	Post weaning
F4 (K88)	F4 (K88)
F5 (K99)	F18
F6 (987P)	
F41	

Table 3 Enterotoxins associated with *Escherichia coli* in pigs

Enterotoxin	Name	Effect
STa	Heat-stable toxin A	Decreases absorption of water and electrolytes
STb	Heat-stable toxin B	Increase fluid secretion by enterocytes
LT	Heat-liable toxin	Increase secretion of Na^+, Cl^- and HCO_3^-
Stx2e	Shiga-like toxin	Increase vascular permeability

2.6 *Haemophilus parasuis*

Haemophilus parasuis (HPS) is a gram-negative bacterium which causes Glässer's disease (fibrinous polyserositis) with at least 15 different serovar groups identified (Kielstein and Rapp-Gabrielson, 1992). There does not appear to be a direct association between serovars and virulence. One of the challenges with HPS is that it tends to be more common in high health status herds. HPS often appears after co-mingling (mixing) pigs from different sources (Wiseman et al., 1989). In acute outbreaks, HPS tends to affect the biggest and healthiest pigs in the group. HPS invades endothelial cells and causes apoptosis and production of pro-inflammatory cytokines creating accumulation of fibrin (Vanier et al., 2006; Bouchet et al., 2008).

Strategic use of antimicrobials can be used to mitigate the sudden effects of HPS by administering it at times of high risk, especially during stressful events (weaning and animal movement). Unfortunately, after killing the bacteria, the antimicrobials do nothing to the fibrin already produced. As this fibrin dries, it becomes fibrous and can affect heart and lung movements, resulting in a chronically ill pig. Effectiveness of HPS vaccination can be variable (Oliveira et al., 2004; Oh et al., 2013).

There are no zoonotic concerns regarding HPS.

2.7 *Lawsonia intracellularis*

Lawsonia intracellularis is an obligate intracellular bacterium which grows preferentially in intestinal epithelial cells, causing ileitis in pigs. There are three quite distinct clinical presentations of the disease:

1. the traditional porcine intestinal adenomatosis (thickening of intestine);
2. the more chronic form of enteritis with fibrinonecrotic membrane; and
3. the peracute haemorrhagic form resulting in sudden death.

The importance of this disease is its continuous effect on decreasing feed efficiency and average daily gain (McOrist et al., 1997; McOrist, 2005). The peracute form usually affects market-ready pigs, resulting in sudden death with significant consequential economic losses.

Prevention involves the use of vaccines and/or strategic pulsing with a variety of antimicrobials (McOrist et al., 1999; Hammer, 2004; Bak and Rathkjen, 2009).

There are no zoonotic concerns regarding *L. intracellularis* infection in pigs.

2.8 *Pasteurella multocida*

Pasteurella multocida is a gram-negative rod or coccobacillus which can cause pneumonia and atrophic rhinitis in pigs. Most pig isolates are either serotype A or D, whereas serotypes B, C and E are found in cattle, reindeer and water buffalo. Most *P. multocida* Type A have a predilection to lung tissue while Type D are usually involved in progressive atrophic rhinitis along with *Bordetella bronchiseptica*, although either type can be found in the other's preferred tissue (Carter, 1955; Pijoan et al., 1983; Rimler and Rhoads, 1987).

P. multocida is the most common bacterial infection found in PRDC and it is the primary target for antimicrobial therapy. There are toxigenic and non-toxigenic strains of *P. multocida*. It is interesting to note that *P. multocida* by itself cannot cause pneumonia even when heavily inoculated. This suggests that there is a need for a primary co-infection to enable the establishment of *P. multocida* (Brockmeier et al., 2001). Vaccines containing *P. multocida* toxin have been effective against progressive atrophic rhinitis in pigs. *P. multocida* has no food safety concerns but has the potential for being zoonotic.

2.9 *Salmonella* spp.

Salmonella spp. are gram-negative rods known to infect a broad range of hosts. The two most important salmonella in pigs are *S. choleraesuis* (pigs only) and *S. typhimurium* (humans and pigs).

Clinical signs of salmonellosis can be variable in pigs depending on the strain. Both *S. choleraesuis* and *S. typhimurium* cause diarrhoea in pigs while *S. choleraesuis* more often tends to be systemic causing cyanosis of the skin as well as an interstitial pneumonia (Schwartz, 1997; Foley and Lynne, 2008; Carlson et al., 2012). There are over 200 virulence factors that have been identified with *Salmonella*, but few have been fully characterized (Carlson et al., 2012).

Vaccination is quite effective in helping prevent disease and antimicrobial use helps with treatment. As a gram-negative enteric pathogen, drug resistance (via plasmids) is common (Barnes and Sorensen, 1975; Schwartz, 1997). In addition to vaccination and antimicrobial therapy, bio-security with heavy emphasis on cleanliness to minimize faecal-oral exposure is important.

Salmonella is of zoonotic concern impacting on food safety.

3 The most common bacterial pathogens in pig production: gram-positive bacteria

3.1 *Clostridium* spp.

Clostridium spp. are anaerobic gram-positive spore-forming rods with several different species causing different diseases in pigs. *Clostridium* spp. cause disease via the different toxins they produce. In this chapter we will only discuss the two *Clostridium* spp. of greatest concern in pigs.

3.1.1 *Clostridium difficile*

In pigs, the clinical signs usually appear in the first few hours or days of life. The disease is believed to be caused by two toxins (Toxins A and B) and the administration of equine-origin antitoxins can mitigate the effects (Ramirez et al., 2014). Antimicrobial use does not appear to affect the severity of the disease in neonatal pigs (Arruda et al., 2013), which makes sense as the microflora is barely established in newborn piglets immediately after birth. Interestingly, although *C. difficile* infections are not seen in older pigs (more than 7 days of age), work by Arruda et al. (2013) has shown pigs are still susceptible at 10 days of age. There are currently no effective pig vaccines against *C. difficile*, which can be found in the faeces of most mammals.

In humans, *C. difficile* infections can be very serious or even deadly with antimicrobial-associated diarrhoea (Bartlet et al., 1978). In humans it can result in simple diarrhoea, colitis, pseudomembranous colitis, ileus, toxic mega colon and even bowel perforation (Kelly et al., 1984). However, there is currently no data to directly link *C. difficile* infection in pigs to zoonotic issues.

3.1.2 *Clostridium perfringens*

There are currently five different toxinotypes (Table 4), with only toxinotypes A and C affecting pigs. Enteritis by *C. perfringens* Type C has been well characterized and pre-farrowing vaccination programmes have been effective in controlling this disease quite well. On the other hand, *C. perfringens* Type A is still a bit of an enigma. Traditionally, Type A infections have only been associated with alpha toxin production. Unfortunately, as seen in Table 3, all other *C. perfringens* toxinotypes also produce this same toxin. Several research labs have suggested a role for a beta2 toxin with this disease (Bueschel et al., 2003; Waters et al., 2003), while others more recently question its role (Faranz et al., 2013). Field vaccination with *C. perfringens* Type A toxoid does not appear to be as effective as Type C vaccination.

There are no zoonotic concerns regarding *C. perfringens* in pigs. Human food poisoning with *C. perfringens* is mostly associated with consumption of beef, poultry or gravies.

Table 4 Five *Clostridium perfringens* toxinotypes and their respective toxin and animals they can infect

Clostridium	Animals affected						Toxins				
perfringens	Pigs	Sheep	Goats	Poultry	Cattle	Horses	Alpha	Beta	Beta2	Epsilon	Iota
Type A	X	X		X	X		X		X		
Type B		X			X	X	X	X		X	
Type C	X	X	X	X	X	X	X	X			
Type D		X	X		X		X			X	
Type E		?			X		X				X

3.2 Tuberculosis

Tuberculosis continues to be responsible for significant economic losses for pig producers in many countries while others such as the United States have practically eliminated it from their pig population (Thoen, 2012). Pigs are susceptible to *Mycobacterium avium* complex (MAC) and *M. tuberculosis* complex amongst other mycobacterial species (Thoen et al., 1975).

Pigs often acquire MAC when reared on ground contaminated by poultry (Schalk et al., 1935) and sometimes even sawdust (Schliesser and Weber, 1973). Most cases in pigs are asymptomatic or non-specific and therefore only diagnosed at slaughter (Thoen, 2012). Slaughter inspection specifically looks for granulomatous lesions in lungs or lymph nodes (granulomatous lymphadenitis). However, current European Union guidelines on pig meat inspection discourage the palpation/incision of such post-mortem lesions during routine slaughter in an effort to minimize bacterial cross-contamination (EUFSA, 2011)

As with most diseases that cause granulomas, the use of antimicrobials is not recommended due to long duration of treatment and poor prognosis. There are no effective vaccines available for pigs.

Although MAC can be a significant zoonotic and food safety risk for humans, especially those with immunocompromised immune systems (e.g., elderly, AIDS), pigs and pork have not been implicated as an exposure risk for human infection (Arasteh et al., 2000; Thoen, 2012).

3.3 *Mycoplasma* spp.

Mycoplasmas are a type of bacteria that lack cell walls. There are several mycoplasmas of importance in pigs including *Mycoplasma hyopneumoniae* (Mhyop), *M. hyorhinis* (Mhyor), *M. hyosynoviae* (Mhyos) and *M. suis*. Because of the large number of different mycoplasmas, it is important to be specific in name and not just refer to them as simply 'mycoplasma'.

There are no zoonotic concerns regarding any of the *Mycoplasmas in* pigs.

3.3.1 *Mycoplasma hyopneumoniae*

Mhyop is the aetiologic agent for enzootic pneumonia in pigs, one of the most significant bacterial respiratory pig pathogens worldwide. The strains of Mhyop are antigenically diverse (Frey et al., 1992; Thacker and Minion, 2012). Mhyop is difficult to grow in most laboratories.

Transmission of Mhyop occurs via nose-to-nose contact especially from sow to pig (Calsamiglia and Pijoan, 2000) but can also occur via aerosol up to 3.2 km (Goodwin, 1985) and 9.2 km (Otke et al., 2010). The organism attaches to ciliary epithelium of the respiratory tract and grows slowly. Protein P97 is involved in adherence (Zhang et al., 1994). Mhyop also alters the function of macrophages (Caruso and Ross, 1990) as well as other parts of the immune system (Thacker and Minion, 2012). Mhyop is an important potentiate of other respiratory pathogens in association with PRDC.

Vaccination of growing pigs can be considered 'standard' in today's pig production. Antimicrobials can also be used strategically to mitigate Mhyop as well as other bacterial co-infections.

3.3.2 *Mycoplasma hyorhinis*

Mhyor is associated with polyserositis and arthritis in three- to ten-week-old pigs. Current interest in this pathogen has increased due to welfare concerns of lameness in growing pigs. Little is known regarding the virulence and pathogenesis of Mhyor (Thacker and Minion, 2012). With the advent of newer PCR technology, many samples are now being tested for Mhyor, possibly creating a false sense of increased prevalence in recent years. Bacterial culture for Mhyor can be quite easy but requires special media (Ross and Whittlestone, 1983).

There are several antimicrobials which are used to treat Mhyor infection, although efficacy is quite variable (Thacker and Minion, 2012). It is suspected that part of the problem is late diagnosis of the disease. As with any joint infection, early detection is key.

3.3.3 *Mycoplasma hyosynoviae*

Mhyos is mostly associated with arthritis, and is very similar to Mhyor except that Mhyos tends to affect older pigs (3–5 months of age). As is the case with Mhyor, the Mhyos bacteria can be found in the tonsils of infected and 'normal' pigs (Thacker and Minion, 2012). Bacterial culture for Mhyos can be quite straightforward but requires special media (Friis et al., 1992).

3.3.4 *Mycoplasma suis*

Eperythrozoon suis, now renamed *Mycoplasma suis*, infects red blood cells of pigs causing anaemia (moderate to severe), respiratory distress (Doyle, 1932) and possible reproductive problems. The organism can live in the cytoplasm as well as in membrane-bound vacuoles of erythrocytes (Groebel et al., 2009) which can make it inaccessible to many antimicrobials. Bacterial culture for *M suis* is not yet possible so diagnosis is currently done via PCR.

3.4 *Staphylococcus* spp.

Staphylococcus spp. are gram-positive cocci that are regarded as normal bacterial flora of adult pig skin (Frana, 2012). There are two primary *Staphylococcus* of importance in pigs: *S. hyicus* and methicillin-resistant *S. aureus* (MRSA).

3.4.1 *Staphylococcus aureus*

Staphylococcus aureus can often be isolated from the skin of pigs as well as from septicaemia, mastitis, metritis and metritis infections. Although *S. aureus* rarely causes disease, recent attention to a specific type of *S. aureus* known as MRSA has stimulated interest in this bacteria. In particular, a unique MRSA known as ST398 was first associated with pigs in Europe (Armad-Leferve et al., 2005). This same sequence type does not appear to be as prevalent or important in pig production in the United States (Frana et al., 2013).

MRSA appears to be asymptomatic in pigs and is not considered to be a herd problem. Although on-farm antimicrobial use is suspected in MRSA, no studies have been able to demonstrate this association, which brings into question the ethical legitimacy of using stigmatization as a direct means to achieve public health outcomes (Plough et al., 2015).

Although MRSA is a significant human health concern, outside Denmark, the role of pigs in MRSA zoonosis does not appear to be significant. MRSA is not a food safety concern.

3.4.2 *Staphylococcus hyicus*

Staphylococcus hyicus is the causative agent for greasy pig disease or exudative epidermitis. This condition has worldwide distribution and presents as a skin infection (pyoderma) in young pigs (nursery age or younger). Although *S. hyicus* is commonly found in pig skin, under the right conditions the bacterium will establish itself in the epidermis via an abrasion in the skin. In severe cases the loss of fluids and electrolytes can lead to dehydration and death. There are several exfoliative toxins that have been identified and are considered the most important virulence factors for greasy pig disease (Amtsberg, 1979).

High humidity in pens, as well as a high number of young gilts farrowing, contributes to higher incidence or acute outbreaks of the disease. Injectable antimicrobials are commonly used to treat affected pigs along with topical treatments (spraying/soaking) which may involve the use of disinfectants (Frana, 2012). Pig farmers often use autogenous vaccines against *S. hyicus* with variable effectiveness.

There are no zoonotic concerns regarding *S. hyicus* in pigs.

3.5 *Streptococcus suis*

Although there are many other *Streptococcus* spp. that infect pigs, we will only discuss *S. suis*. *Streptococcus suis* is a gram-positive encapsulated coccus that can frequently be found in the tonsils, intestines and genital tract of healthy pigs (Gottschalk, 2012). There are 35 serotypes based on capsular polysaccharide but serotype 2 is the most common and most important one in pigs because of its zoonotic potential.

There are a large number of virulence factors identified for *S. suis* but none are fully understood due to the complexity of multiple factors (Braums and Valentin-Weigand, 2009). *Streptococcus suis* infection in pigs can be variable, in part due to variations in virulence and may include septicaemia, central nervous signs (meningitis), arthritis, pneumonia, vegetative valvular endocarditis, rhinitis and abortions (Sanford and Tilker, 1982).

Vaccination against *S. suis* is often ineffective. Beta-lactams and macrolide antimicrobials are commonly used to prevent and treat *S. suis* infections.

In South-East Asia, *S. suis* is the most common cause of bacterial meningitis in humans and therefore has been identified as a serious emerging public health threat (Wertheim, 2009). *S. suis* poses a zoonotic concern including a food safety concern in countries with particular cultural practices and preferences such as drinking uncooked blood from infected pigs and eating organs such as the uterus.

4 The most common viral pathogens in pig production

As a general rule it is helpful to know if the viral pathogen is a DNA or an RNA virus as well as whether it is an enveloped or non-enveloped virus. Compared with DNA viruses, RNA viruses mutate often as they do not have the necessary proofing mechanism when they replicate. Thus there can be great variability between strains and developing vaccines can be more challenging. Non-enveloped viruses tend to be much more resistant to inactivation than enveloped viruses. This means that non-enveloped viruses tend to persist longer in the environment. Clearly there are always exceptions to the rule, but these guiding principles can be very useful when learning about a new virus and its possible behaviour regarding transmission between pigs. Many of these important viruses are listed by the World Organization for Animal Health (OIE) as notifiable diseases due to their importance regarding animal and human health as well as international trade.

4.1 African swine fever virus

ASFv is an enveloped DNA virus from the family *Asfarvideae* that causes a highly contagious and haemorrhagic disease in pigs of all ages. ASF is an OIE-listed disease with important international trade consequences. The disease is endemic in sub-Saharan Africa and has now spread to several Eastern European countries. There are many different hosts for the virus as well as significant variation in virulence between strains (De Villeret al., 2010).

ASFv can be transmitted by many means including soft ticks and direct contact with contaminated oral and nasal secretions (Colgrove et al., 1969), consumption of contaminated feed and possibly short distances via aerosol (CFSPH, 2015a). The virus can survive long periods in cured meats (Mebus et al., 1993). ASFv can infect multiple tissues but their primary cells for replication include monocytes and macrophages (Malmquist and Hays, 1960; Minguez et al., 1988). Although ASFv does not induce neutralizing antibodies

(De Boer, 1967), protective immunity against homologous but not heterologous re-infection still occurs (Ruiz Gonzalvo et al., 1981).

Currently there are no good vaccines against ASFv and treatment is not recommended as disease eradication should be the goal.

ASFv is not zoonotic but the severity of the disease causes significant food security and sustainability concerns.

4.2 Aujeszky's disease virus

Aujeszky's disease, also known as pseudorabies, is caused by the herpes virus, which is an enveloped DNA virus whose only natural host is pigs. Aujeszky's disease presents with central nervous system indicators, reproductive problems including abortions, respiratory illness and mortality. All other mammals, except humans, are end hosts for the virus, resulting in close to 100% mortality in these species. Its distribution is worldwide, with variations in virulence between strains and is an OIE-reportable disease.

Aujeszky's disease virus is primarily transmitted between pigs via direct and indirect contact including long-distance aerosol (Christen et al., 1990). Even after recovery, pigs remain infected for the rest of their life (common amongst herpes virus infections) and stress can reactivate viral shedding, helping to spread the disease to other pigs (Wittmann and Rziha, 1989).

Gene-deleted vaccines can be used with DIVA (differentiate vaccinated from infected) capabilities. They are quite effective at protecting against viraemia and clinical signs but unfortunately they do not prevent latent infections. Aujeszky's disease is not of zoonotic concern.

4.3 Classical swine fever virus

Classical swine fever (CSF), also known as hog cholera, is caused by an enveloped RNA Pestivirus that causes generalized systemic disease indistinguishable from many other common, endemic bacterial and viral pig diseases. CSF is an OIE-reportable disease. CSF is endemic in parts of Asia, South and Central America and some Caribbean islands (CFSPH, 2015b).

CSF virus is highly contagious and can cause septicaemia, anorexia, constipation, diarrhoea, lethargy and abortions, to mention just a few. The variable clinical signs and virulence of the virus is dependent on many factors including variation in strains (asymptomatic to high mortality) (Depner et al., 1997; Moenning et al., 2003).

As with natural infections, a combination of cell-mediated immunity and neutralizing antibodies appear to be important in producing sterilizing immunity (Pirou et al., 2003). There are several vaccines, especially live or modified live, that provide good protection against disease including some oral vaccines for wild boars (CFSPH, 2015b). CSF disease is not of zoonotic concern.

4.4 Coronaviruses

There are several different coronaviruses of importance in pigs. For the most part, they have quite similar clinical presentation, primarily causing diarrhoea with similar treatment and control but immunologically are very distinct from each other (i.e. no cross-protection between these different viruses). They are enveloped RNA viruses with spike proteins that give them a crown-like appearance under an electron microscope. None of the pig coronaviruses are of zoonotic concern.

4.4.1 Deltacoronavirus

With the introduction of porcine epidemic diarrhoea virus (PEDv) into the United States in early 2013 a new porcine deltacoronavirus (PDCov) was identified in faeces from pigs with diarrhoea (Wang et al., 2014). Work done by June et al. (2015) demonstrated that this virus is capable of causing disease in gnotobiotic pigs. Anecdotal field evidence suggests PDCov can cause diarrhoea, but is significantly milder and shorter in duration than PEDv. Currently there are no vaccines for PDCov.

4.4.2 Porcine epidemic diarrhoea virus

The importance of PEDv has re-surfaced after its introduction into the United States in May 2013. This introduction highlighted the possible role feed and feed ingredients can play in disease transmission – an area previously ignored most of the time. The clinical presentation for the disease is almost exactly the same as that of transmissible gastroenteritis virus (TGEv). The challenge of any new disease introduced into a naïve population was quite apparent. Disease spread easily from place to place and the infectious dose of the virus was quite low as data to be published in peer-review journals will show. Recent research suggest faecal viral shedding occurs within one day of exposure, peaks around 7 days and can continue for over 28 days (Magstadt et al., 2014). Interestingly, faecal consistency in the same study was clinically normal at 10 days post-inoculation, even though PEDv shedding was still occurring.

Clinical disease is more severe in piglets less than 3 weeks of age, with piglets less than 10 days old usually experiencing 100% mortality in an acute outbreak. Clinical signs can be short in duration to asymptomatic in older pigs. Immunity to PEDv appears to be short term (probably less than 4 months) and both colostral and lactogenic immunity are important in protecting baby pigs (Thomas, 2014). Vaccines can be helpful in herds with previous field exposure to PEDv but do not appear to be as effective when used in naïve animals (Thomas, 2014; Schelkopf et al., 2015). Although currently there are at least three different reported isolates in the United States, research suggests there is still good cross-protection between these different isolates (Zhang et al., 2015). Current vaccines against PEDv appear to be effective when used in animals previously exposed to live virus (Jung and Saif, 2015).

4.4.3 Transmissible gastroenteritis virus

The occurrence of TGEv has significantly decreased over the past decade as the appearance and widespread prevalence of the TGEv respiratory mutant (porcine respiratory coronavirus) spread throughout the United States suggesting possible cross-protection between these two related pathogens (Yaeger et al., 2002).

Clinical presentation, treatment and control options are the same for TGEv and PEDv. Vaccines against TGEv are not very effective and produce partial immunity of short duration (Saif et al., 1994).

4.5 Foot-and-mouth disease virus

Foot-and-mouth disease virus (FMDv) is a member of the *Picornaviridae*, a family of non-enveloped RNA viruses of significant international importance. It appears that FMDv is one of the best recognized and dreaded livestock diseases causing severe vesicular lesions in cloven-hooved animals and is at the top of the OIE-reportable diseases. FMDv is endemic in large areas of Africa, Asia, the Middle East and South America.

Seven serotypes (O, A, C, SAT 1, SAT 2, SAT 3 and Asia 1) which are important for vaccination are recognized. Serotype O is the most common serotype worldwide (CFSPH, 2014). Foot-and-mouth disease (FMD) can be transmitted via direct, indirect and aerosol means, with all secretions and excretions from infected animals containing the infectious virus (Alexandersen et al., 2012). Unlike ruminants, pigs do not become carriers or harbour FMDv for more than 28 days. FMDv is quite resistant and can remain infectious in the environment, cured meats and dairy products for several weeks (Bachrach, 1968; Cottral, 1969; CFSPH, 2014).

Vesicular lesions from FMD are clinically indistinguishable from any other vesicular disease including swine vesicular disease, vesicular exanthema of swine, vesicular stomatitis and Seneca virus A. Mortality in adults tends to be low, but animals have difficulty eating and moving around, resulting in a welfare concern.

Vaccination requires matching the proper serotype as there is no cross-protection between all seven serotypes. Most FMD vaccination has been focused on cattle and not pigs. There is also a wide range of strains within each serotype, thus complicating vaccine efficacy (Kitching et al., 1989). This requirement to match the strain and serotype of FMDv makes it difficult to react to an emergency outbreak. Vaccine-induced protection is short, lasting only about 4–6 months necessitating at least two doses per year (Domenech et al., 2010). Most of the largest producers are free of FMD and do not vaccinate.

The first OIE/Food and Agriculture Organization of the United Nations Global Conference on FMD led to the development of a Global FMD Control Strategy. This effort is focused on improving FMD control in regions where the disease is still endemic through the use of a Progressive Control Pathways (PCP) tool and is supported by many countries including the European Commission for the Control of Foot-and-Mouth Disease (EUFMD). This PCP offers a structured five-stage approach to FMD control, allowing FMD-endemic countries to become more successful in achieving FMD-free status strategically (OIE and FAO, 2012).

FMDv is not zoonotic, but the severity of the disease and international trade restrictions as an OIE-listed disease causes significant food security and sustainability concerns.

4.6 Influenza A virus in swine

Influenza A viruses in swine (IAv-S) are members of the family *Orthomyxoviridae*, which are enveloped viruses with segmented RNA, and can cause respiratory infections in most mammals. The segmented genome of IAv-S facilitates the exchange of gene segments between different IAv-S which may infect the same cell. This rearrangement of genes can generate new strains of the virus. Virus replication is limited to the upper and lower respiratory tract (Van Reeth et al., 2012).

The primary IAv-S are H1N1 and H3N2. Within each of these influenza types there are many different groupings of strains and cross-protection between strains, even within one type, can be quite variable. Infection occurs primarily via direct nose-to-nose contact between pigs and between people and pigs but more rarely between pigs and people (Nelson, 2014).

Vaccination continues to be the best means for prevention (Van Reeth et al., 2012). With such large diversity in field strains, matching the correct vaccine isolates to field isolates is extremely challenging, especially in livestock where vaccine regulations limit how quickly vaccine isolates can be changed. It is important to note that maternal antibodies do interfere with vaccine efficacy.

IAv-S are of zoonotic importance but are not a food safety concern. The biggest public health fear is the rearrangement of IAv-S into a novel strain infecting a naïve human population as was the case in the 2009 influenza pandemic.

4.7 Porcine circovirus type 2 virus

Porcine circovirus 2 viruses (PCV2v) belong to the family *Circoviridae* and are non-enveloped DNA viruses of global importance in pigs. PCV2v causes a variety of systemic diseases in pigs including wasting, pneumonia, late-term abortions, stillbirths, porcine dermatitis, nephropathy syndrome and diarrhoea (Harding and Clark, 1997).

PCV2v causes immunosuppression in pigs, making them vulnerable to a wide variety of infections (Chianini et al., 2003). The characteristic case definition for post-weaning multi-wasting syndrome (PMWS) requires three components:

1. lymphoid depletion,
2. large number of PCV2v in the lesion, and
3. clinical signs of wasting with a doubling of mortality (Sorden, 2000).

PCV2-infected animals develop good neutralizing antibodies in 10–28 days (Pogtanichniy et al., 2000; Fort et al., 2007). Vaccines are extremely effective in preventing PCV2-associated disease and is a standard part of any vaccination programme for growing pigs.

PCV2v is not of zoonotic concern, but the severity of the disease causes significant food security and sustainability concerns.

4.8 Porcine reproductive and respiratory syndrome virus

Porcine reproductive and respiratory syndrome virus (PRRSv) is an enveloped RNA virus from the family *Arterivirdae*, which affects pigs. PRRSv is present in most pig-producing regions, although there are a few countries with significant pig production where the virus has never been documented. Porcine reproductive and respiratory syndrome (PRRS) is an OIE-reportable disease.

One of the most significant characteristics of PRRSv is its ability to mutate. PRRSv has been reported to have a mutation rate multiple times that of the human immunodeficiency virus. This high mutation rate results in the development of quasi-species (a grouping of more than one genetic sequence related to a common mutation in an animal at the same time) (Rowland et al., 1999; Goldberg et al., 2003). This quasi-species is important because when genetically sequencing a sample from an infected animal, it is actually obtaining the consensus sequence of the different PRRSv viruses present. The high mutation rate also makes it difficult to find a single vaccine isolate that will provide broad cross-protection.

PRRSv infects macrophages, especially pulmonary macrophages (Thanawongnuwech et al., 1997; Duan et al., 1997). Targeting pulmonary macrophages results in an increased susceptibility to secondary infections in pigs. Protective immunity against PRRSv is slow to develop (4–6 weeks), so frequently requires closure of the herd for around 200+ days (Torremorell et al., 2002). The slow protective immunity along with low transmissibility of the virus has generated sub-populations of animals with different levels of immunity (Dee et al., 1996).

There are many vaccines available in the market with none able to provide universal protection against all strains. With the diversity of PRRSv, cross-protection becomes difficult to predict. Even today with sequencing technology, this genetic information only helps with epidemiological investigations but cannot be used in any way to predict cross-protection. New technology has identified genetic resistance in some pigs (Rowland et al., 2012) as well as the creation in 2015 of the first genetically engineered pigs completely resistant to PRRSv (Basi, 2015).

PRRSv is not of zoonotic concern, but the severity of the disease causes significant food security and sustainability concerns.

4.9 Rotaviruses

Rotaviruses are non-enveloped RNA viruses of the family *Reoviridae* and are a major cause of diarrhoea in neonatal and young pigs. There are four different porcine rotavirus serogroups identified (A, B, C and E). It was not until the last five to ten years that PCR technology has enabled the detection of serogroups other than type A (Médici et al., 2011). This new technology has resulted in increased detection and awareness of rotavirus B and C as well as the concept of co-infections with more than one rotavirus at a time.

Rotaviruses are highly prevalent throughout the world with some countries demonstrating up to 100% sero-prevalence in adult pigs (Chang et al., 2012).

Rotaviruses replicate predominately in villous epithelium in the small intestine (Buller and Moxley, 1988) as well as the large intestine (Theil et al., 1978) causing villous atrophy and diarrhoea. The high prevalence of rotavirus in the field suggests piglets are constantly being exposed to the virus and are likely to have reduced performance due to infections. Rotavirus infections are well recognized in neonatal pigs but are often ignored post-weaning.

Colostrum and lactogenic immunity play an important role in helping protect neonatal piglets from clinical disease (Saif, 1999; Wagstrom et al., 2000). Currently there are only vaccines against rotavirus type A infections.

Porcine rotaviruses are not considered to be of zoonotic concern.

5 The most common parasitic pathogens in pig production

Modern pig production has moved pigs indoors and thus limited their exposure to many internal and external parasites. Unfortunately, there is a general misconception that indoor pigs do not have internal parasites. Although many indoor facilities are clean and disinfected between groups of pigs, eggs from internal parasites are quite resistant. Personal experience suggests that the discontinued use of anthelmintic in indoor production has allowed some internal parasites to slowly propagate, increasing the exposure dose of growing and breeding pigs.

5.1 Ascarids

There are several ascarids that can infect pigs but *Ascaris suum* (pig round worm) is the most common and most important one. The life cycle is direct, taking four weeks for eggs passed in faeces to develop infectivity. These ascarid eggs are highly resistant to desiccation allowing them to build up in outdoor as well as indoor facilities (Barrett, 1976).

As infective eggs are consumed by the pig, the larvae will hatch in the intestine and travel through the intestinal wall to the liver, causing the traditional milk spots (scars) The larvae then travel on to the lungs causing verminous pneumonia before being re-swallowed completing the life cycle by developing into adults in the small intestine. This whole migration process causes internal damage reducing pig growth and production. *Ascaris suum* are zoonotic but are not a food safety concern.

5.2 Cysticerci

The adult tapeworm (*Taenia solium*) produces larvae which are then ingested by pigs and hatch into cysticercus (*Cysticercus cellulosae*). They are found in the skeletal and cardiac muscle of infected pigs and are referred to as 'measly pork' or 'pork measles'.

Pigs do not appear to have many clinical signs when infected but they can serve as a source of infection for humans (Greve, 2012). Pigs cannot re-infect themselves, infection requires ingestion of human tapeworm eggs. Pigs must not be allowed to ingest human faeces.

Cysticercosis is not a concern for modern pig production farms due to bio-security practices. In 2010 the World Health Organization added cysticercosis as a neglected tropical disease of zoonotic food safety concern, especially in undeveloped countries with free-roaming pigs (WHO, 2016). In humans cysticerci have a predilection for the central nervous system, making this a serious disease.

5.3 Coccidia

Coccidia are obligate intracellular protozoan with two main coccidia in pigs. Pigs can be infected with both *Isospora suis* or *Eimeria* spp.

5.3.1 *Isospora suis*

Isospora suis is the primary protozoal disease of neonatal pigs while *Eimeria* spp. are rarely identified. There appears to be a seasonal incidence of this disease as warmer temperatures and higher humidity favours sporulation of *I. suis* (Stuart and Lindsay, 1986). *I. suis* is not affected by the use of traditional coccidiostats as employed in other livestock (decoquinate, Amprolium, sulphas and ionophores). Prevention is focused on increased sanitation.

I. suis is not zoonotic.

5.3.2 *Eimeria* spp.

Eimeria spp. can infect pigs but are rarely identified in them (Lindsay et al., 1987). There are many different species of *Eimeria* which can be found in pigs worldwide. Although clinical disease in pigs is rare there have been sporadic reports of clinical diarrhoea in pigs of different ages (Hill et al., 1985). There are no studies on treatment or control options for *Eimeria* spp. Prevention should be focused on increased sanitation. *Eimeria* spp. are not zoonotic.

5.4 *Sarcoptes scabiei*

Sarcoptic mange (*Sarcoptes scabiei*) is the most important external parasite of pigs globally as the mite creates a highly pruritic condition affecting average daily gain, feed efficiency and even reproductive performance in herds (Kessler et al., 2003). The intense itching causes significant property damage with consequential financial losses for the pig farmer.

Eradication programmes can be effective because *S. scabiei* lives and completes its entire life cycle on the skin of pigs and environmental contamination is fairly trivial (Smith, 1986). Mange is not zoonotic or of food safety concern.

5.5 *Trichuris suis*

The pig whip worm *Trichuris suis* occurs primarily in the cecum of pigs and can cause diarrhoea with or without blood and mucous, affecting growth rate and feed efficiency. Its clinical significance is that many anthelmintics used for roundworms are ineffective against whipworms.

5.6 *Trichinella spiralis*

Trichinae in pigs is usually caused by *Trichinella spiralis*, which have minimal effect on pigs, but have significant health effects on people. Garbage feeding, as well as pig access to infected rodent carcasses or other dead pigs, is the primary means for transmission. Raising pigs indoors with limited access to wildlife along with aggressive rodent control has practically eliminated this disease from commercial pigs in the United States (Greve, 2012). This elimination of Trichinae has also allowed for new lower cooking recommendations for pork in the United Sates (from 71°C to 63°C), enabling people to enjoy a tastier (less dry) pork chop (USDA, 2011). *Trichinella spiralis* is of great zoonotic and food safety concern.

6 Case studies

The evolution and complexity of pig disease can be challenging. Knowledge of diseases continues to change as production practices and pathogens change, requiring veterinarians, nutritionists and animal scientists to be constantly attentive while monitoring the health and well-being of pigs.

PCV2v serves as a perfect case study to demonstrate these points and the complexity of evolution in knowledge and the pathogen itself. PCV2v was first identified in the late 1990s. This new pathogen appeared to be causing post-weaning wasting in several European countries while in the United States, most pigs tested positive for antibodies to the disease but otherwise were unaffected. The development and availability of new diagnostic techniques (antibody detection via ELISA) left veterinarians and pig farmers unsure of how to interpret the results, some even publically mocked the new discovery calling it a 'Circus' virus. Then in 2007–8 there was a genotype shift from PCV2a to PCV2b, which resulted in a highly pathogenic strain causing post-weaning wasting in pigs in North America (Carman et al., 2008). This small mutation had now changed the virus from a 'routine commensal' to a devastating, highly virulent, systemic wasting disease with herd mortalities at times in excess of 50%. In the same pen one would find pigs starting to waste right next to healthy-looking pigs (Fig. 2). All the pigs were eating, yet those infected pigs would quickly waste until they were euthanized or they would die. There was no halting the process. Blood samples would show that most pigs (healthy and wasting ones) had antibodies and virus in their blood. It was no longer a matter of just knowing they were positive or negative. A new definition had to be developed to help clarify which pigs had PMWS and which did not. A diagnosis of PMWS now required three components:

1 lymphoid depletion,
2 large number of PCV2v in the lesion, and
3 clinical signs of wasting with a doubling in mortality (Sorden, 2000).

Then the miracle of vaccination came. Although pigs were being affected by PCV2b, the new vaccine used a PCV2a strain. Initially pig farmers and veterinarians were unsure about vaccinating their pigs against the old 'less pathogenic' strain when it was the new variant causing the high mortalities. The structure of the pig industry in the United States facilitated the spread of the PCV2b strain over the entire continental United States within just a few months. There was a new disease with new technology (diagnostics and vaccine), new knowledge and a new industry structure. The disease triad discussed in the introduction of this chapter (Fig. 1) was in full effect. Fortunately, it was quickly realized the new PCV2 vaccine truly was a miracle. The killed bacterin reduced mortalities from more than 25 to between 4 and 6% instantly; vaccinated pigs were now protected. Academically, it appeared the vaccine efficacy sounded too good to be true, but it was. Even herds that did not have the clinical disease and had 'normal' productivity who started using the PCV2 vaccine noted slight improvements in the overall health of the herd and lower mortalities.

Figure 2 Size variation appearing in a group of field pigs infected with the new PCV2b variant starting to cause wasting in some of these pigs while others appear to be perfectly normal.

7 Summary

Hopefully through reading this chapter the complexity of the various diseases has been emphasized. Diseases do not occur in a vacuum. They are impacted directly by a large cohort of factors including environment, nutrition, animal husbandry, genetics and co-infections amongst others.

Ultimately scientists, veterinarians, pig farmers, nutritionists, researchers and pig lovers are all striving to improve the health and well-being of the pigs raised so as to provide a more sustainable, abundant, wholesome, safe, economical and delicious protein for mankind.

8 Future trends

The future of sustainable pig production is positive. Vast amounts of knowledge are being gained rapidly. New technologies in diagnostic surveillance such as spatial-temporal pen sampling with oral fluids, metagenomics and microbiota in pig health are already being developed. Caution must be exercised when applying these new technologies to ensure a better understanding of what is known as well as what is unknown. A point of information overload is being reached as well as at times over-interpretation of the information. To ensure farmers stay focused on the ultimate goal, it is critical to collaborate with others who have expertise in different fields while embracing the clinical significance and implementation of new discoveries.

9 Where to look for further information

There are several sources available for additional information on swine diseases. *Diseases of Swine*, which is currently in its 10th edition, is recognized as the most comprehensive and authoritative textbook on swine diseases. Additionally the following websites can be consulted for information: The Pig Site (http://www.thepigsite.com/diseaseinfo/), Pig333 (https://www.pig333.com/pig-diseases/), The Merck Veterinary Manual (http://www.merckvetmanual.com/) and American Association of Swine Veterinarians Swine Disease Manual (https://vetmed.iastate.edu/vdpam/about/food-supply/swine/swine-disease-manual).

10 References

Alexandersen, S., Knowles, N. J., Dekker, A., Belsham, G. J., Zhang, Z. and Koenen, F. (2012) Picornaviruses. In: Zimmerman, J. J., Karriker, L. A., Ramirez, A., Schwartz, K. J. and Stevenson, G. W. (eds), *Diseases of* Swine, 10th ed. Ames: Wiley-Blackwell Publishing, pp 592.

Amtsberg, G. (1979) Determination of exfoliation triggering substances in cultures of *Staphylococcus hyicus* in swine and *Staphylococcus epidermidis* biotype 2 in cattle. *Zentralbl Veterinarmed B.* 26(4):257–72. German.

Arasteh, K. N., Cordes, C., Ewers, M., Simon, V., Dietz, E., Futh, U. M., Brockmeyer, N. H. and L'Age, P. (2000) Human immunodeficiency virus-related nontuberculous mycobacterial infection: incidence, survival analysis and associated risk factors. *Eur. J. Med. Res.* 5:424–30.

Armand-Lefevre, L., Ruimy, R. and Andremont, A. (2005) Clonal comparison of *Staphylococcus aureus* isolates from healthy pig farmers, human controls, and pigs. *Emerg. Infect. Dis.* 11(5):711–4.

Arruda, P. H., Madson, D. M., Ramirez, A., Rowe, E., Lizer, J. T. and Songer, J. G. (2013) Effect of age, dose and antibiotic therapy on the development of *Clostridium difficile* infection in neonatal piglets. *Anaerobe*. 22:104–10.

Bachrach, H. L. (1968) Foot-and-mouth disease. *Ann. Rev. Microbiol.* 22:201–44.

Bak, H. and Rathkjen, P. H. (2009) Reduced use of antimicrobials after vaccination of pigs against porcine proliferative enteropathy in a Danish SPF herd. *Acta Vet. Scand.* 51:1.

Barnes, D. M. and Sorensen, D. K. (1975) Salmonellosis. In: Dunne, H. W. and Leman, A. D. (eds), *Diseases of Swine*, 4th ed. Ames: Iowa State University Press, pp. 560–61.

Barrett, J. (1976) Studies on the induction of permeability in *Ascaris lumbricoides* eggs. *Parasitology* 73:109–21.

Bartlett, J. G., Chang, T. W., Gurwith, M., Gorbach, S. L. and Onderdonk, A. B. (1978) Antibiotic-associated pseudomembranous colitis due to toxin-producing clostridia. *N. Engl. J. Med.* 298(10):531–4.

Basi, C. (2015) Pigs that are resistant to incurable disease developed at University of Missouri. *New Bureau – University of Missouri.* [Online] 8 December. Available from: http://munews.missouri.edu/news-releases/2015/1208-pigs-that-are-resistant-to-incurable-disease-developed-at-university-of-missouri/ [Accessed 12 December 2015].

Baums, C. G. and Valentin-Weigand, P. (2009) Surface-associated and secreted factors of *Streptococcus suis* in epidemiology, pathogenesis and vaccine development. *Anim. Health Res. Rev.* 10(1):65–83.

Beier, D. and Gross, R. (2008) The BvgS/BvgA phosphorelay system of pathogenic *Bordetellae*: structure, function and evolution. *Adv. Exp. Med. Biol.* 631:149–60.

Bouchet, B., Vanier, G., Jacques, M. and Gottschalk, M. (2008) Interactions of *Haemophilus parasuis* and its LOS with Porcine Brain Microvascular Endothelial Cells. *Vet. Res.* 39:42.

Brochmeier, S. L., Halbur, P. G. and Thacker, E. L. (2002) Porcine Respiratory Disease Complex. In: Brogden, K. A. and Guthmiller, J. M. (eds). *Polymicrobial Diseases.* ASM Press:Washington DC.

Brochmeier, S. L., Register, K. B., Nicholson, T. L. and Loving, C. L. (2012) Bordetellosis. In: Zimmerman, J. J., Karriker, L. A., Ramirez, A., Schwartz, K. J. and Stevenson, G. W. (eds), *Diseases of Swine*, 10th ed. Ames: Wiley-Blackwell Publishing.

Brockmeier, S. L., Palmer, M. V., Bolin, S. R. and Rimler, R. B. (2001) Effects of intranasal inoculation with *Bordetella bronchiseptica*, porcine reproductive and respiratory syndrome virus, or a combination of both organisms on subsequent infection with *Pasteurella multocida* in pigs. *Am. J. Vet. Res.* 62(4):521–5.

Bueschel, D. M., Jost, B. H., Billington, S. J., Trinh, H. T. and Songer, J. G. (2003) Prevalence of cpb2, encoding beta2 toxin, in *Clostridium perfringens* field isolates: correlation of genotype with phenotype. *Vet. Microbiol.* 94:121.

Buller, C. R. and Moxley, R. A. (1988) Natural infection of porcine ileal dome M cells with rotavirus and enteric adenovirus. *Vet. Pathol.* 25:516–17.

Burrough, E. R., Strait, E. L., Kinyon, J. M., Bower, L. P., Madson, D. M., Wilberts, B. L., Schwartz, K. J., Frana, T. S. and Songer, J. G. Comparative virulence of clinical *Brachyspira spp.* isolates in inoculated pigs. *J. Vet. Diagn. Invest.* 24(6):1025–34.

Calsamiglia, M. and Pijoan, C. (2000) Colonisation state and colostral immunity to *Mycoplasma hyopneumoniae* of different parity sows. *Vet. Rec.* 146:530–2.

Carlson, S. A., Barnhill, A. E. and Griffith, R. W. (2012) Salmonellosis. In: Zimmerman, J. J., Karriker, L. A., Ramirez, A., Schwartz, K. J. and Stevenson, G. W. (eds), *Diseases of Swine*, 10th ed. Ames: Wiley-Blackwell Publishing.

Carman, S., Cai, H. Y., DeLay, J., Youssef, S. A., McEwen, B. J., Gagnon, C. A., Tremblay, D., Hazlett, M., Lusis, P., Fairles, J., Alexander, H. S. and van Dreumel, T. (2008) The emergence of a new strain of porcine circovirus-2 in Ontario and Quebec swine and its association with severe porcine circovirus associated disease–2004–2006. *Can. J. Vet. Res.* 72:259–68.

Carter, G. R. (1955) Studies on *Pasteurella multocida*: A hemagglutination test for the identification of serological types. *Am. J. Vet. Res.* 16:481–4.

CFSPH - Center for Food Security and Public Health. (2005) Tanea infections. [Online] Available from: http://www.cfsph.iastate.edu/Factsheets/pdfs/taenia.pdf. [Accessed 31 May 2016].

CFSPH - Center for Food Security and Public Health. (2014) *Foot and mouth disease*. [Online] Available from: http://www.cfsph.iastate.edu/Factsheets/pdfs/foot_and_mouth_disease.pdf. [Accessed: 17th December 2015].

CFSPH - Center for Food Security and Public Health. (2015a) *African Swine Fever*. [Online] Available from: http://www.cfsph.iastate.edu/Factsheets/pdfs/african_swine_fever.pdf. [Accessed: 10th January 2016].

CFSPH - Center for Food Security and Public Health. (2015b) *Classical Swine Fever*. [Online] Available from: http://www.cfsph.iastate.edu/Factsheets/pdfs/classical_swine_fever.pdf. [Accessed: 22nd January 2016].

Chander, Y., Primus, A., Oliveira, S. and Gebhart, C. J. (2012) Phenotypic and molecular characterization of a novel strongly hemolytic *Brachyspira* species, provisionally designated '*Brachyspira hampsonii*'. *J. Vet. Diagn. Invest.* 24(5):903–10.

Chang, K.-O., Saif, L. J. and Kim, Y. (2012) Reoviruses (Rotaviruses and Reoviruses). In: Zimmerman, J. J., Karriker, L. A., Ramirez, A., Schwartz, K. J. and Stevenson, G. W. (eds), *Diseases of Swine*, 10th ed. Ames: Wiley-Blackwell Publishing.

Chianini, F., Majo, N., Segales, J., Dominguez, J. and Domingo, M. (2003) Immunohistochemical characterization of PCV2 associate lesions in lymphoid and non-lymphoid tissues of pigs with natural postweaning multisystemic wasting syndrome (PMWS). *Vet. Immunol. Immunopathol.* 94:63–75.

Chiers, K., De Waele, T., Pasmans, F., Ducatelle, R. and Haesebrouck, F. (2010) Virulence factors of *Actinobacillus pleuropneumoniae* involved in colonization, persistence and induction of lesions in its porcine host. *Vet. Res.* 41(5):65.

Christensen, L. S., Mousing, J., Mortensen, S., Soerensen, K. J., Strandbygaard, S. B., Henriksen, C. A. and Andersen, J. B. (1990) Evidence of long distance airborne transmission of Aujeszky's disease (pseudorabies) virus. *Vet. Rec.* 127:471–4.

Clota, J., Foix, A., March, R., Riera, P. and Costa, L. (1996) Caracterización serológica de cepas de *Actinobacillus pleuropneumoniae* aisladas en Espaa. *Med. Vet.* 13:17–22.

Colgrove, G. S., Haelterman, E. O. and Coggins, L. (1969) Pathogenesis of African swine fever in young pigs. *Am. J. Vet. Res.* 30(8):1343–59.

Cottral, G. E. (1969) Persistence of foot-and-mouth disease virus in animals, their products and the environment. *Bull.Off. Int. Epizoot.* 70:549–68.

De Boer, C. V. (1967) Studies to determine neutralizing antibody in sera from animals recovered from African swine fever and laboratory animals inoculated with African swine fever virus with adjuvants. *Arch Gesamte Virusforsch.* 20:164–79.

de Jong, M. F. and Nielsen, J. P. (1990) Definition of progressive atrophic rhinitis. *Vet. Rec.* 126:93.

De Villier, E. P., Gallardo, C., Arias, M., Da Silva, M., Upton, C., Martin, R. and Bishop, R. P. (2010) Phylogenomic analysis of 11 complete African swine fever virus genome sequences. *Virol.* 400(1):128–36.

Dee, S. A., Joo, H. S., Henry, S., Tokach, L., Park, K., Molitor, T. and Pijoan, C. (1996) Detecting subpopulations after PRRS virus infection in large breeding herds using multiple serologic tests. *J. Swine Health Prod.* 4(4):181–4.

Depner, K. R., Hinrichs, U., Bickhardt, K., Greiser-Wilke, I., Pohlenz, J., Moennig, V. and Liess, B. (1997) Influence of breed-related factors on the course of classical swine fever virus infection. *Vet. Rec.* 140:506–7.

Desrosiers, R. (2004) Epidemiology, diagnosis and control of swine diseases. Howard Dunne Memorial Lecture. In *Proc. 35th Annu. Meet Am. Assoc. Swine Pract.*, pp. 9–37.

Domenech, J., Lubroth, J. and Sumption, K. (2010) Immune protection in animals: the examples of rinderpest and foot-and-mouth disease. *J. Comp. Pathol.* 142 Suppl 1:S120–4.

Doyle, L. (1932) A rickettsia-like or anaplasmos-like disease in swine. *J. Am. Vet. Med. Assoc.* 8:668–71.

Duan, X., Nauwynck, H. J. and Pensaert, M. B. (1997) Virus quantification and identification of cellular targets in the lungs and lymphoid tissues of pigs at different time intervals after inoculation with porcine reproductive and respiratory syndrome virus (PRRSV). *Vet. Microbiol.* 56(1–2):9–19.

European Union Food Safety Authority. (2011) Scientific Opinion on the public health hazards to be covered by inspection of meat (swine). *EFSA Journal.* 9(10):2351.

Fairbrother, J. M. and Nadeau, E. (2006) *Escherichia coli*: on-farm contamination of animals. *Rev. Sci. Tech.* 25(2):555–69.

Farez, S. and Morley, R. S. (1997) Potential animal health hazards of pork and pork products. *Rev. Sci. Tech.* 16(1):65–78.

Farzan, A., Kircanski, J., DeLay, J., Soltes, G., Songer, J. G., Friendship, R. and Prescott, J. F. (2013) An investigation into the association between cpb2-encoding *Clostridium perfringens* type A and diarrhea in neonatal piglets. *Can. J. Vet. Res.* 77(1):45–53.

Foley, S. L., Lynne, A. M. and Nayak, R. Salmonella challenges: prevalence in swine and poultry and potential pathogenicity of such isolates. *J. Anim. Sci.* 2008 April; 86(14 Suppl):E149–62. Epub 2 October 2007. Review. PubMed PMID: 17911227.

Fort, M., Olvera, A., Sibila, M., Segalés, J. and Mateu, E. (2007) Detection of neutralizing antibodies in postweaning multisystemic wasting syndrome (PMWS)-affected and non-PMWS-affected pigs. *Vet. Microbiol.* 125:244–55.

Frana, T. S. (2012) Staphylococcosis. In: Zimmerman, J. J., Karriker, L. A., Ramirez, A., Schwartz, K. J. and Stevenson, G. W. (eds) *Diseases of Swine*, 10th ed. Ames: Wiley-Blackwell Publishing.

Frana, T. S., Beahm, A. R., Hanson, B. M., Kinyon, J. M., Layman, L. L., Karriker, L. A., Ramirez, A. and Smith, T. C. (2013) Isolation and characterization of methicillin-resistant *Staphylococcus aureus* from pork farms and visiting veterinary students. *PLoS One.* 8(1):e53738.

Frey, J., Haldimann, A. and Nicolet, J. (1992) Chromosomal heterogeneity of various *Mycoplasma hyopneumoniae* field strains. *Int. J. Syst. Bacteriol.* 42:275–80.

Friis, N. F., Hansen, K. K., Schirmer, A. L. and Aabo, S. (1992) *Mycoplasma hyosynoviae* in joints with arthritis in abattoir baconers. *Acta Vet. Scand.* 33:425–9.

Goldberg, T. L., Lowe, J. F., Milburn, S. M. and Firkins, L. D. (2003) Quasispecies variation of porcine reproductive and respiratory syndrome virus during natural infection. *Virology.* 20; 317(2):197–207.

Goodwin, R. F. (1985) Apparent reinfection of enzootic-pneumonia-free pig herds: search for possible causes. *Vet. Rec.* 116:690–4.

Gottschalk, M. (2012) Actinobacillosis. In: Zimmerman, J. J., Karriker, L. A., Ramirez, A., Schwartz, K. J. and Stevenson, G. W. (eds), *Diseases of Swine*, 10th ed. Ames: Wiley-Blackwell Publishing.

Gottschalk, M. (2012) Streptococcosis. In: Zimmerman, J. J., Karriker, L. A., Ramirez, A., Schwartz, K. J. and Stevenson, G. W. (eds), *Diseases of Swine*, 10th ed. Ames: Wiley-Blackwell Publishing.

Gottschalk, M., Broes, A. and Fittipaldi, N. (2003) Recent developments on *Actinobacillus pleuropneumoniae*. In *Proc. 34th Annu. Meet Am. Assoc. Swine Pract*, pp. 387–93.

Greve, J. H. (2012) Internal parasites: Helminths. In: Zimmerman, J. J., Karriker, L. A., Ramirez, A., Schwartz, K. J. and Stevenson, G. W. (eds), *Diseases of Swine*, 10th ed. Ames: Wiley-Blackwell Publishing.

Groebel, K., Hoelzle, K., Wittenbrink, M. M., Ziegler, U. and Hoelzle, L. E. (2009) *Mycoplasma suis* invades porcine erythrocytes. *Infect. Immun.* 77:576–84.

Hammer, J. M. (2004) The temporal relationship of fecal shedding of *Lawsonia intracellularis* and seroconversion in field cases. *J. Swine Health Prod.* 12: 29–33.

Hampson, D. (2012) Brachyspiral Colitis. In: Zimmerman, J. J., Karriker, L. A., Ramirez, A., Schwartz, K. J. and Stevenson, G. W. (eds), *Diseases of Swine*, 10th ed. Ames: Wiley-Blackwell Publishing.

Harding, J. C. and Clark, E. G. (1997) Recognizing and diagnosing postweaning multisystemic wasting syndrome (PMWS). *J. Swine Health Prod.* 5:201–3.

Hill, J. E., Lomax, L. G., Lindsay, D. S. and Lynn, B. S. (1985) Coccidosis caused by *Eimeria scabra* in a finishing hog. *J. Am. Vet. Med. Assoc.* 186(9):981–3.

Jung, K., Hu, H. and Saif, L. J. (2016) Porcine deltacoronavirus induces apoptosis in swine testicular and LLC porcine kidney cell lines in vitro but not in infected intestinal enterocytes in vivo. *Vet. Microbiol.* 182:57–63.

Jung, K. and Saif, L. J. (2015) Porcine epidemic diarrhea virus infection: Etiology, epidemiology, pathogenesis and immunoprophylaxis. *Vet. J.* 204(2):134–43.

Kelly, A. R., Jones, R. J., Gillick, J. C. and Sims, L. D. (1984) Outbreak of botulism in horses. *Eq. Vet. J.* 16:519–21.

Kessler, E., Matthes, H. F., Schein, E. and Wendt, M. (2003) Detection of antibodies in sera of weaned pigs after contact infection with Sarcoptes scabiei var. suis and after treatment with an antiparasitic agent by three different indirect ELISAs. *Vet. Parasitol.* 114(1):63–73.

Kielstein, P. and Rapp-Gabrielson, V. J. (1992) Designation of 15 serovars of *Haemophilus parasuis* on the basis of immunodiffusion using heat-stable antigen extracts. *J. Clin. Microbiol.* 30(4):862–5.

Kitching, R. P., Knowles, N. J., Samuel, A. R. and Donaldson, A. I. (1989) Development of foot-and-mouth disease virus strain characterisation--a review. *Trop. Anim. Health Prod.* 21:153–66.

Kucerova, Z., Jaglic, Z., Ondriasova, R. and Nedbalcova, K. (2005) Serotype distribution of *Actinobacillus pleuropneumoniae* isolated from porcine pleuropneumonia in the Czech Republic during period 2003–2004. *Vet. Med. – Czech.* 50:355–60.

Leiser, O. P., Corn, J. L., Schmit, B. S., Keim, P. S. and Foster, J. T. (2013) Feral swine brucellosis in the United States and prospective genomic techniques for disease epidemiology. *Vet. Microbiol.* 166(1–2):1–10.

Leuenberger, R., Boujon, P., Thür, B., Miserez, R., Garin-Bastuji, B., Rüfenacht, J. and Stärk, K. D. (2007) Prevalence of classical swine fever, Aujeszky's disease and brucellosis in a population of wild boar in Switzerland. *Vet. Rec.* 160(11):362–8.

Li, G., Chen, Q., Harmon, K. M., Yoon, K. J., Schwartz, K. J. and Hoogland, M. J. (2014) Full-length genome sequence of porcine deltacoronavirus strain USA/IA/2014/8734. *Genome Announc.* 2:e00278–14.

Lindsay, D. S., Blagburn, B. L. and Boosinger, T. R. (1987) Experimental *Eimeria debliecki* infections in nursing and weaned pigs. *Vet. Parasitol.* 25(1):39–45.

Magstadt, D., Madson, D., Arruda, P., Hoag, H., Sun, D., Pillatzki, A., Stevenson, G., Wilberts, B., Brodie, J., Harmon, K., Chong, W. and Main, R. (2014) Porcine epidemic diarrhea: Pathogenesis and viremia in weaned and neonatal pigs. *In Proc 22nd Annual Swine Disease Conference for Swine Practitioners*, pp. 34–7.McOrist, S. (2005) Defining the full cost of endemic porcine proliferative enteropathy. *Vet. J.* 170(1):8–9.

McOrist, S., Smith, S. H. and Green, L. J. (1997) Estimate of direct financial losses due to porcine proliferative enteropathy. *Vet. Rec.* 140: 579–81.

McOrist, S., Smith, S. H. and Klein, T. (1999) Monitored control programme for proliferative enteropathy on British pig farms. *Vet. Rec.* 144: 202–4.

Mebus, C. A., House, C., Ruiz Gonzalvo, F., Pineda, J. M., Tapiador, J., Pire, J. J., Bergada J., Yedloutschnig, R. J., Sahu, S., Becerra, V. and Sanchez-Vizcaino, J. M. (1993) Survival of foot-and-mouth disease, African swine fever, and hog cholera viruses in Spanish serrano cured hams and Iberian cured hams, shoulders and loins. *Food Micro.* 10:133–43.

Médici, K. C., Barry, A. F., Alfieri, A. F. and Alfieri, A. A. (2011) Porcine rotavirus groups A, B, and C identified by polymerase chain reaction in a fecal sample collection with inconclusive results by polyacrylamide gel electrophoresis. *J. Swine Health Prod.* 19(3):146–50.

Mittal, K. R., Higgins, R., Larivière, S. and Nadeau M. (1992) Serological characterization of *Actinobacillus pleuropneumoniae* strains isolated from pigs in Quebec. *Vet. Microbiol.* 32(2):135–48.

Moennig, V., Floegel-Niesmann, G. and Greiser-Wilke, I. (2003) Clinical signs and epidemiology of classical swine fever: A review of new knowledge. *Vet. J.* 165:11–20.

Nelson, M. I. and Vincent, A. L. (2014) Reverse zoonosis of influenza to swine: new perspectives on the human-animal interface. *Trends Microbiol.* 23(3):142–53.

Oh, Y., Han, K., Seo, H. W., Park, C. and Chae, C. (2013) Program of vaccination and antibiotic treatment to control polyserositis caused by Haemophilus parasuis under field conditions. *Can. J. Vet. Res.* 77(3): 183–90.

Oliveira, S., Pijoan, C. and Morrison, R. (2004) Evaluation of *Haemophilus parasuis* control in the nursery using vaccination and controlled exposure. *J. Swine Health Prod.* 12(3):123–8.

Olsen, S. C., Garin-Bastuji, B., Blasco, J. M., Nicola, A. M, and Samartino, L. (2012) Brucellosis In: Zimmerman, J. J., Karriker, L. A., Ramirez, A., Schwartz, K. J. and Stevenson, G. W. (eds), *Diseases of Swine*, 10th ed. Ames: Wiley-Blackwell Publishing.

Otake, S., Dee, S., Corzo, C., Oliveira, S. and Deen, J. Long-distance airborne transport of infectious PRRSV and *Mycoplasma hyopneumoniae* from a swine population infected with multiple viral variants. *Vet. Microbiol.* 145(3–4):198–208.

Pappas, G., Akritidis, N., Bosilkoviski, M. and Tsianos, E. (2005) Brucellosis. *N. Engl. J. Med.* 352:2325–36.

Pijoan, C., Morrison, R. B. and Hilley, H. D. (1983) Serotyping of *Pasteurella multocida* isolated from swine lungs collected at slaughter. *J. Clin. Microbiol.* 17:1074–6.

Piriou, L., Chevallier, S., Hutet, E., Charley, B., Le Potier, M. F. and Albina, E. (2003) Humoral and cell-mediated immune responses of d/d histocompatible pigs against classical swine fever (CSF) virus. *Vet. Res.* 34:389–404.

Ploug, T., Holm, S. and Gjerris, M. (2015) The stigmatization dilemma in public health policy-the case of MRSA in Denmark. *BMC Public Health.* 15:640.

Pogranichny, R. M., Yoon, K. J., Harms, P. A., Swenson, S. L., Zimmerman, J. J. and Sorden, S. D. (2000) Characterization of immune response of young pigs to porcine circovirus type 2 infection. *Viral. Immunol.* 13:143–53.

Ramirez, A., Rowe, E. W., Arruda, P. H. and Madson, D. (2014) Use of equine-origin antitoxins in piglets prior to exposure to mitigate the effects of *Clostridium difficile* infection – a pilot study. *J. Swine Health Prod.* 2014, 22(1):29–32.

Rimler, R. B. and Rhoades, K. R. (1987) Serogroup F, a new capsule serogroup of *Pasteurella multocida*. *J. Clin. Microbiol.* 25:615–18.

Ross, R. F. and Whittlestone, P. (1983) Recovery of, identification of, and serological response to porcine mycoplasmas. In: Tully, J. G. and Razin, S. (eds), *Methods in Mycoplasmology*. New York: Academic Press.

Rowland, R. R., Lunney, J. and Dekkers, J. (2012) Control of porcine reproductive and respiratory syndrome (PRRS) through genetic improvements in disease resistance and tolerance. *Front. Genet.* 3:260.

Rowland, R. R., Steffen, M., Ackerman, T. and Benfield, D. A. (1999) The evolution of porcine reproductive and respiratory syndrome virus: quasispecies and emergence of a virus subpopulation during infection of pigs with VR-2332. *Virology*. 5, 259(2):262–6.

Ruiz Gonzalvo, F., Carnero, M. E. and Bruyel, V. (1981) Immunological responses of pigs to partially attenuated ASF and their resistance to virulent homologous and heterologous viruses. In: Wilkinson, P. J. (ed.), *FAO/CEC Expert Consultation in ASF Research*. Rome, pp. 206–16.

Saif, L. J. (1999) Comparative pathogenesis of enteric viral infections of swine. *Adv. Exp. Med. Biol.* 473:47–59.

Saif, L. J., van Cott, J. L. and Brim, T. A. (1994) Immunity to transmissible gastroenteritis virus and porcine respiratory coronavirus infections in swine. *Vet. Immunol. Immunopathol.* 43(1–3):89–97.

Sanford, S. E. and Tilker, M. E. (1982) *Streptococcus suis* type II-associated diseases in swine: observations of a one-year study. *J. Am. Vet. Med. Assoc.* 181(7):673–6.

Schalk, A. F., Roderick, L. M., Foust, H. L. and Harshfield, G. S. (1935) Avian Tuberculosis: Collected Studies. *ND Agric. Exp. Stn. Tech. Bull.* 279.

Schelkopf, A. C., Magstadt, D. R., Arruda, B. L. Arruda, P. H. E., Zimmerman, J. J. Wetzell, T., Dee, S. A. and Madson, D. M. (2015) Lactogenic Protection against porcine epidemic diarrhea virus in piglets following homologous challenge. *In Proc 23rd Annual Swine Disease Conference for Swine Practitioners*, pp. 41–4.

Schliesser, T. and Weber, A. (1973) Untersuchungen über die Tenazitzt von Mykobakterien der Gruppe III nach Runyon in Sagemehleinstreu. *Zentralbl Veterinärmed (B)* 20:710–14.

Schwartz, K. J. 1997. Salmonellosis. Proc AASP, Quebec City, Quebec.

Sjölund, M., de la Fuente, A., Fossum, C. and Wallgren, P. (2009) Responses of pigs to a re-challenge with *Actinobacillus pleuropneumoniae* after being treated with different antimicrobials following their initial exposure. *Vet Rec.* 164:550–5.

Smith, H. J. (1986) Transmission of *Sarcoptes scabiei* in Swine by Fomites. *Can. Vet J.* 27(6):252–54.

Sorden, S. D. (2000) Update on porcine circovirus and postweaning multisystemic wasting syndrome. *J. Swine Health Prod.* 8:133–6.

Stuart, B. P. and Lindsay, D. S. (1986) Coccidiosis in swine. *Vet. Clin. North Am. Food Anim. Pract.* 2(2):455–68.

Thacker, E. J. and Minion, F. C. (2012) Mycoplasmosis. In: Zimmerman, J. J., Karriker, L. A., Ramirez, A., Schwartz, K. J. and Stevenson, G. W. (eds), *Diseases of Swine*, 10th ed. Ames: Wiley-Blackwell Publishing.

Thanawongnuwech, R., Thacker, E. L. and Halbur, P. G. (1997) Effect of porcine reproductive and respiratory syndrome virus (PRRSV) (isolate ATCC VR-2385) infection on bactericidal activity of porcine pulmonary intravascular macrophages (PIMs): in vitro comparisons with pulmonary alveolar macrophages (PAMs). *Vet. Immunol. Immunopathol.* 59(3–4):323–35.

Theil, K. W., Bohl, E. H., Cross, R. F., Kohler, E. M. and Agnes, A. G. (1978) Pathogenesis of porcine rotaviral infection in experimentally inoculated gnotobiotic pigs. *Am. J. Vet. Res.* 39:213–20.

Thoen, C. O., Jarnagin, J. L. and Richards, W. D. (1975) Isolation and identification of mycobacteria from porcine tissues: A three-year summary. *Am. J. Vet. Res.* 36:1383–6.

Thoen, C. O. (2012) Tuberculosis. In: Zimmerman, J. J., Karriker, L. A., Ramirez, A., Schwartz, K. J. and Stevenson, G. W. (eds), *Diseases of Swine*, 10th ed. Ames: Wiley-Blackwell Publishing.

Thomas, P. (2014) Field experiences using porcine epidemic diarrhea virus (PEDv) vaccine in herds experiencing endemic disease. *In Proc 22nd Annual Swine Disease Conference for Swine Practitioners*, pp. 38–42.

Torremorrell, M., Moore, C. and Christianson, W. T. (2002) Establishment of a herd negative for porcine reproductive and respiratory syndrome virus (PRRSV) from PRRSV-positive sources. *J. Swine Health Prod.* 10(4):153–60.

United States Department of Agriculture. (2011) *News Release.* [Online] Available from: http://www.fsis.usda.gov/wps/portal/fsis/newsroom/!ut/p/a0/04_Sj9CPykssy0xPLMnMz0vMAfGjzOINAg3MDC2dDbwMDIHQ08842MTDy8_YwNtMvyDbUREAzbjixQ!!/?1dmy&current=true&urile=wcm%3Apath%3A%2FFSIS-Archives-Content%2Finternet%2Fmain%2Fnewsroom%2Fnews-releases-statements-and-transcripts%2Fnews-release-archives-by-year%2Farchives%2FCT_Index919. [Accessed: 31st May 2016].

Vanier, G., Szczotka, A., Friedl, P., Lacouture, S., Jacques, M. and Gottschalk, M. (2006) *Haemophilus parasuis* invades porcine brain microvascular endothelial cells. *Microbiology.* 152:135–42.

Wagstrom, E. A., Yoon, K. J. and Zimmerman, J. J. (2000) Immune components in porcine mammary secretions. *Viral. Immunol.* 13:383–97.

Wang, L., Byrum, B. and Zhang, Y. (2014) Detection and genetic characterization of deltacoronavirus in pigs, Ohio, USA, 2014. *Emerg. Infect Dis.* 20:1227–30.

Waters, M., Savoie, A., Garmory, H. S., Bueschel, D., Popoff, M. R., Songer, J. G., Titball, R. W., McClane, B. A. and Sarker, M. R. (2003) Genotyping and phenotyping of beta2-toxigenic *Clostridium perfringens* fecal isolates associated with gastrointestinal diseases in piglets. *J. Clin. Microbiol.* 41:3584–91.

Wertheim, H. F., Nghia, H. D., Taylor, W. and Schultsz, C. (2009) *Streptococcus suis*: an emerging human pathogen. *Clin. Infect. Dis.* 48(5):617–25.

Wiseman, B., Harris, D. L., Glock, R. D. and Wilkins, L. (1989) Management of seedstock that is negative for *Haemophilus parasuis*. In *Proc 20th Annu Meet Am Assoc Swine Pract.*, pp. 23–5.

Wittmann, G. and Rziha, H. J. (1989) Aujeszky's disease (pseudorabies) in pigs. In: Wittmann, G. (ed.), *Herpesvirus diseases of cattle, horses and pigs.* Boston: Kluwer Academic Publishers.

World Health Organization. (2016) *Taeniasis/cysticercosis.* [Online] Available from: http://www.who.int/mediacentre/factsheets/fs376/en/. [Accessed: 26th May 2016].

World Organization for Animal Health and Food and Agriculture Organization of the United Nations. (2012) The Global Foot and Mouth Disease Control Strategy – Strengthening animal health systems through improved control of major disease. [Online] Available from: http://www.oie.int/doc/ged/D11886.PDF. [Accessed: 26th May 2016].

Yaeger, M., Funk, N. and Hoffman, L. (2002) A survey of agents associated with neonatal diarrhea in Iowa swine including *Clostridium difficile* and porcine reproductive and respiratory syndrome virus. *J. Vet. Diagn. Invest.* 14(4):281–7.

Zhang, J., Chen, Q., Thomas, J. and Gauger, P. (2015) Characterization of pathogenicity and cross-protective immunity of U. S. PEDVs. *In Proc 23rd Annual Swine Disease Conference for Swine Practitioners*, pp. 45–9.

Zhang, Q., Young, T. F. and Ross, R. F. (1994) Microtiter plate adherence assay and receptor analogs for *Mycoplasma hyopneumoniae. Infect. Immun.* 62:1616–22.

Chapter 2

Changing patterns of disease affecting pigs: Porcine Reproductive and Respiratory Syndrome (PRRS) and Porcine Epidemic Diarrhoea (PED)

Carla Correia-Gomes, Scotland's Rural College, UK

1 Introduction

Any disease in humans or animals has its own unique epidemiological pattern, that is, behaviour (occurrence, disappearance, re-occurrence as well as distribution of outbreaks and cases in space and time) (Blaha, 2000). The epidemiological behaviour is determined by the biological properties of the causative agent (such as pathogenicity and infectivity), the characteristics of the pathogen–host interaction (such as immune response, shedding and transmission pattern) and socio-economic conditions (such as structure of the industry, animal movements and herd size) (Thrusfield, 2005). The pig industry, like any other livestock sector, is not immune to disease, and recent examples of outbreaks that have affected the pig industry are the African Swine Fever outbreaks in Eastern European countries (Gogin et al., 2013), the Porcine Epidemic Diarrhoea (PED) outbreak in the United States (US) in 2013 (Stevenson et al., 2013) and the Porcine Circovirus Associated Disease (PCVAD) outbreak in Ontario, Canada, in 2004 (Carman et al., 2008). Some of these outbreaks have led to epidemics and then went on to become endemic in some countries, changing their disease epidemiological behaviour over time and space. Overall, in the last three decades, the pig industry worldwide has experienced several major disease epidemics

http://dx.doi.org/10.19103/AS.2017.0013.15

– all caused by viruses: Swine Influenza, Porcine Circovirus (PCV), Porcine Reproductive and Respiratory Syndrome virus (PRRSv) and Porcine Epidemic Diarrhoea virus (PEDv). Some of these viruses are not host-specific and can spread to other hosts, including humans (H1N1 influenza); others are highly host-specific (PRRSv and PEDv). The latter agents also share remarkable features exhibiting rapid rates of mutation and appear to have been associated with pigs for years to decades before highly pathogenic disease syndromes were manifested (Davies, 2015). In this chapter we focus on these two agents (PRRSv and PEDv) due to their importance in pig production worldwide and their evolution over the years. We also discuss the agent, transmission, clinical presentation and evolution and spread of these two viruses in the pig population both worldwide, and more specifically, the United Kingdom, in detail. This will provide readers with an overview of the complexity of these two agents and how that influences their clinical presentation and evolution over time and space. However, this is not an exhaustive literature review and as these agents are two of the most widely studied viruses, current and future research may update some of the findings reported here. As PED outbreaks have only been seen quite recently in the Western world, the US epidemic of 2013 and the current situation in Europe (including surveillance) are summarised in more detail.

2 Porcine Reproductive and Respiratory Syndrome (PRRS)

PRRSv is the major cause of reproductive and respiratory problems in pigs worldwide (Holtkamp et al., 2013). It was first described in the late 1980s as a mystery disease causing reproductive failure and respiratory disease in North America (Collins, 1991; Keffaber, 1989). A few years later a syndrome with similar clinical signs was observed in Western Europe (Terpstra et al., 1991; Wensvoort et al., 1991). Then the disease spread and appeared in Germany, the Netherlands, and then the rest of Western Europe (Wensvoort et al., 1991). In 1991, a large-scale laboratory investigation was conducted to search for the aetiological agent and it was found to be a virus (Wensvoort et al., 1991). In 1992 it was named PRRSv and its syndrome PRRS. Genomic sequence analysis revealed significant genetic differences between viruses isolated in Europe and North America, with only 60% nucleotide identity (Allende et al., 1999). These genomic differences confirmed the presence of two distinct genotypes, known today as Type 1 for the European isolates and Type 2 for the North American isolates. Over time the pattern of PRRS has changed from sudden outbreaks that spread rapidly through the pig population (i.e. epidemic stage), to frequent and constant occurrences (i.e. endemic stage). Today, PRRSv is endemic in all pork-producing countries with both genotypes distributed worldwide (Shi et al., 2010a).

2.1 The agent

PRRSv is a small, enveloped, single-stranded positive-sense RNA virus belonging to the Arteriviridae family, within the order Nidovirales (Benfield et al., 1992; Conzelmann et al., 1993; Wensvoort et al., 1991). The PRRSv genome is approximately 15 kb in length with a 5′cap and a 3′polyadenylated tail and it encodes at least ten open reading frames (ORFs) (Firth et al., 2011; Johnson et al., 2011; Snijder and Meulenberg, 1998; Wu et al., 2001). ORF1a and ORF1b cover the first three quarters of the PRRSv genome and encode two long non-structural polyproteins, pp1a and pp1ab, where the latter is expressed after a ribosomal frame shift (Snijder and Meulenberg, 1998). pp1a and pp1ab are proteolytically cleaved into at least 12 non-structural proteins (Snijder and Meulenberg, 1998; Fang and Snijder, 2010). ORFs 2–5 encode the membrane glyco-proteins, GP2–GP5, while ORF6 and ORF7 encode a non-glycosylated membrane protein (M) and the nucleocapsid (N) protein, respectively. Two small genes, ORF2b fully embedded in ORF2 and ORF5a partially overlapping ORF5, encode the non-glycosylated protein E and ORF5, a protein (Wu et al., 2001; Johnson et al., 2011; Firth et al., 2011). Within a production system, PRRSv infection predominantly occurs as a subclinical infection, participating as a co-factor in various polymicrobial disease syndromes, such as Porcine Respiratory Disease Complex and PCVAD (Chand et al., 2012). A common example of this is the noticeable increase in severity of enzootic pneumonia in grower/finisher pigs when they become infected with PRRSv. Although PRRSv is considered to be among the most genetically labile viruses, it is highly host-specific and has not been found to replicate in any species other than pigs (Butler et al., 2014). The genetic liability of PRRSv appears to play a central role in the difficulties involved in controlling the disease due to the regular emergence of genetic variants with limited heterologous immunity, which limits the vaccination success (Davies, 2015; Kappes and Faaberg, 2015; Perez et al., 2015; Wang et al., 2015). The PRRSv compromises the cellular immune response and damages mucosal surfaces. Primary virus replication takes place in local macrophages from where the virus rapidly spreads to lymphoid organs and lungs. Other tissues may also be infected, but not as commonly as these (Dietze et al., 2011). In the environment, PRRSv favours moist and cold conditions, at or below 20°C, with a pH range of 5.5–6.5 (Dietze et al., 2011).

PRRSv is differentiated into two genetically distinct genotypes: Type 1 (or European genotype) and Type 2 (or North American genotype). Type 1 is further divided into subtypes 1 (pan European), 2 and 3 (East European), while Type 2 is divided into nine lineages (Dietze et al., 2011) (Table 1). Lineages 1, 5, 8 and 9 are further divided into several sublineages (Jantafong et al., 2015), which are not presented in Table 1.

2.2 Transmission

PRRSv has been recovered from a variety of porcine secretions and excretions including blood, semen, saliva, faeces, aerosols, milk and colostrum (Rossow et al., 1994; Swenson et al., 1994; Wagstrom et al., 2001; Wills et al., 1997). Transmission of PRRSv primarily occurs via direct contact between the infected and naïve pigs (Rossow, 1998), including via semen from infected boars (Christopher-Hennings et al., 1995, Yaeger et al., 1993).

2.3 Clinical presentation

The clinical presentation and clinical signs of PRRS vary greatly between herds. Infection with PRRSv shows two different sets of clinical signs: reproductive and respiratory syndromes (Keffaber, 1989; Wensvoort et al., 1991). The clinical presentation of PRRSv infections depends on the age of the pigs at the stage of infection and on the pregnancy status and trimester of gestation of the infected sow/gilt (Rossow, 1998). The clinical presentation also varies with the type of PRRSv. Animals infected with mildly virulent isolates or Type 1 PRRSv showed transient pyrexia, dyspnoea and tachypnea, whereas those infected with highly virulent isolates (Type 2 PRRSv) showed induced laboured breathing, pyrexia, lethargy and anorexia. Furthermore, studies have reported that the impact on reproductive

Table 1 Division for Type 2 and Type 1 of PRRS into lineages and clades, respectively, based on phylogenetic analysis (using ORF5 sequences), with international distribution. The lineages and clades marked with an asterisk contain modified-live virus vaccine strains, with the name in brackets. This table was constructed based on Jantanfog et al. (2015) and Shi et al. (2010a,b)

PRRS type		Lineage	International distribution (not exhaustive)
Type 2		1	USA, Canada, Thailand
		2	USA, Canada
		3	Hong Kong, Taiwan, China
		4	Japan
		5* (Ingelvac MLV)	USA, Austria, Canada, Denmark, China, Japan, Poland, South Korea, Thailand
		6	USA
		7* (PrimePac)	USA
		8* (Ingelvac ATP)	USA, China, Canada, Italy, Thailand
		9	USA, Italy, Japan, Mexico
PRSS type		Clade	International distribution (not exhaustive)
Type 1	Subtype I	A* (Porcilis vaccine)	Pan Europe, USA, Thailand
		B	Austria, Italy, Belgium
		C	Spain, Germany, South Korea, Denmark
		D* (Pyrsvac, Amervac)	Spain, Hungary, Germany, Poland, China, Thailand
		E	Spain, Germany, Poland
		F	Poland
		G	Denmark, China
		H	Poland, Thailand, Germany, Spain, Belgium, Denmark, South Korea
		I	Germany
		J	Italy, Czech Republic
		K	Italy
		L	Italy
	Subtype II		Russia, Lithuania, Belarus
	Subtype III		Belarus

performance may be isolate-dependent (Halbur et al., 1995a,b; Mengeling et al., 1996). In sows, the usual clinical signs are inappetence, anorexia and reproductive disorders such as abortion, premature birth, birth of dead or weak piglets and foetal death with or without mummification. A less frequently observed sign is transient blue discolouration of the ears, abdomen or vulva (Terpstra et al., 1991). PRRSv infection in weaned pigs is characterised by fever, pneumonia, lethargy and failure to thrive (Rossow, 1998).

Gross lesions observed following PRRSv infection vary widely and may be dependent on the virus isolate, genetics of the infected pig and stress factors (e.g. environment and health status of the pig herd). Lung lesions vary from none to diffuse consolidation and are commonly complicated by lesions resulting from concurrent bacterial infections (Rossow, 1998). The pig's adaptive immune response against PRRSv is characterised by being delayed and defective (Beura et al., 2010). Following a natural infection, it takes at least three months to reach peak immunity levels and it does not appear to be solid enough to prevent reinfection, especially if the reinfection is caused by antigenically heterologous PRRSv strains (Murtaugh et al., 2002; Zuckermann et al., 2007). Once the virus has entered a herd, it tends to remain.

2.4 PRRS virus origin

The origin of PRRSv is still a mystery. Sophisticated tools of nucleotide sequencing and phylogenetic analysis have been used to investigate the phylogeny and evolution of PRRS viruses. Although estimates of time to most recent common ancestor vary widely, most studies point to divergence of the prototype viruses at least one century before the emergence of the clinical syndromes (Forsberg, 2005). A recent study estimated that Type 1 PRRS viruses found worldwide diverged approximately 100 years ago and appear to have been endemic in several populations without recognition of any clinical diseases (Nguyen et al., 2014). Furthermore, due to the genetic diversity of the Type 1 viruses, it was found that the Western European epidemic was caused by transfer of several lineages from the original reservoir (Forsberg et al., 2001). The likely origin of transfer was Eastern Europe, as Eastern European pigs harboured highly diverse Type 1 PRRSv lineages, which was shown to diversify before the Western Europe Type 1 PRRSv lineages (Stadejek et al., 2002). Thus, a plausible hypothesis is that Type 1 PRRSv originated in Eastern European pigs and was transmitted to Western European pigs following the increased East–West European pig trade that began in the late 1980s (Stadejek et al., 2002).

2.5 Evolution over time

PRRSv is endemic in the majority of pork-producing countries and both genotypes are distributed worldwide (Shi et al., 2010a). A Type 1 PRRSv isolate was identified in a pig herd in the United States in 1999 (Fang et al., 2004; Ropp et al., 2004) and has been introduced to Canada (Dewey et al., 2000), South Korea (Lee et al., 2010), China (Jiang et al., 2000) and Thailand (Thanawongnuwech et al., 2004). Type 2 PRRSv was introduced in Europe in 1996 probably through the use of a live attenuated vaccine that reverted back to virulence in Danish pig herds (Botner et al., 1997; Madsen et al., 1998). Subsequently, Type 2 PRRSv has been reported sporadically in the rest of Europe (Stadejek et al., 2013). Type 2 PRRSv also circulates throughout most of Asia (An et al., 2007; Feng et al., 2008). At first it was believed that Type 2 PRRS viruses were more genetically diverse, while Type 1 PRRS viruses exhibited a lower degree of variation (Meng et al., 1995; Kapur et al., 1996; Suarez et al., 1996). This perception changed following extensive sampling of Type 1 viruses, which revealed an even greater diversity among European isolated viruses than North American isolated viruses (Stadejek et al., 2002, 2006, 2008). Country-specific clusters were formed by viruses isolated in Great Britain, Italy and Denmark (Forsberg et al., 2002, Frossard et al., 2013). Further sampling and sequencing revealed even more diversity among the Type 1 PRRSv isolated in Europe, particularly those isolated in Eastern Europe which showed high divergence and diversity compared to Western European isolates. The European genotype was then further separated into three subtypes (1–3) (Table 1). The poor geographic structuring in the phylogeny of Type 1 is evident in Table 1, as each clade is associated with multiple countries (except for Italian lineages). This indicates that frequent inter-country virus flow occurs in Western Europe.

Type 1 PRRSv has been found in five countries outside Europe and so far only subtype 1 has been detected (Shi et al., 2010a). The diversity of Type 1 PRRSv isolated outside Europe suggests multiple introductions of PRRSv into these countries, confirmed by the appearance of viruses from the same country in different clades (Shi et al., 2010a) (Table 1). The Type 2 viruses isolated in North America showed the greatest diversity as they were distributed throughout seven of the nine lineages. Lineages 3 and 4 only included Asian isolated viruses. Within the seven lineages constituting the North American viruses, Asian and European isolated viruses were also represented. Lineage 5 is the cluster with most countries represented, including Denmark. This lineage also harboured the Ingelvac MLV vaccine, which is the preferred vaccine against Type 2 PRRSv in most European countries (Stadejek et al., 2013) and the Type 2 protogenotype VR-2332 (Collins et al., 1992). The European interest for the diversity of Type 2 PRRSv isolated in Europe has been minor, since to date, the majority of Type 2 PRRSv circulating in the European pig herds were vaccine-like (>98% identical) (Greiser-Wilke et al., 2010; Stadejek et al., 2013). However, in Hungary two Type 2 PRRSv isolates showed major divergence compared to the vaccine strain on the basis of partial ORF5 sequences and the origin of these viruses could not be explained (Balka et al., 2008). From time to time highly virulent PRRSv strains evolve, which results in poor welfare for the pigs and huge economic losses for the pig industry (Gauger et al., 2012, Han et al., 2006, Karniychuk et al., 2010, Neumann et al., 2005).

In China in 2006, a severe 'high fever' disease occurred in several pig farms and subsequently spread to almost half of China (Tian et al., 2007). The epidemic persisted for several months partly due to the limited knowledge of the aetiological agent (Tian et al., 2007). The aetiological agent was found to be PRRSv from extensive and systematic investigations, and polymerase chain reaction (PCR) and immunohistochemistry experiments (Tian et al., 2007). Today the atypical PRRSv strain is referred to as highly pathogenic PRRSv (HP-PRRSv). Shortly after the epidemic of HP-PRRSv in China, this strain rapidly spread to pigs in Southeast Asia and nearby countries (An et al., 2011). Phylogenetic analysis of HP-PRRSv ORF5 sequences in these countries revealed that they were very similar to typical Chinese PRRSv isolates (Tian et al., 2007) and they were clustered on Type 2, lineage 8, sublineage 8.7 (Jantafong et al., 2015). In the phylogenetic tree, HP-PRRSVs were divided into four clades, namely A, B, JXA1-like and GXFCH08-like (Jantanfog et al., 2015). Clade A, the so-called Southeast Asian clade, comprises HP-PRRSV isolates from various countries in Southeast Asia, including Thailand, Lao PDR, Cambodia and Vietnam. HP-PRRSV isolates in this clade are closely related to SX2009, the Chinese PRRSv isolated in 2009 (Zhu et al., 2011). The SX2009-like PRRSv was introduced into Vietnam and Thailand in 2010 and has become the predominant strain in these regions since then. The majority of the viruses in clade B are HP-PRRS viruses isolated from different provinces in Thailand from 2010 to 2013. The viruses in clade B are closely related to 09HEN1, the HP-PRRSv that emerged in 2009 in Henan, which is located in Central China (Zhu et al., 2011). The JXA1-like clade consists of HP-PRRS viruses from Thailand, the Philippines and China and is closely related to the JXA1 strain, a Chinese HP-PRRSv prototype isolated in 2006 (Tian et al., 2007). The last clade, called GXFCH08-like, comprises HP-PRRSv isolates from Cambodia, China, Vietnam and Thailand that are closely related to GXFCH08, a Chinese

HP-PRRSv that appeared in 2008. The two Southeast Asian HP-PRRS viruses, including NA/TH/NMA031/2008 (KF698665) from Thailand, are grouped in the GXFCH08-like clade.

In the winter of 2007, a Belarusian farm confirmed positive for PRRSv experienced abortions; birth of mummified, dead and weak piglets; high mortality rate before weaning; respiratory disorders and mortality (up to 70%) in growing pigs (Karniychuk et al., 2010). PRRSv was isolated from lung tissues obtained from weak-born piglets, and from the ORF5 and ORF7 sequences, the virus was classified as belonging to PRRSv Type 1 subtype 3 and the virus strain was designated 'Lena' (Karniychuk et al., 2010). The complete genome of Lena has been sequenced and alignments of Lena with the Type 1 subtype 1 reference sequence showed variations over the entire genome (Van Doorsselaere et al., 2012). Three experimental infection studies with subtype 3 PRRS viruses have been conducted (Karniychuk et al., 2010; Morgan et al., 2013; Weesendorp et al., 2013). Two of these studies were conducted using the Lena strain and Type 1 PRRS viruses belonging to subtype 1 for comparison (Karniychuk et al., 2010; Van Doorsselaere et al., 2012; Weesendorp et al., 2013). The study by Karniychuk (2010) showed that Lena induced high fever, lasting for several weeks, viremia detectable for four weeks post-infection, and gross lesions such as fibrinous pleuropneumonia, pericarditis and peritonitis. The second infection study with Lena showed less severe clinical signs compared to the first study; however, the immune response from pigs inoculated with Lena showed differences compared to Type 1 subtype 1 PRRS viruses (Weesendorp et al., 2013). The third experimental infection study with a Type 1 subtype 3 PRRSv SU1-bel (but distinct from Lena) was performed (Morgan et al., 2013), and the results indicated increased clinical and pathological effects of the SU1-bel strain caused by an enhanced inflammatory immune response (Morgan et al., 2013). From these studies it is evident that Type 1 subtype 3 PRRSv is able to induce severe disease in pigs (Karniychuk et al., 2010; Morgan et al., 2013; Weesendorp et al., 2013).

3 PRRS virus in the United Kingdom

Surveillance in Great Britain has been done through the Veterinary Investigation Diagnosis Analysis (VIDA) system since 1999. This system collates data from voluntary submissions from farmers and their vets to Animal and Plant Health Agency (APHA) Regional Laboratories and Scotland's Rural College Disease Surveillance Centres (Gibbens et al., 2008). Carcass and non-carcass submissions are submitted through private veterinary surgeons for laboratory testing and diagnostic investigation. Individual results are sent back to the private veterinary surgeon and the data contribute to surveillance for the detection of new and emerging threats, including significant changes in endemic disease trends. The trends are reported back regularly to the industry and academics (APHA, 2016b). From 2004 to 2015 the PRRS in pigs as percentage diagnosable submissions has increased (Fig. 1). Most of the PRRS cases were systemic disease due to PRRS, followed by pneumonia associated with PRRS (APHA, 2016c). However these values are an underestimation of the real number of cases as many producers and veterinarians only submit a low percentage of cases with disease. At the seasonal level a rise in PRRS cases in winter is usually observed in Great Britain (APHA, 2016b). The same system recently reported that sequences of ORF5 were obtained from 29 PRRS viruses detected in the 2015–16 Great Britain diagnostic submissions. All strains were Type 1 with no Type 2 ever having been detected in Great Britain pigs. The change in diversity follows the expected

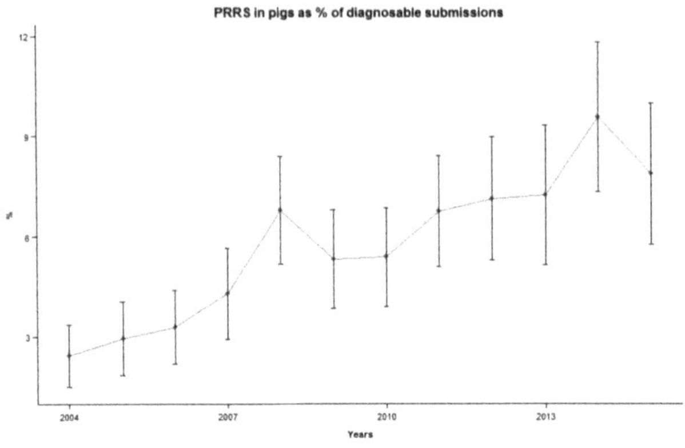

Figure 1 PRRS in pigs in GB as percentage of diagnosed submissions from 2004 to 2015 (data source: APHA, VIDA data set).

trend with ever-increasing branch lengths being seen as the existing viruses continue to evolve in pigs. According to APHA no entirely new branches have appeared in the last year, suggesting no new incursions of 'foreign' virus strains (APHA, 2016a).

Surveys to estimate the prevalence of the disease in Great Britain or the United Kingdom are rare. A cross-sectional study of 103 British pig herds was conducted in 2003–04 to investigate the between- and within-herd variability of PRRSv antibodies (Evans et al., 2008). Fifty pigs were tested per farm. Thirty-five herds (34.0%) were seronegative, 41 (39.8%) were seropositive and 27 (26.2%) were vaccinated. The authors reported that herds were more likely to be seronegative if they had <250 sows [odds ratio (OR) 3.86 (95% confidence interval (CI): 1.46–10.19)] and if the nearest pig herd was ≥2 miles away [OR 3.42 (95% CI: 1.29–9.12)]. Sixteen seropositive herds had only seropositive adult pigs. A UK-wide survey in 2013 tested several pigs at slaughter to PRRSv, *Salmonella spp.*, *Yersinia spp.* and *Toxoplasma gondii* (Powell et al., 2016). Most of the pigs were from farms in England (81.7%), followed by Northern Ireland (13.4%), Scotland (4.5%) and Wales (0.3%). The majority of samples were from pigs aged <12 months and from finishing only or farrow-to-finish farms. A total of 621 plasma samples were tested for antibodies to PRRSv, of which 362 were seropositive giving a seroprevalence of antibodies to PRRSv of 58.3% (95% CI: 53.1–63.4). The lowest seroprevalence was found in pigs aged >12months (32.1%, 95% CI: 15.0–49.3) and the highest in pigs aged <6months (68.5%, 95% CI: 54.5–82.5) (P = 0.002). All these results confirm that PRRS is endemic in the United Kingdom.

4 Porcine Epidemic Diarrhoea (PED)

PEDv causes acute diarrhoea, vomiting, dehydration and high mortality in neonatal piglets, resulting in significant economic losses (Jung and Saif, 2015). Unlike PRRSv, which was previously an unknown agent, PEDv is a familiar pig pathogen that has increased in geographical range and clinical severity (Davies, 2015).

PEDv first appeared in the United Kingdom (Wood, 1977) and Belgium (Pensaert and de Bouck, 1978) in the early 1970s. The virus was classified in the family Coronaviridae

(Pensaert and de Bouck, 1978). Subsequently, in the 1980s and 1990s, PEDv was identified as a cause of severe epidemics in Japan and South Korea (Kweon et al., 1993; Takahashi et al., 1983). Despite extensive application of PEDv vaccines, PED has remained endemic in South Korea (Park et al., 2013). During the 1980s and 1990s in Europe, outbreaks of PED appeared infrequently, but the virus continued to spread and persisted in an endemic form in the pig population. Subsequent serological surveys showed a low to moderate prevalence of PEDv in European pigs (Carvajal et al., 1995; Van Reeth and Pensaert, 1994). The prevalence of PEDv in European pigs then declined greatly for unknown reasons. Outbreaks of PED were observed only sporadically in Europe: in the Netherlands in 1989–91 (Pijpers et al., 1993); in Hungary in 1995 (Nagy et al., 1996) and in England in 1998 (Pritchard et al., 1999). However, a typical epidemic outbreak of PED was identified in Italy in 2005–06 (Martelli et al., 2008). In Thailand in 2007–08, several outbreaks of severe PED were reported with Thai PEDv isolates in the same clade phylogenetically as the Chinese strain JS-2004-2 (Puranaveja et al., 2009). In China in 2010–12, severe PED outbreaks in seropositive pigs were reported in different regions (Li et al., 2012; Sun et al., 2012; Wang et al., 2013). For almost two decades since PEDv first emerged in China, many pig herds have been vaccinated with the prototype strain CV777-inactivated or related vaccines. However, the moderate to high mortality of suckling piglets in vaccinated herds indicates the low effectiveness of the CV777 vaccines (Li et al., 2012). The PED outbreaks in China, in 2010–12, were caused by both classical and new PEDv variant strains that differ genetically from the prototype CV777 (Wang et al., 2013).

4.1 The agent

PED viruses are enveloped viruses with a large, capped and polyadenylated RNA genome of approximately 28 000 nucleotides (Song and Park, 2012). PED viruses are members of the genera *Alphacoronavirus* in the family Coronaviridae of the order Nidovirales (Jung and Saif, 2015). They belong to the same genus as other coronaviruses which infect pigs [Transmissible Gastroenteritis Virus (TGEv) and its respiratory variant, Porcine Respiratory Coronavirus], dogs (canine coronavirus), cats (feline infectious peritonitis virus), humans (human coronavirus 229E or human coronavirus NL-63) and bats (Vijaykrishna et al., 2007). They have an electron-dense core with central halo and club-shaped projections of approximately 20 nm length radiating from the core (Pensaert et al., 1981). The virus is ether- and chloroform-sensitive and does not hemagglutinate erythrocytes from a wide variety of species (Pensaert, 1999). PEDv is stable under low temperatures, but adversely affected by high temperatures. It survives between pH 5.0 and 9.0 at 4°C while only between pH 6.5 and 7.5 at 37°C. It can survive for at least 28 days in slurry at 4°C, 7 days in contaminated dry feed at 25°C or 14 days in contaminated wet feed at 25°C (EFSA, 2014). These data indicate that PEDv will be inactivated by acidic or alkaline disinfectants if they are applied for a certain period at a higher temperature (>37°C) (Jung and Saif, 2015).

4.2 Transmission

The faecal–oral route is the main means of PEDv transmission, although the possibility of airborne transmission has also been suggested (Alonso et al., 2014). Viral shedding in faeces starts on post-infection day one or two and continues for a period of 7–10 days (Carvajal et al., 1995; Song et al., 2006), although it can extend up to 36 weeks in some animals (Crawford et al., 2015; Madson et al., 2014). The transmission of the infection is facilitated by the high viral load in faeces from infected animals (Jung et al., 2014; Sun et al., 2014) as well as by the minimum infectious dose required to infect naïve pigs (EFSA, 2014). Its resistance in the environment favours the indirect transmission by different faeces-contaminated fomites such as transport vehicles (Lowe et al., 2014), feed (Dee et al., 2014), clothing or footwear. Another possible reservoir for PEDv includes carriers, such as older pigs with asymptomatic infection, in which the virus spreads subclinically. Previous studies showed a low to moderate detection rate (23–41%) of PEDv RNA in milk samples of affected, lactating sows (Li et al., 2012; Sun et al., 2012), suggesting that sow milk might be a potential route for the vertical transmission of PEDv. However infectious PEDv in these samples has not been demonstrated and contamination with faecal material cannot be excluded (Carvajal et al., 2015). The role that vectors play in the transmission of PEDv has also been investigated. So far, there has been no evidence of PEDv replication in non-porcine hosts, including rodents (Kamau et al., 2010). Another study demonstrated a significant detection rate of PEDv RNA in acute serum samples (55–100%) of experimentally infected piglets or naturally infected grower pigs (Jung et al., 2014). Whether pork plasma used as a feed additive could be a transmission source of PEDv remains questionable, since discrepant results were reported in two different infection studies that investigated whether spray-dried porcine plasma that had tested positive for PEDv RNA was infectious in seronegative pigs (Opriessnig et al., 2014; Pasick et al., 2014).

5 Epidemiological presentation of PED

Two epidemiological presentations of PED have been described on farms: epidemic outbreaks or endemic presentation (this is normally due to concomitant infections with bacteria). Detailed clinical disease due to epidemic PED outbreaks was documented in several countries (Martelli et al., 2008; Pensaert and de Bouck, 1978; Puranaveja et al., 2009; Stevenson et al., 2013; Sueyoshi et al., 1995; Wood, 1977). The clinical outbreaks on seronegative farms were characterised by a sudden epidemic of severe diarrhoea and/or vomiting, accompanied by anorexia and significantly reduced appetite, in pigs of all ages. The severity of clinical signs and mortality appeared to be inversely related to the age of the pigs (Shibata et al., 2000). In weaner to finisher pigs and pregnant sows, clinical signs are self-limiting within 5–10 days after the onset of disease and are not as severe as those of nursing piglets under two weeks of age (Martelli et al., 2008; Puranaveja et al., 2009). When pregnant sows become immune after exposure to the virus, they protect their offspring by lactogenic immunity. The interval between onset and cessation of the disease is generally 3–4 weeks (Puranaveja et al., 2009); however, clinical signs mainly develop in the seronegative lactating sows and their suckling piglets. In farrowing herds, morbidity can approach 100% in piglets, but varies in sows. Mortality of piglets less than two weeks of age can exceed 95% (50% on average) at 3–5 days after onset of severe watery diarrhoea and/or vomiting. While in weaned pigs mortality rates are typically only 1–3% (Martelli et al., 2008; Stevenson et al., 2013). No mortality associated with PED is usually observed among adult pigs (Carvajal et al., 2015). Field observations on epidemic PED in the United States in 2013 and the United Kingdom in 1976–77, as well as experimental findings (Pensaert and de Bouck, 1978), suggest that the incubation period of PEDv before clinical signs are detected varied, ranging from 1 to 7 days (Stevenson et al., 2013) or 5 to 8 days (Wood, 1977), respectively. The endemic presentation of PED affects mainly post-weaning piglets that have lost their lactogenic immunity as well newly introduced seronegative gilts (Carvajal et al., 2015). It causes severe diarrhoea in seronegative gilts and 6–10-week-old pigs, and was mild or absent in nursing and weaning pigs with no mortality (Pijpers et al., 1993). Another typical endemic PED has been manifested in Korean pig farms. Korean farms have employed live or inactivated PEDv vaccines using three Korean strains, DR13, KPEDV-9 and SM98-1, or a Japanese strain P-5V. Studies reported that recent prevalent Korean PEDv field isolates are closely related to Chinese strains and differ genetically from the four vaccine strains used in Korea and the prototype CV777 (Park et al., 2013). This divergence of historic vaccine and recent field PEDv strains may contribute to the reduced efficacy of the vaccines, causing difficulty with eradication of PEDv from pig farms with endemic PED in the Korean pig population. Like endemic TGE, PEDv-related mortality and morbidity of nursing piglets passively immunised are lower than those seen in seronegative pigs (Bohl et al., 1978). Apart from differences in the virulence of PEDv strains, many other parameters including management, immune status of the population and herd sanitary status could also explain variations in the clinical outcome of PED outbreaks (Stevenson et al., 2013). Endemic PED is manifested mainly in weaned pigs (Pijpers et al., 1993), and the severity of clinical disease in nursing piglets may be exacerbated by co-infections of other enteropathogens, such as *Escherichia coli* (Turgeon et al., 1980, Wang et al., 2013), or viruses including PCV type 2 (PCV2), TGEv, rotavirus (Hirai et al., 2001; Jung et al., 2006; Wang et al., 2013) and more recently porcine delta coronavirus (PDCov) and mammalian orthoreovirus 3 (MRv3). The latter two viruses have been detected in faecal samples collected from PEDv-positive farms in the United States. PDCov has been associated with mild to moderate diarrhoea in experimentally inoculated naïve suckling piglets (Chen et al., 2015), while MRv3 caused severe diarrhoea with 100% mortality in 3-day-old piglets (Thimmasandra Narayanappa et al., 2015).

5.1 PEDv epidemic in the United States in 2013–2014

An outbreak of PEDv infection occurred in the United States in Iowa in April 2013 and within one year had spread all over the country and to neighbouring countries, such as Canada and Mexico (Vlasova et al., 2014). Within one year, PED outbreaks had also occurred across the Pacific Ocean in Korea and Japan (Cho et al., 2014; Lin et al., 2014). The US PEDv strains identified during the initial outbreak in 2013 were closely related genetically to the Chinese strains (China/2012/AH2012) reported in 2011–12 (Chen et al., 2014; Huang et al., 2013), indicating the emergence of AH2012-like Chinese PEDv strains in the United States. The US-like PEDv strains were also found in diarrhoeic piglets in South Korea and Taiwan during late 2013 and early 2014 (Cho et al., 2014; Lin et al., 2014), although whether Chinese or US PEDv strains could have been transmitted to pigs in South Korea and Taiwan is unknown. Within a year since the first outbreak, other novel US PEDv strains (OH/OH851) with multiple deletions and insertions in their S gene, which clustered closely with Chinese strain HBQX-2010 or CH/ZMZDY/11, rather than AH2012, were found to possess low nucleotide identity in their 5′-end S1 region (first 1170 nucleotides) and high nucleotide identity in the remaining S gene, compared to the major US PEDv strains (Vlasova et al., 2014; Wang et al., 2014). Possible recombination events involving strain(s) from China may have contributed to a rapid evolution of US PEDv strains and the emergence of multiple variants, complicating the molecular epidemiology of US PEDv strains (Tian et al., 2014). Remarkably, another PEDv variant, which has a large 197 amino acid (aa) deletion in the N-terminal portion of the S protein of major cell-cultured

US PEDv strains, such as PEDv strain TC-PC22A (GenBank accession number KM392224), has emerged only one year after the first outbreak (Oka et al., 2014). Another PEDv variant with a large (204 aa) deletion at positions 713–916 of the S protein was identified among Korean PEDv strains (Park et al., 2014). Recently a root cause investigation was carried out to provide a possible answer to the question of how PED had entered the United States (Scott et al., 2016). The most plausible scenario identified was that flexible intermediate bulk containers ('feed totes') used to transport bulk feed served as fomites for movement of PED viruses, as they were frequently reused, and were not likely to have been cleaned or disinfected in a manner conducive to eliminating viruses.

In 2016, at the end of the third year of the PEDv outbreak, most US pig herds had been infected and an endemic disease phase with much lower, but still chronic losses is observed. Current data from the United States show that there has been a steady number of cases recorded; however, the numbers are lower than when it first emerged in 2013 (AASV, 2017). In Canada, a small number of cases on both breeding and finishing units were diagnosed in May and June 2016 (Anonymous, 2016).

5.2 Current situation in Europe at the end of 2016

Recent PED outbreaks have been reported in Europe in Germany (Hanke et al., 2015), Italy (Alborali et al., 2014), the Netherlands (Van der Woof et al., 2015), Austria (Steinrigl et al., 2015), Belgium (Theuns et al., 2015), France (Grasland et al., 2015), Portugal (Mesquita et al., 2015), Spain (Laranjo et al., 2015), Slovenia (Toplak et al., 2015) and Ukraine (Dastjerdi et al., 2015). In Europe PED is only notifiable in Finland, France, Ireland, Sweden and United Kingdom (EFSA, 2016). The European Food Standards Authority (EFSA) has set up a network to monitor PED. This network, representing several countries, voluntarily reported 245 case reports where pig herds met the PED case definition and 71 pig herds with PCR confirmation of PEDv-genome (EFSA, 2016). Based on this network the first confirmed PED case report in Europe was in March 2014 and most of the case reports were reported in the beginning of 2015. This could be due to an increasing number of clinical cases, increased vigilance for signs or changes in laboratory protocols and/or the range of tests applied to samples. Between May and August 2015, the number of PED case reports decreased, and then a slight increase was observed in September 2015. As the disease is not notifiable to the European Union (EU), any temporal distribution should only be considered to be indicative as there are likely to be infected herds which have not been reported or tested. Forty-three per cent of case reports were finisher herds and 31% were grower–producer herds; all pig herds were raised indoors with only 3% having outdoor access (EFSA, 2016). As clinical signs are more noticeable in piglets, it is surprising that the majority of herds for which confirmed cases were reported are finishers; however, this production system is the dominant type of pig production in the reporting countries; moreover, frequent movements of animals both on and off farm in rearing herds may increase the risk of PEDv introduction. Based on the low number of case reports that have been submitted for the 2014–15 period, the number of PED-affected herds in the EU is very low (EFSA, 2016).

Phylogenetic analysis of full-length genome sequences from EU member states indicated that the reference INDEL strain USA/OH851/2014 shared greater than 99% sequence identity with the German, French, Belgian and Spanish strains (Grasland et al., 2015; Hanke et al., 2015; Stadler et al., 2015; Theuns et al., 2015). The German INDEL strain GER/L00719/2014 isolated from a fattening pig farm in May 2014 (Hanke et al., 2015) had 99.9% identity with strain KR011756 from France (Grasland et al., 2015). The Belgian strain (BEL/15V010/2015) had 99.9% identity with L00721 also isolated in Germany in 2014 and 97% identity with the older European strain CV777 (Theuns et al., 2015). In Italy, comparative analysis of partial sequences of the RdRp and M genes and the total spike (S1) gene of the circulating strains responsible for the outbreaks in 2014 showed a high identity with the S-INDEL strain USA/OH851/2014 from the United States (98.7, 99.8 and 99.3–99.5%, respectively) and with the strains detected in Germany (100, 100 and 99.7%, respectively) (Boniotti et al., 2016). This differs from the PEDv strains previously circulating from 2007 to 2012 in Italy, which showed a high genetic variability (EFSA, 2016). In the Netherlands, a partial sequence analysis of 11 PCR products from 11 farms showed that all viruses were > 99.5% similar to the German isolate and the US INDEL USA/OH851/2014 strain (EFSA, 2016). In Austria, sequencing of the complete or nearly complete S gene (>4 kb) obtained from affected farms showed over 99.5% sequence identity to other recent PEDv strains from Western and Central Europe (Steinrigl et al., 2015). In Portugal, the analyses of amplified sequences of the S gene shared a very high (99.0%) identity with the INDEL strain USA/OH851/2014 and were identical (100%) to the strains reported in Germany (Mesquita et al., 2015). However the strain (Ukraine/Poltava01/2014) isolated from an outbreak in a farm in Poltrava in the Ukraine in 2014 had 99.8% sequence identity to strains USA/Kansas29/2013 and USA/Colorado30/2013; therefore it was identified as a non-IDEL strain similar to the strains isolated during the US 2013 outbreaks, and had only shown 98.5% sequence identity with the German strains and 96.5% with the CV777 strain (Dastjerdi et al., 2015). This Ukraine strain is, therefore, distinct from the strains currently circulating in the rest of Europe and poses a risk to the pig population naïve to this novel PEDv variant. Recently, two studies reported the detection of previously undescribed recombinant TGEv-PEDv virus strains circulating in samples from 2009 in Italy (Boniotti et al., 2016) and in 2012 in Germany with clinical disease reported to be similar to cases of PED in both countries (Akimkin et al., 2016). The recombinant probably originated in a country in which both PEDv and TGEv are endemic (of which Italy is one) and reflected the potential for natural recombination among coronaviruses. In countries where PED, TGE or both of these diseases are reportable, correct diagnosis and reporting might be difficult (Akimkin et al., 2016).

5.3 Surveillance in Europe

Austria, Belgium, Denmark, Estonia, Finland, France, Germany, Ireland, Italy, the Netherlands, Norway, Sweden and United Kingdom have surveillance and/or monitoring activities in place for PED. Among these countries, Denmark, Finland, Ireland, Norway and United Kingdom did not confirm any PED cases in their country's pigs between June 2013 and September 2015 (EFSA, 2016). Activities are based on active or passive surveillance around serological testing or PCR testing at farm or slaughterhouse level. Clinical investigations of suspected samples have been adopted by several countries (Austria, Belgium, Denmark, Estonia, Finland, France, Ireland, Italy, the Netherlands and United Kingdom). Sweden had an active surveillance programme for PED between 1993 and 2005, but is not now performing surveillance for PED. Denmark had an active surveillance programme using an in-house ELISA to detect PEDv-specific antibodies which tested approximately 2500 pig sera each year between 2000 and 2006, but no positive results were reported during this period. In the Netherlands, in 2014, a retrospective serological survey of 838 serum samples from sows obtained from herds and slaughterhouses resulted in two positive samples as determined by an indirect ELISA and subsequent confirmation using a virus neutralisation test. In Belgium, in 2014, sow serum samples from 12 herds randomly selected from each province were tested using an immunoperoxidase monolayer assay method and no samples tested positive. France conducted a prevalence survey in 2014 of serum samples from 300 breeding sows in 30 herds using an indirect ELISA test. The test performance parameters were estimated from pig farm samples from Canada. After correcting for herd clustering and test sensitivity and specificity the prevalence of seropositive sows was 3.6% with a 95% CI of 1.55–6.47% (EFSA, 2016). There is uncertainty about the sensitivity and specificity of existing antibody tests when used for the assessment of exposure to emerging PEDv strains. Many tests were developed using strains circulating in Europe in the 1970s and 1980s [for example, CV777 isolated from a pig breeding farm in Belgium in 1977 (Pensaert and de Bouck, 1978)] and the ability to detect new strains from outside Europe requires further investigation (EFSA, 2016).

5.4 PEDv in the United Kingdom

PEDv has not been detected in UK pigs since 2002 and TGE has not been detected in UK pigs since 1999 (APHA, 2016a). PED has been a notifiable disease in England since December 2015 [The Specified Diseases (Notification) (Amendment) (England) Order 2015 No. 2023) and in Scotland since March 2016 (The Specified Diseases (Notification) Amendment (Scotland) Order 2016 No. 41]. A similar approach has been followed by both countries: suspect or confirmed disease must be reported and control of disease will be industry-led. The new law makes it mandatory for any suspected case of PED in Scotland and England to be notified to the relevant authority. There have been no suspect PED cases in either country since it was made notifiable. Between June 2013 and July 2016, 441 submissions from outbreaks of diarrhoea have tested negative for PEDv by PCR (APHA, 2016b). In 2013, UK conducted a seroprevalence survey for PED using serum samples obtained at the slaughterhouse (Cheney et al., 2014). The results showed that a small proportion (9%) of the pigs sampled at the slaughterhouse were seropositive to PEDv using a blocking ELISA. The blocking ELISA test used for this survey (van Nieuwstadt and Zetstra, 1991) is known to be capable of detecting antibodies to virulent US and historic Great Britain PEDv strains; however, the authors (Cheney et al., 2014) identified some uncertainties with respect to the antibodies and false-positive results have been reported to occur.

6 Conclusion

The unexpected appearance and evolution of highly infectious agents that can cause explosive outbreaks with high mortality rates raise questions about the complex issues behind such outbreaks and more critically, what can be done to prevent similar events. Recently, PRRS and PED demonstrated how vulnerable the world pig industry is to incursions of a new pathogen and recrudescence or re-emergence of a known pathogen due to extensive commercial trading of live animals, their products, feed and fomites. In today's globalised context the risk of pathogen transfer cannot be reduced to zero and future adverse events are inevitable. Exposure to hazards, however, can be limited if the industry adopts effective measures to increase their biosecurity levels, within and between herds.

7 Where to look for further information

There are a number of online documents and review papers that provide further reading about PRRS and PED such as:

EFSA Scientific Opinion about PRRS ('Assessment of listing and categorisation of animal diseases within the framework of the Animal Health Law (Regulation (EU) No 2016/429): porcine reproductive and respiratory syndrome (PRRS)')
http://onlinelibrary.wiley.com/doi/10.2903/j.efsa.2017.4949/epdf

OIE Disease Information Sheets about PRRS:
http://www.oie.int/fileadmin/Home/eng/Media_Center/docs/pdf/Disease_cards/PPRS-EN.pdf

The National Pork Board's new Porcine Reproductive and Respiratory Syndrome (PRRS) virus research booklet contains Checkoff-funded PRRS research from 1997 to 2016 that can help producers, veterinarians and researchers who would like to learn more about how to control this costly virus:
http://www.pork.org/wp-content/uploads/2017/05/prrsbook2016.pdf

The European Food Safety Authority (EFSA) has recently published a scientific report about epidemiological data on PED ('Collection and review of updated scientific epidemiological data on porcine epidemic diarrhoea'):
https://www.efsa.europa.eu/en/efsajournal/pub/4375

A good introduction to PED is also provided by Carvajal et al (2015) and Jung and Saif (2005) in their peer-reviewed papers (see References for full details). The American Association of Swine Veterinarians (AASV) provides a good update of PED case report over time as also other relevant information for related virus:
https://www.aasv.org/aasv%20website/Resources/Diseases/PorcineEpidemicDiarrhea.php

Key conferences:

- The NA PRRS Symposium is an annual conference for everyone interested in PRRS. The scope of the meeting is further expanded to include emerging and foreign animal diseases, such as Seneca Valley virus (SVV), PEDv, PCVAD, African swine fever virus, classical swine fever virus, and other high-consequence diseases of swine.
- The International Pig Veterinary Society holds biennial congresses that address PRRS, PED and other diseases of relevance to the swine industry.
- The European Symposium of Porcine Health Management is an annual conference for swine veterinarians to meet and interact and includes PRRS- and PED-related topics.

8 Acknowledgements

The author would like to acknowledge Susanna Williamson and APHA for providing Great Britain data for Fig. 1, Jo Baughan for the literature search and proofreading the chapter and finally Rita Ribeiro and Prof. George Gunn for their suggestions on how to improve the chapter.

9 References

AASV (2017), '*Porcine Epidemic Diarrhea Virus (PEDv), What's New This Week?*', https://www.aasv.org/Resources/PEDv/PEDvWhatsNew.php. Accessed 16 August 2017.

Alborali, G. L., Boniotti, B. and Lavazza, A. (2014), 'Surveillance and control of PED coronavirus in pig in Italy', In: *International Conference on Swine Enteric Coronavirus Diseases, Chicago, USA*, https://www.aphis.usda.gov/animal_health/animal_dis_spec/swine/downloads/meeting/presentations/24%20-%2020%20-%20Alborali.pdf. Accessed 09 November 2016.

Allende, R., Lewis, T. L., Lu, Z., Rock, D. L., Kutish, G. F., Ali, A., Doster, A. R. and Osorio, F. A. (1999), 'North American and European porcine reproductive and respiratory syndrome viruses differ in non-structural protein coding regions', *J. Gen. Virol.* 80, 307–15.

Alonso, C., Goede, D. P., Morrison, R. B., Davies, P. R., Rovira, A., Marthaler, D. G. and Torremorell, M. (2014), 'Evidence of infectivity of airborne porcine epidemic diarrhea virus and detection of airborne viral RNA at long distances from infected herds', *Vet. Res.* 45, 73. https://veterinaryresearch.biomedcentral.com/articles/10.1186/s13567-014-0073-z.

Akimkin, V., Beer, M., Blome, S., Hanke, D., Höper, D., Jenckel, M. and Pohlmann, A. (2016), 'New chimeric porcine coronavirus in swine feces, Germany, 2012', *Emerg. Infect. Dis.* 22(7):1314–15.

An, T. Q., Zhou, Y. J., Liu, G. Q., Tian, Z. J., Li, J., Qiu, H. J. and Tong, G. Z. (2007), 'Genetic diversity and phylogenetic analysis of glycoprotein 5 of PRRSV isolates in mainland China from 1996 to 2006: Coexistence of two NA-subgenotypes with great diversity', *Vet. Microbiol.* 123, 43–52.

An, T. Q., Tian, Z. J., Leng, C. L., Peng, J. M. and Tong, G. Z. (2011), 'Highly pathogenic porcine reproductive and respiratory syndrome virus, Asia', *Emerg. Infect. Dis.* 17, 1782–4.

Anonymous (2016), '*Ontario Pork: Porcine Epidemic Diarrhea Virus (PED)*', http://www.ontariopork.on.ca/Producers/Herd-Health#PED%20. Accessed on 09 November 2016.

APHA (2016a), '*GB Emerging Threats Quarterly Report Pig Diseases Vol 20: Q1 January to March 2016*', Available at: https://www.gov.uk/government/uploads/system/uploads/attachment_data/file/526240/pub-survrep-p0116.pdf.

APHA (2016b), '*GB Emerging Threats Quarterly Report Pig Diseases Vol 20: Q2 April to June 2016*', Available at: https://www.gov.uk/government/uploads/system/uploads/attachment_data/file/549113/pub-survrep-p0216.pdf.

APHA (2016c), '*Veterinary Investigation Diagnosis Analysis (VIDA) Report, 2015: Yearly Trends 2008 to 2015: Pigs*', Available at: https://www.gov.uk/government/statistics/veterinary-investigation-diagnosis-analysis-vida-report-2015. Accessed at 30 October 2016.

Balka, G., Hornyak, A., Balint, A., Kiss, I., Kecskemeti, S., Bakonyi, T. and Rusvai, M. (2008), 'Genetic diversity of porcine reproductive and respiratory syndrome virus strains circulating in Hungarian swine herds', *Vet. Microbiol.* 127, 128–35.

Benfield, D. A., Nelson, E., Collins, J. E., Harris, L., Goyal, S. M., Robison, D., Christianson, W. T., Morrison, R. B., Gorcyca, D. and Chladek, D. (1992), 'Characterization of swine infertility and respiratory syndrome (SIRS) virus (isolate ATCC VR-2332)', *J. Vet. Diagn. Invest.* 4, 127–33.

Beura, L. K., Sarkar, S. N., Kwon, B., Subramaniam, S., Jones, C., Pattnaik, A. K. and Osorio, F. A. (2010), 'Porcine reproductive and respiratory syndrome virus nonstructural protein 1beta modulates host innate immune response by antagonizing IRF3 activation', *J. Virol.* 84, 1574–84.

Blaha, T. (2000), 'The 'colourful' epidemiology of PRSS', *Vet Res.* 31, 77–83.

Bohl, E. H., Kohler, E. M., Saif, L. J., Cross, R. F., Agnes, A. G. and Theil, K. W. (1978), 'Rotavirus as a cause of diarrhea in pigs', *J. Am. Vet. Med. Assoc.* 172, 458–63.

Boniotti, M. B., Papetti, A., Lavazza, A., Alborali, G., Sozzi, E., Chiapponi, C., Faccini, S., Bonilauri, P., Cordioli, P. and Marthaler, D., (2016), 'Porcine epidemic diarrhea virus and discovery of a recombinant swine enteric coronavirus, Italy', *Emerg. Infect. Dis.* 22, 1, 83–7.

Botner, A., Strandbygaard, B., Sorensen, K. J., Have, P., Madsen, K. G., Madsen, E. S. and Alexandersen, S. (1997), 'Appearance of acute PRRS-like symptoms in sow herds after vaccination with a modified live PRRS vaccine', *Vet. Rec.* 141, 497–9.

Butler, J. E., Lager, K. M., Golde, W., Faaberg, K. S., Sinkora, M., Loving, C. and Zhang, Y. I. (2014), 'Porcine reproductive and respiratory syndrome (PRRS): an immunedysregulatory pandemic', *Immunol. Res.* 59, 81–108.

Carman, S., Cai, H. Y., DeLay, J., Youssef, S. A., McEwen, B. J., Gagnon, C. A., Tremblay, D., Hazlett, M., Lusis, P., Fairles, J., Alexander, H. S. and van Dreumel, T. (2008), 'The emergence of a new strain of porcine circovirus-2 in Ontario and Quebec swine and its association with severe porcine circovirus associated disease–2004–2006', *Can J. Vet. Res.* 72 (3), 259–68.

Carvajal, A., Lanza, I., Diego, R., Rubio, P. and Carmenes, P. (1995), 'Evaluation of a blocking ELISA using monoclonal antibodies for the detection of porcine epidemic diarrhea virus and its antibodies', *J. Vet. Diagn. Invest.* 7, 60–4.

Carvajal, A., Argüello, H., Martínez-Lobo, F. J., Costillas, S., Miranda, R., de Nova, P. J. G. and Rubio, P. (2015), 'Porcine epidemic diarrhoea: New insights into an old disease', *Porcine Health Manag.*, 1, 12, https://porcinehealthmanagement.biomedcentral.com/articles/10.1186/s40813-015-0007-9.

Chand, R. J., Trible, B. R. and Rowland, R. R. (2012), 'Pathogenesis of porcine reproductive and respiratory syndrome virus', *Curr. Opin. Virol.* 2, 256–63.

Chen, Q., Li, G., Stasko, J., Thomas, J. T., Stensland, W. R., Pillatzki, A. E., Gauger, P. C., Schwartz, K. J., Madson, D., Yoon, K. J., Stevenson, G. W., Burrough, E. R., Harmon, K. M., Main, R. G. and Zhang J. (2014), 'Isolation and characterization of porcine epidemic diarrhea viruses associated with the 2013 disease outbreak among swine in the United States', *J. Clin. Microbiol.* 52, 234–43.

Chen, Q., Gauger, P., Stafne, M., Thomas, J., Arruda, P., Burrough, E., Madson, D., Brodie, J., Magstadt, D., Derscheid, R., Welch and M., Zhang, J. (2015),'Pathogenicity and pathogenesis of a United States porcine deltacoronavirus cell culture isolate in 5-day-old neonatal piglets', *Virology* 482, 51–9.

Cheney, T., Powell, L., Steinbach, F. and Williamson, S. (2014), '*Study of Porcine Epidemic Diarrhoea Virus in UK Pigs at Slaughter in 2013*', Animal Health and Veterinary Laboratories Agency, UK, http://development.bpex.org.uk/?R-and-D/Pig-Health/pedv.aspx.

Cho, Y. Y., Lim, S. I., Kim, Y. K., Song, J. Y., Lee, J. B. and An, D. J. (2014), 'Complete genome sequence of K14JB01, a novel variant strain of porcine epidemic diarrhea virus in South Korea', *Genome Announc.* 2(3), e00505–14.

Christopher-Hennings, J., Nelson, E. A., Nelson, J. K., Hines, R. J., Swenson, S. L., Hill, H. T., Zimmerman, J. J., Katz, J. B., Yaeger, M. J. and Chase, C. C. (1995), 'Detection of porcine reproductive and respiratory syndrome virus in boar semen by PCR', *J. Clin. Microbiol.* 33, 1730–4.

Collins, J. E. (1991), 'Diagnostic note; new recognized respiratory syndrome in North American swine herds', *AASP Newsl.* 3, 7–11.

Collins, J. E., Benfield, D. A., Christianson, W. T., Harris, L., Hennings, J. C., Shaw, D. P., Goyal, S. M., McCullough, S., Morrison, R. B. and Joo, H. S. (1992), 'Isolation of swine infertility and respiratory syndrome virus (isolate ATCC VR-2332) in North America and experimental reproduction of the disease in gnotobiotic pigs', *J. Vet. Diagn. Invest.* 4, 117–26.

Conzelmann, K. K., Visser, N., Van Woensel, P. and Thiel, H. J. (1993), 'Molecular characterization of porcine reproductive and respiratory syndrome virus, a member of the arterivirus group', *Virology* 193, 329–39.

Crawford, K., Lager, K., Miller, L., Opriessnig, T., Gerber, P. and Hesse, R. (2015), 'Evaluation of porcine epidemic diarrhea virus transmission and the immune response in growing pigs', *Vet Res.*, 46, 49.

Dastjerdi, A., Carr, J., Ellis, R. J., Steinbach, F. and Williamson S. (2015), 'Porcine epidemic diarrhea virus among farmed pigs, Ukraine', *Emerg. Infect. Dis.* 21(12):2235–7.

Davies, P. R. (2015), 'The dilemma of rare events: Porcine epidemic diarrhea virus in North America', *Prev. Vet. Med.* 122, 235–41.

Dee, S., Clement, T., Schelkopf, A., Nerem, J., Knudsen, D., Christopher-Hennings, J. and Nelson, E. (2014), 'An evaluation of contaminated complete feed as a vehicle for porcine epidemic diarrhea virus infection of naive pigs following consumption via natural feeding behavior: Proof of concept', *BMC Vet. Res.* 10, 176. https://bmcvetres.biomedcentral.com/articles/10.1186/s12917-014-0176-9

Dewey, C., Charbonneau, G., Carman, S., Hamel, A., Nayar, G., Friendship, R., Eernisse, K. and Swenson, S. (2000), 'Lelystad-like strain of porcine reproductive and respiratory syndrome virus (PRRSV) identified in Canadian swine', *Can. Vet. J.* 41, 493–4.

Dietze, K., Pinto J., Wainwright, S. and Hamilton, C. (2011), 'Porcine reproductive and respiratory syndrome (PRRS): Virulence jumps and persistent circulation in Southeast Asia', In: *Focus On.* Food and Agriculture Organization of the United Nations, Rome, No. 5, p. 8.

EFSA AHAW Panel (EFSA Panel on Animal Health and Welfare) (2014), 'Scientific Opinion on porcine epidemic diarrhoea and emerging pig deltacoronavirus', *EFSA J.* 12, 3877.

EFSA (2016), 'Collection and review of updated scientific epidemiological data on porcine epidemic diarrhoea', *EFSA J.* 14(2), 4375 p. 52, http://www.efsa.europa.eu/en/efsajournal/pub/4375.

Evans, C. M., Medley, G. F. and Green, L. E. (2008), 'Porcine reproductive and respiratory syndrome virus (PRRSV) in GB pig herds: Farm characteristics associated with heterogeneity in seroprevalence', *BMC Vet. Res.*, 4, 48, http://bmcvetres.biomedcentral.com/articles/10.1186/1746-6148-4-48

Fang, Y., Kim, D. Y., Ropp, S., Steen, P., Christopher-Hennings, J., Nelson, E. A. and Rowland, R. R. (2004), 'Heterogeneity in Nsp2 of European-like porcine reproductive and respiratory syndrome viruses isolated in the United States', *Virus Res.* 100, 229–35.

Fang, Y. and Snijder, E. J. (2010), 'The PRRSV replicase: Exploring the multifuncionality of an intriguing set of non-structural proteins', *Virus Res.* 154, 61–76.

Feng, Y., Zhao, T., Nguyen, T., Inui, K., Ma, Y., Nguyen, T. H., Nguyen, V. C., Liu, D., Bui, Q. A., To, L. T., Wang, C., Tian, K. and Gao, G. F. (2008), 'Porcine respiratory and reproductive syndrome virus variants, Vietnam and China, 2007', *Emerg. Infect. Dis.* 14, 1774–6.

Firth, A. E., Zevenhoven-Dobbe, J. C., Wills, N. M., Go, Y. Y., Balasuriya, U. B., Atkins, J. F., Snijder, E. J. and Posthuma, C. C. (2011), 'Discovery of a small arterivirus gene that overlaps the GP5 coding sequence and is important for virus production', *J. Gen.Virol.* 92, 1097–106.

Forsberg, R., Oleksiewicz, M. B., Petersen, A. M., Hein, J., Botner, A. and Storgaard, T. (2001), 'A molecular clock dates the common ancestor of European-type porcine reproductive and respiratory syndrome virus at more than 10 years before the emergence of disease', *Virology* 289, 174–9.

Forsberg, R., Storgaard, T., Nielsen, H. S., Oleksiewicz, M. B., Cordioli, P., Sala, G., Hein, J. and Botner, A. (2002), 'The genetic diversity of European type PRRSV is similar to that of the North American type but is geographically skewed within Europe', *Virology* 299, 38–47.

Forsberg, R. (2005), 'Divergence time of porcine reproductive and respiratory syndrome virus subtypes', *Mol. Biol. Evol.* 22, 2131–4.

Frossard, J. P., Hughes, G. J., Westcott, D. G., Naidu, B., Williamson, S., Woodger, N. G., Steinbach, F. and Drew, T. W. (2013), 'Porcine reproductive and respiratory syndrome virus: Genetic diversity of recent British isolates', *Vet. Microbiol.* 162, 507–18.

Gauger, P. C., Faaberg, K. S., Guo, B., Kappes, M. A. and Opriessnig, T. (2012), 'Genetic and phenotypic characterization of a 2006 United States porcine reproductive and respiratory virus isolate associated with high morbidity and mortality in the field', *Virus Res.* 163, 98–107.

Gibbens, J. C., Robertson, S., Willmington, J., Milne, S. A., Rya, N. J. B., Wilesmith, J. W., Cook, A. J. and David, G. P. (2008), 'Use of laboratory data to reduce the time taken to detect new diseases: VIDA to FarmFile', *Vet. Rec.*, 162(24):771–6.

Gogin, A., Gerasimov, V., Malogolovkin, A. and Kolbasov, D. (2013), 'African swine fever in the North Caucasus region and the Russian Federation in years 2007–2012', *Virus Res.* 173, 198–203.

Grasland, B., Bigault, L., Bernard, C., Quenault, H., Toulouse, O., Fablet, C., Rose, N., Touzain, F. and Blanchard, Y. (2015), 'Complete genome sequence of a porcine epidemic diarrhea S gene Indel strain isolated in France in December 2014', *Genome Announc.* 3, e00535–15.

Greiser-Wilke, I., Fiebig, K., Drexler, C. and Grosse Beilage, E. (2010), 'Genetic diversity of Porcine reproductive and respiratory syndrome virus (PRRSV) in selected herds in a pig-dense region of North-Western Germany', *Vet. Microbiol.* 143, 213–23.

Halbur, P. G., Miller, L. D., Paul, P. S., Meng, X. J., Huffman, E. L. and Andrews, J. J. (1995a), 'Immunohistochemical identification of porcine reproductive and respiratory syndrome virus (PRRSV) antigen in the heart and lymphoid system of three-week-old colostrum-deprived pigs', *Vet. Pathol.* 32, 200–4.

Halbur, P. G., Paul, P. S., Frey, M. L., Landgraf, J., Eernisse, K., Meng, X. J., Lum, M. A., Andrews, J. J. and Rathje, J. A. (1995b), 'Comparison of the pathogenicity of two US porcine reproductive and respiratory syndrome virus isolates with that of the Lelystad virus', *Vet. Pathol.* 32, 648–60.

Han, J., Wang, Y. and Faaberg, K. S. (2006), 'Complete genome analysis of RFLP 184 isolates of porcine reproductive and respiratory syndrome virus', *Virus Res.* 122, 175–82.

Hanke, D., Jenckel, M., Petrov, A., Ritzmann, M., Stadler, J., Akimkin, V., Blome, S., Pohlmann, A., Schirrmeier, H., Beer, M. and Höper, D. (2015), 'Comparison of porcine epidemic diarrhea viruses from Germany and the United States, 2014', *Emerg. Infect. Dis.* 21, 493–6.

Hirai, T., Nunoya, T., Ihara, T., Kusanagi, K. and Shibuya, K. (2001), 'Dual infection with PCV-2 and porcine epidemic diarrhoea virus in neonatal piglets', *Vet. Rec.* 148, 482–4.

Holtkamp, D. J., Kliebenstein, J. B., Neumann, E. J., Zimmerman, J. J., Rotto, H., Yoder,T. K., Wang, C., Yeske, P., Mowrer, C. and Haley, C. (2013), 'Assessment of the economic impact of porcine reproductive and respiratory syndrome virus on U.S. pork producers', *Swine Health Prod.* 21, 72–84.

Huang, Y. W., Dickerman, A. W., Pineyro, P., Li, L., Fang, L., Kiehne, R., Opriessnig, T. and Meng, X. J. (2013), 'Origin, evolution, and genotyping of emergent porcine epidemic diarrhea virus strains in the United States', *MBio* 4, e00737–13.

Jantafong, T., Sangtong, P., Saenglub, W., Mungkundar, C., Romlamduan, N., Lekchareonsuk, C. and Lekcharoensuk, P. (2015), 'Genetic diversity of porcine reproductive and respiratory syndrome virus in Thailand and Southeast Asia from 2008 to 2013', *Vet. Microbiol.* 176(3–4), 229–38.

Jiang, P., Chen, P. Y., Dong, Y. Y., Cai, J. L., Cai, B. X. and Jiang, Z. H. (2000), 'Isolation and genome characterization of porcine reproductive and respiratory syndrome virus in P.R. China', *J. Vet. Diagn. Invest.* 12, 156–8.

Johnson, C. R., Griggs, T. F., Gnanandarajah, J. and Murtaugh, M. P. (2011), 'Novel structural protein in porcine reproductive and respiratory syndrome virus encoded by an alternative ORF5 present in all arteriviruses', *J. Gen. Virol.* 92, 1107–16.

Jung, K., Ha, Y., Ha, S. K., Kim, J., Choi, C., Park, H. K., Kim, S. H. and Chae, C. (2006), 'Identification of porcine circovirus type 2 in retrospective cases of pigs naturally infected with porcine epidemic diarrhoea virus', *Vet. J.* 171, 166–8.

Jung, K., Wang, Q., Scheuer, K. A., Lu, Z., Zhang, Y. and Saif, L. J. (2014), 'Pathology of US porcine epidemic diarrhea virus strain PC21A in gnotobiotic pigs', *Emerg. Infect. Dis.* 20, 662–5.

Jung, K. and Saif, L. J. (2015), 'Porcine epidemic diarrhea virus infection: Etiology, epidemiology, pathogenesis and immunoprophylaxis', *Vet. J.* 204, 134–43.

Kamau, N., Park, J., Park, J., Hyun, B., Yang, D., Song, J. and Shin, H. J. (2010), 'Susceptibility of mice to porcine epidemic diarrhoea virus', *J. Anim. Vet. Adv.* 9:3114–16.

Kappes, M. A. and Faaberg, K. S. (2015), 'PRRSV structure, replication and recombination: Origin of phenotype and genotype diversity', *Virology* 479–480, 475–86.

Kapur, V., Elam, M. R., Pawlovich, T. M. and Murtaugh, M. P. (1996), 'Genetic variation in porcine reproductive and respiratory syndrome virus isolates in the Midwestern United States', *J. Gen. Virol.* 77, 1271–6.

Karniychuk, U. U., Geldhof, M., Vanhee, M., Van Doorsselaere, J., Saveleva, T. A. and Nauwynck, H. J. (2010), 'Pathogenesis and antigenic characterization of a new East European subtype 3 porcine reproductive and respiratory syndrome virus isolate', *BMC Vet. Res.* 6, 30, http://bmcvetres.biomedcentral.com/articles/10.1186/1746-6148-6-30.

Keffaber, K. K. (1989), 'Reproductive failure of unknown etiology', *Am. Assoc. Swine Pract. Newsl.* 1, 1–10

Kweon, C. H., Kwon, B. J., Jung, T. S., Kee, Y. J., Hur, D. H., Hwang, E. K., Rhee, J. C. and An, S. H. (1993), 'Isolation of porcine epidemic diarrhea virus (PEDV) infection in Korea', *Korean J. Vet. Res.* 33, 249–54.

Laranjo, M., Allepuz, A., Pleguezuelos, P., Lopez, S., Casal, J., Segales, J., Martın, M. and Pujols, J. (2015), 'Estudio epidemiologico prospectivo en granjas con brotes recientes de diarrea epidemica porcina', 95p. Unpublished report.

Lee, C., Kim, H., Kang, B., Yeom, M., Han, S., Moon, H., Park, S., Kim, H., Song, D. and Park, B. (2010), 'Prevalence and phylogenetic analysis of the isolated type I porcine reproductive and respiratory syndrome virus from 2007 to 2008 in Korea', *Virus Genes.* 40, 225–30.

Li, W., Li, H., Liu, Y., Pan, Y., Deng, F., Song, Y., Tang, X. and He, Q. (2012), 'New variants of porcine epidemic diarrhea virus, China, 2011', *Emerg. Infect. Dis.* 18, 1350–3.

Lin, C. N., Chung, W. B., Chang, S. W., Wen, C. C., Liu, H., Chien, C. H. and Chiou, M. T. (2014), 'US-like strain of porcine epidemic diarrhea virus outbreaks in Taiwan, 2013–2014', *J. Vet. Med. Sci.* 76, 1297–9.

Lowe, J., Gauger, P., Harmon, K., Zhang, J., Connor, J., Yeske, P., Loula, T., Levis, I., Dufresne, L. and Main, R. (2014), 'Role of transportation in spread of porcine epidemic diarrhea virus infection, United States', *Emerg. Infect. Dis.* 20, 872–4.

Madsen, K. G., Hansen, C. M., Madsen, E. S., Strandbygaard, B., Botner, A. and Sorensen, K. J. (1998), 'Sequence analysis of porcine reproductive and respiratory syndrome virus of the American type collected from Danish swine herds', *Arch. Virol.* 143, 1683–700.

Madson, D. M., Magstadt, D. R., Arruda, P. H., Hoang, H., Sun, D., Bower, L. P., Bhandari, M., Burrough, E. R., Gauger, P. C., Pillatzki, A. E., Stevenson, G. W., Wilberts, B. L., Brodie, J., Harmon, K. M., Wang, C., Main, R. G., Zhang, J. and Yoon, K. J. (2014), 'Pathogenesis of porcine epidemic diarrhoea virus isolate (US/Iowa/18984/2013) in 3-week-old weaned pigs', *Vet Microbiol.*, 174, 60–8.

Martelli, P., Lavazza, A., Nigrelli, A. D., Merialdi, G., Alborali, L. G. and Pensaert, M. B. (2008), 'Epidemic of diarrhoea caused by porcine epidemic diarrhoea virus in Italy', *Vet. Rec.* 162, 307–10.

Mengeling, W. L., Vorwald, A. C., Lager, K. M. and Brockmeier, S. L. (1996), 'Comparison among strains of porcine reproductive and respiratory syndrome virus for their ability to cause reproductive failure', *Am. J. Vet. Res.* 57, 834–9.

Meng, X. J., Paul, P. S., Halbur, P. G. and Lum, M. A. (1995), 'Phylogenetic analyses of the putative M (ORF 6) and N (ORF 7) genes of porcine reproductive and respiratory syndrome virus (PRRSV): Implication for the existence of two genotypes of PRRSV in the U.S.A. and Europe', *Arch. Virol.* 140, 745–55.

Mesquita, J. R., Hakze-van der Honing, R., Almeida, A., Lourenço, M., van der Poel, W. H. M. and Nascimento, M. S. J. (2015), 'Outbreak of porcine epidemic diarrhea virus in Portugal, 2015', *Transbound Emerg. Dis.* 62, 586–8, doi:10.1111/tbed.12409.

Morgan, S. B., Graham, S. P., Salguero, F. J., Sanchez Cordon, P. J., Mokhtar, H., Rebel, J. M., Weesendorp, E., Bodman-Smith, K. B., Steinbach, F. and Frossard, J. P. (2013), 'Increased pathogenicity of European porcine reproductive and respiratory syndrome virus is associated with enhanced adaptive responses and viral clearance', *Vet. Microbiol.* 163, 13–22.

Murtaugh, M. P., Xiao, Z. and Zuckermann, F. (2002), 'Immunological responses of swine to porcine reproductive and respiratory syndrome virus infection', *Viral Immunol.* 15, 533–47.

Nagy, B., Nagy, G., Meder, M. and Mocsari, E. (1996), 'Enterotoxigenic Escherichia coli, rotavirus, porcine epidemic diarrhoea virus, adenovirus and calici-like virus in porcine postweaning diarrhoea in Hungary', *Acta Vet. Hung.* 44, 9–19.

Neumann, E. J., Kliebenstein, J. B., Johnson, C. D., Mabry, J. W., Bush, E. J., Seitzinger, A. H., Green, A. L. and Zimmerman, J. J. (2005), 'Assessment of the economic impact of porcine reproductive and respiratory syndrome on swine production in the United States', *J. Am. Vet. Med. Assoc.* 227, 385–92.

Nguyen, V. G., Kim, H. K., Moon, H. J., Park, S. J., Chung, H. C., Choi, M. K. and Park, B. K. (2014), 'A bayesian phylogeographical analysis of type 1 porcine reproductive and respiratory syndrome virus (PRRSV)', *Transbound Emerg. Dis.* 61, 537–45.

Oka, T., Saif, L. J., Marthaler, D., Esseili, M. A., Meulia, T., Lin, C. M., Vlasova, A. N., Jung, K., Zhang, Y. and Wang, Q. (2014), 'Cell culture isolation and sequence analysis of genetically diverse US porcine epidemic diarrhea virus strains including a novel strain with a large deletion in the spike gene', *Vet. Microbiol.* 173, 258–69.

Opriessnig, T., Xiao, C. T., Gerber, P. F., Zhang, J. and Halbur, P. G. (2014), 'Porcine epidemic diarrhea virus RNA present in commercial spray-dried porcine plasma is not infectious to naive pigs', *PLoS ONE* 9, e104766. http://journals.plos.org/plosone/article?id=10.1371/journal.pone.0104766

Park, S. J., Song, D. S. and Park, B. K. (2013), 'Molecular epidemiology and phylogenetic analysis of porcine epidemic diarrhea virus (PEDV) field isolates in Korea', *Arch. Virol.* 158, 1533–41.

Park, S., Kim, S., Song, D. and Park, B. (2014), 'Novel porcine epidemic diarrhea virus variant with large genomic deletion, South Korea', *Emerg. Infect. Dis.* 20, 2089–92.

Pasick, J., Berhane, Y., Ojkic, D., Maxie, G., Embury-Hyatt, C., Swekla, K., Handel, K., Fairles, J. and Alexandersen, S. (2014), 'Investigation into the role of potentially contaminated feed as a source of the first-detected outbreaks of porcine epidemic diarrhea in Canada', *Transbound Emerg. Dis.* 61, 397–410.

Pensaert, M. and de Bouck, P. (1978), 'A new coronavirus-like particle associated with diarrhea in swine', *Arch. Virol.* 58, 243–7.

Pensaert, M. B., Debouck, P. and Reynolds, D. J. (1981), 'An immunoelectron microscopic and immunofluorescent study on the antigenic relationship between the coronavirus-like agent, CV 777, and several coronaviruses', *Arch. Virol.* 68, 45–52.

Pensaert, M. B. (1999), 'Porcine epidemic diarrhea', In: Straw, B. E., D'Allaire, S., Mengeling, W. L. and Taylor, D. J. (Eds) *Diseases of Swine*, 8th ed. Iowa State University Press, Ames, Iowa, pp. 179–85.

Perez, A., Davies, P., Goodell, C., Holtkamp, D., Mondaca, E., Poljak, Z., Tousignant, S., Valdes-Donoso, P., Zimmerman, J. and Morrison, R. (2015), 'Review of literature for epidemiology and control of porcine reproductive 1 and respiratory syndrome virus (PRRSV) in North America: Lessons learned and knowledge gaps', *J. Am. Vet. Med. Assoc.* 246, 1304–17.

Pijpers, A., van Nieuwstadt, A. P., Terpstra, C. and Verheijden, J. H. (1993), 'Porcine epidemic diarrhoea virus as a cause of persistent diarrhoea in a herd of breeding and finishing pigs', *Vet. Rec.* 132, 129–31.

Powell, L. F., Cheney, T. E. A., Williamson, S., Guy, E., Smith, R. P. and Davies, R. H. (2016), 'A prevalence study of Salmonella spp., Yersinia spp., Toxoplasma gondii and porcine reproductive and respiratory syndrome virus in UK pigs at slaughter', *Epidemiol. Infect.* 144, 1538–49.

Pritchard, G. C., Paton, D. J., Wibberley, G. and Ibata, G. (1999), 'Transmissible gastroenteritis and porcine epidemic diarrhoea in Britain', *Vet. Rec.* 144, 616–18.

Puranaveja, S., Poolperm, P., Lertwatcharasarakul, P., Kesdaengsakonwut, S., Boonsoongnern, A., Urairong, K., Kitikoon, P., Choojai, P., Kedkovid, R., Teankum, K. and Thanawongnuwech, R. (2009), 'Chinese-like strain of porcine epidemic diarrhea virus, Thailand', *Emerg. Infect. Dis.* 15, 1112–15.

Ropp, S. L., Wees, C. E., Fang, Y., Nelson, E. A., Rossow, K. D., Bien, M., Arndt, B., Preszler, S., Steen, P., Christopher-Hennings, J., Collins, J. E., Benfield, D. A. and Faaberg, K. S. (2004), 'Characterization of emerging European-like porcine reproductive and respiratory syndrome virus isolates in the United States', *J. Virol.* 78, 3684–703.

Rossow, K. D., Bautista, E. M., Goyal, S. M., Molitor, T. W., Murtaugh, M. P., Morrison, R. B., Benfield, D. A. and Collins, J. E. (1994), 'Experimental porcine reproductive and respiratory syndrome virus infection in one-, four-, and 10-week-old pigs', *J. Vet. Diagn. Invest.* 6, 3–12.

Rossow, K. D. (1998), 'Porcine reproductive and respiratory syndrome', *Vet. Pathol.* 35, 1–20.

Scott, A., McCluskey, B., Brown-Reid, M., Grear, D., Pitcher, P., Ramos, G., Spencer, D. and Singrey, A. (2016), 'Porcine epidemic diarrhea virus introduction into the United States: Root cause investigation', *Prev. Vet. Med.* 123, 192–201. doi:10.1016/j.prevetmed.2015.11.013.

Shi, M., Lam, T. T., Hon, C. C., Hui, R. K., Faaberg, K. S., Wennblom, T., Murtaugh, M. P., Stadejek, T. and Leung, F. C. (2010a), 'Molecular epidemiology of PRRSV: A phylogenetic perspective', *Virus Res.* 154, 7–17.

Shi, M., Lam, T. T., Hon, C. C., Murtaugh, M. P., Davies, P. R., Hui, R. K., Li, J., Wong, L. T., Yip, C. W., Jiang, J. W. and Leung, F. C. (2010b), 'Phylogeny-based evolutionary, demographical, and geographical dissection of North American type 2 porcine reproductive and respiratory syndrome viruses', *J. Virol.* 84, 8700–11.

Shibata, I., Tsuda, T., Mori, M., Ono, M., Sueyoshi, M. and Uruno, K. (2000), 'Isolation of porcine epidemic diarrhea virus in porcine cell cultures and experimental infection of pigs of different ages', *Vet. Microbiol.* 72, 173–82.

Snijder, E. J. and Meulenberg, J. J. (1998), 'The molecular biology of arteriviruses', *J. Gen.Virol.* 79, 961–79.

Song, D., Leem, S. S., Yang, J. S., Moon, H. J., Ohc, J. S., Ha, G. W., Jang, Y. S. and Park, B. K. (2006), 'Use of an internal control in a quantitative RT-PCR assay for quantitation of porcine epidemic diarrhoea virus shedding in pigs', *J. Virol Methods* 133, 27–33.

Song, D. and Park, B. (2012), 'Porcine epidemic diarrhoea virus: A comprehensive review of molecular epidemiology, diagnosis, and vaccines', *Virus Genes* 44, 167–75.

Stadejek, T., Stankevicius, A., Storgaard, T., Oleksiewicz, M. B., Belak, S., Drew, T. W. and Pejsak, Z. (2002), 'Identification of radically different variants of porcine reproductive and respiratory syndrome virus in Eastern Europe: towards a common ancestor for European and American viruses', *J. Gen. Virol.* 83, 1861–73.

Stadejek, T., Oleksiewicz, M. B., Potapchuk, D. and Podgorska, K. (2006), 'Porcine reproductive and respiratory syndrome virus strains of exceptional diversity in eastern Europe support the definition of new genetic subtypes', *J. Gen. Virol.* 87, 1835–41.

Stadejek, T., Oleksiewicz, M. B., Scherbakov, A. V., Timina, A. M., Krabbe, J. S., Chabros, K. and Potapchuk, D. (2008), 'Definition of subtypes in the European genotype of porcine reproductive and respiratory syndrome virus: Nucleocapsid characteristics and geographical distribution in Europe', *Arch. Virol.* 153, 1479–88.

Stadejek, T., Stankevicius, A., Murtaugh, M. P. and Oleksiewicz, M. B. (2013), 'Molecular evolution of PRRSV in Europe: Current state of play', *Vet. Microbiol.* 165(1–2), 21–8.

Stadler, J., Zoels, S., Fux, R., Hanke, D., Pohlmann, A., Blome, S., Weissenböck, H., Weissenbacher-Lang, C., Ritzmann, M. and Ladinig, A. (2015), 'Emergence of porcine epidemic diarrhea virus in southern Germany', *BMC Vet. Res.* 11, 142.

Steinrigl, A., Revilla Fernandez, S., Stoiber, F., Pikalo, J., Sattler, T. and Schmoll, F. (2015), 'First detection, clinical presentation and phylogenetic characterization of Porcine epidemic diarrhea virus in Austria', *BMC Vet. Res.* 11, 310.

Stevenson, G. W., Hoang, H., Schwartz, K. J., Burrough, E. B., Sun, D., Madson, D., Cooper, V. L., Pillatzki, A., Gauger, P., Schmitt, B. J., Koster, L. G., Killian, M. L. and Yoon, K. J. (2013), 'Emergence of Porcine epidemic diarrhea virus in the United States: Clinical signs, lesions, and viral genomic sequences', *J. Vet. Diagn. Invest.* 25, 649–54.

Suarez, P., Zardoya, R., Martin, M. J., Prieto, C., Dopazo, J., Solana, A. and Castro, J. M. (1996), 'Phylogenetic relationships of european strains of porcine reproductive and respiratory syndrome virus (PRRSV) inferred from DNA sequences of putative ORF-5 and ORF-7 genes', *Virus Res.* 42, 159–65.

Sueyoshi, M., Tsuda, T., Yamazaki, K., Yoshida, K., Nakazawa, M., Sato, K.Minami, T., Iwashita, K., Watanabe, M., Suzuki, Y., et al. (1995), 'An immunohistochemical investigation of porcine epidemic diarrhoea', *J. Comp. Pathol.* 113, 59–67.

Sun, R., Leng, Z., Dekun, C. and Song, C. (2014), 'Multiple factors contribute to persistent porcine epidemic diarrhoea infection in the field: An investigation on porcine epidemic diarrhoea repeated outbreaks in the same herd', *J. Anim. Vet. Adv.* 13, 410–15.

Sun, R. Q., Cai, R. J., Chen, Y. Q., Liang, P. S., Chen, D. K. and Song, C. X. (2012), 'Outbreak of porcine epidemic diarrhea in suckling piglets, China', *Emerg. Infect. Dis.* 18, 161–3.

Swenson, S. L., Hill, H. T., Zimmerman, J. J., Evans, L. E., Landgraf, J. G., Wills, R. W., Sanderson, T. P., McGinley, M. J., Brevik, A. K. and Ciszewski, D. K. (1994), 'Excretion of porcine reproductive and respiratory syndrome virus in semen after experimentally induced infection in boars', *J. Am. Vet. Med. Assoc.* 204, 1943–8.

Takahashi, K., Okada, K. and Ohshima, K. (1983), 'An outbreak of swine diarrhoea of a new type associated with coronavirus-like particles in Japan', *Nihon Juigaku Zasshi.* 45, 829–32.

Terpstra, C., Wensvoort, G. and Pol, J. M. (1991), 'Experimental reproduction of porcine epidemic abortion and respiratory syndrome (mystery swine disease) by infection with Lelystad virus: Koch's postulates fulfilled', *Vet. Q.* 13, 131–6.

Thanawongnuwech, R., Amonsin, A., Tatsanakit, A. and Damrongwatanapokin, S. (2004), 'Genetics and geographical variation of porcine reproductive and respiratory syndrome virus (PRRSV) in Thailand', *Vet. Microbiol.* 101, 9–21.

Theuns, S., Conceição-Neto, N., Christiaens, I., Zeller, M., Desmarets, L. M. B., Roukaerts, I. D. M., Acar, D. D., Heylen, E., Matthijnssens, J. and Nauwynck, H. J. (2015), 'Complete genome sequence of a porcine epidemic diarrhea virus from a novel outbreak in Belgium, January 2015', *Genome Announc.* 3, e00506–15. doi:10.1128/genomeA.00506-15.

Thimmasandra Narayanappa, A., Sooryanarain, H., Deventhiran, J., Cao, D., Ammayappan Venkatachalam, A., Kambiranda, D., LeRoith, T., Heffron, C. L., Lindstrom, N., Hall, K., Jobst, P., Sexton, C., Meng, X. J. and Elankumaran, S. (2015), 'A novel pathogenic mammalian orthoreovirus from diarrheic pigs and swine blood meal in the United States', *MBio.* 6, e00593–15.

Thrusfield, M. (2005), 'Chapter 5: Determinants of disease', In: *Veterinary Epidemiology*, 3rd ed. Blackwell Publishing, Oxford UK, pp. 75–97.

Tian, K., Yu, X., Zhao, T., Feng, Y., Cao, Z., Wang, C., Hu, Y., Chen, X., Hu, D., Tian, X., Liu, D., Zhang, S., Deng, X., Ding, Y., Yang, L., Zhang, Y., Xiao, H., Qiao, M., Wang, B., Hou, L., Wang, X., Yang, X., Kang, L., Sun, M., Jin, P., Wang, S., Kitamura, Y., Yan, J. and Gao, G. F. (2007), 'Emergence of fatal PRRSV variants: Unparalleled outbreaks of atypical PRRS in China and molecular dissection of the unique hallmark', *PLoS One* 2, e526. http://journals.plos.org/plosone/article?id=10.1371/journal.pone.0000526.

Tian, P. F., Jin, Y. L., Xing, G., Qv, L. L., Huang, Y. W. and Zhou, J. Y. (2014), 'Evidence of recombinant strains of porcine epidemic diarrhea virus, United States, 2013', *Emerg. Infect. Dis.* 20, 1735–8.

Toplak, I., Štukelj, M., Rihtarič, D., Hostnik, P. and Grom, J. (2015), 'First detection of porcine epidemic diarrhoea virus in Slovenia, 2015', In: *Abst Proc 10th Int Congress of Veterinary Virology*, Montpellier, France. http://esvv2015.cirad.fr/program-proceedings/proceedings.

Turgeon, D. C., Morin, M., Jolette, J., Higgins, R., Marsolais, G. and DiFranco, E. (1980), 'Coronavirus-like particles associated with diarrhea in baby pigs in Quebec', *Can Vet J.* 21(3), 100–xxiii.

Van der Wolf, P. J., Van Walderen, A., Meertens, M. N., Van Hout, A. J., Duinhof, T. F., Geudeke, M. J., Franssen, J. M., Dortmans, J. C. F. M. and Dikman, R. (2015), 'First case of porcine epidemic diarrhoea (PED) caused by a new variant of PED virus in the Netherlands', In: *Proc. 7th European Symposium of Porcine Health Management 2015*, p. 79.

Van Doorsselaere, J., Brar, M. S., Shi, M., Karniychuk, U., Leung, F. C. and Nauwynck, H. J. (2012), 'Complete genome characterization of a East European Type 1 subtype 3 porcine reproductive and respiratory syndrome virus', *Virus Genes* 44, 51–4.

van Nieuwstadt, A. P. and Zetstra, T. (1991), 'Use of two enzyme-linked immunosorbent assays to monitor antibody responses in swine with experimentally induced infection with porcine epidemic diarrhea virus', *Am. J. Vet. Res.* 52, 1044–50.

Van Reeth, K. and Pensaert, M. (1994), 'Prevalence of infections with enzootic respiratory and enteric viruses in feeder pigs entering fattening herds', *Vet. Rec.* 135, 594–7.

Vijaykrishna, D., Smith, G. J., Zhang, J. X., Peiris, J. S., Chen, H. and Guan, Y. (2007), 'Evolutionary insights into the ecology of coronaviruses', *J. Virol.* 81, 4012–20.

Vlasova, A. N., Marthaler, D., Wang, Q., Culhane, M. R., Rossow, K. D., Rovira, A., Collins, J. and Saif, L. J. (2014), 'Distinct characteristics and complex evolution of PEDV strains, North America, May 2013–February 2014', *Emerg. Infect. Dis.* 20, 1620–8.

Wagstrom, E. A., Chang, C. C., Yoon, K. J. and Zimmerman, J. J. (2001), 'Shedding of porcine reproductive and respiratory syndrome virus in mammary gland secretions of sows', *Am. J. Vet. Res.* 62, 1876–80

Wang, J., Zhao, P., Guo, L., Liu, Y., Du, Y., Ren, S., Li, J., Zhang, Y., Fan, Y., Huang, B., Liu, S and Wu, J. (2013), 'Porcine epidemic diarrhea virus variants with high pathogenicity, China', *Emerg. Infect. Dis.* 19, 2048–9.

Wang, L., Byrum, B. and Zhang, Y. (2014), 'New variant of porcine epidemic diarrhea virus, United States, 2014', *Emerg. Infect. Dis.* 20, 917–19.

Wang, X., Marthaler, D., Rovira, A., Rossow, S. and Murtaugh, M. P. (2015), 'Emergence of a virulent porcine reproductive and respiratory syndrome virus in vaccinated herds in the United States', *Virus Res.* 210, 34–41.

Weesendorp, E., Morgan, S., Stockhofe-Zurwieden, N., Popma-De Graaf, D. J., Graham, S. P. and Rebel, J. M. (2013), 'Comparative analysis of immune responses following experimental infection of pigs with European porcine reproductive and respiratory syndrome virus strains of differing virulence', *Vet. Microbiol.* 163, 1–12.

Wensvoort, G., Terpstra, C., Pol, J. M., ter Laak, E. A., Bloemraad, M., de Kluyver, E. P., Kragten, C., van Buiten, L., den Besten, A. and Wagenaar, F. (1991), 'Mystery swine disease in The Netherlands: The isolation of Lelystad virus', *Vet. Q.* 13, 121–30.

Wills, R. W., Zimmerman, J. J., Yoon, K. J., Swenson, S. L., Hoffman, L. J., McGinley, M. J., Hill, H. T. and Platt, K. B. (1997), 'Porcine reproductive and respiratory syndrome virus: Routes of excretion', *Vet. Microbiol.* 57, 69–81.

Wood, E. N. (1977), 'An apparently new syndrome of porcine epidemic diarrhoea', *Vet. Rec.* 100, 243–4.

Wu, W. H., Fang, Y., Farwell, R., Steffen-Bien, M., Rowland, R. R., Christopher-Hennings, J. and Nelson, E. A. (2001), 'A 10-kDa structural protein of porcine reproductive and respiratory syndrome virus encoded by ORF2b', *Virology* 287, 183–91.

Yaeger, M. J., Prieve, T., Collins, J. E., Christopher-Hennings, J., Nelson, E. and Benfield, D. A. (1993), 'Evidence for the transmission of porcine reproductive and respiratory syndrome (PRRS) virus in boar semen', *Swine Health Prod.* 1, 7–9.

Zhu, L., Zhang, G., Ma, J., He, X., Xie, Q., Bee, Y. and Gong, S. Z. (2011), 'Complete genomic characterization of a Chinese isolate of porcine reproductive and respiratory syndrome virus', *Vet. Microbiol.* 147, 274–82.

Zuckermann, F. A., Garcia, E. A., Luque, I. D., Christopher-Hennings, J., Doster, A., Brito, M. and Osorio, F. (2007), 'Assessment of the efficacy of commercial porcine reproductive and respiratory syndrome virus (PRRSV) vaccines based on measurement of serologic response, frequency of gamma-IFN-producing cells and virological parameters of protection upon challenge', *Vet. Microbiol.* 123, 69–85.

Chapter 3

The influence of gut microbiome on developing immune and metabolic systems in the young pig

Mick Bailey, Emily Porter and Ore Francis, University of Bristol, UK

1 Introduction

The term 'enteric health' is becoming widely used, but is associated with a variety of meanings depending on the context in which it is used. It can be empirically and commercially defined as a physiological state promoting optimum growth over input, where performance provides the measure of enteric health. This is essentially a nutritional or metabolic definition, but the weight of evidence now suggests that growth and health are strongly affected by both the immune system and the types of bacteria resident within the intestine – the 'microbiome'. These three domains – metabolism, the intestinal microbiome and the immune system – interact in complex ways to modify performance in growing pigs and provide important opportunities for interventions to improve pig production in the future. However, there are a number of confounding factors, such as welfare and disease state, that also need to be considered as desired outcomes when attempting to optimise gut health (Bailey, 2015). Thus, optimal enteric health under conventional, intensive husbandry may be different from that in outdoor systems, and very different from that in the expanding smallholder sector. It seems unlikely that a single definition or measure of enteric health status is universally applicable to all production systems, since different conditions during rearing are likely to result in different immune and metabolic systems arising from colonisation with a variety of intestinal bacterial species.

It is clear that the function of the immune system, specifically of the components present in the intestinal mucosa, is on the causal pathway for maintaining normal intestinal function and, as such, is a critical component of enteric health. The mucosal immune system must

http://dx.doi.org/10.19103/AS.2017.0013.16

maintain the ability to respond to pathogens: acute, transient inflammatory responses that impair intestinal digestive and absorptive processes may be acceptable if they result in pathogen clearance and a return to normal digestive function in the long term. Similarly, the mucosal immune system must avoid the expression of damaging immune responses to harmless food antigens and to wholly commensal intestinal bacteria, since these may

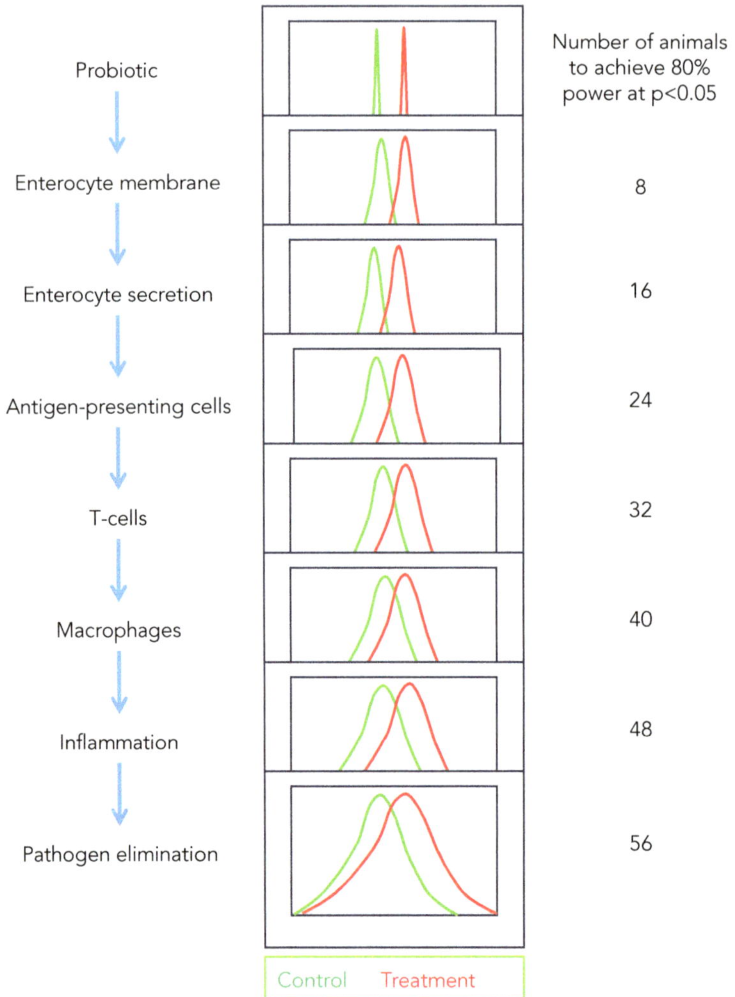

Figure 1 The problem of using proxy measures of the effectiveness of an intervention where the causal link between the intervention (in this case, probiotic administration) and the desired outcome (in this case, eliminating a pathogen) is long. Where inter-individual variation broadens the spread of the effect within a population, the use of a proxy measure (such as stabilising the enterocyte membrane) close to the intervention increases the likelihood of finding a significant effect on that measure, but decreases the likelihood that the proxy measure will be correlated with the desired outcome.

also result in unnecessary inflammation and consequent impairment of digestive function. Finally, it may need to maintain some level of continuous, low-level response in order to limit potential opportunist pathogens or 'pathobionts', and this may result in chronic, low-level effects on digestive and absorptive functions, which may be a necessary compromise to prevent progression to severe damage. Thus, the immune system may contribute to enteric health in two ways: firstly, the expression of appropriate immune responses is necessary in order to maintain optimal digestive and absorptive functions and, secondly, the immune system itself is a nutritional cost, which can divert resources away from growth. The ability of the mucosal immune system to distinguish between pathogens and harmless antigens, then, is a crucial aspect of enteric health, and one that increasing evidence suggests is influenced by the early life experiences of the piglet.

However, there may be an increasing problem in the future when proxy measures are used to determine 'enteric health', such as crypt depth and villus height, or immune response to vaccination, or composition of the intestinal microbiome; while all of these are likely to modulate the conversion of food to body mass, the relative impacts and the extent to which they interact make them potentially unreliable as single measures. Specifically, if we consider the length of a causal chain between an intervention (for example, a probiotic), a particular measure of an effect (e.g. circulating neutrophils, enterocyte brush border peptidases) and a desired outcome (fewer days to finish), the further away from the desired outcome our measure is, the less reliable it becomes as a proxy for 'enteric health' (Fig. 1). Ultimately, it will be necessary to fully understand the interactions between metabolism, microbiome and immunity in order to design rational interventions, but this is still several decades away. At present, we are beginning to use high-throughput analytical techniques coupled with multi-variate analyses and machine learning to identify associations between particular patterns, in order to empirically identify functional interventions.

2 The mucosal immune system

In very simple terms, the main function of the immune system is to recognise and respond to non-self-molecules, while avoiding responses to self-molecules. Multiple mechanisms which allow the immune system to distinguish between self and non-self have been characterised. However, mammals, including pigs and humans, are also exposed to a range of antigenic, non-self-molecules such as food, aerial pollutants, commensal bacteria, in addition to the pathogens and toxins which the immune system protects against, and inappropriate inflammatory responses to these harmless foreign antigens can be as detrimental as the response to self-antigens. The major components of the immune system may escape exposure to such environmental, harmless antigens, but mucosal surfaces will be continuously exposed.

The mucosal immune system refers to that part of the immune system which is associated with any mucosal surface within the body, whether this be the gastrointestinal tract, the respiratory tract or the urogenital tract. In the simplest terms, it may be defined as the immunological components which are associated with these surfaces. However, a more functional definition includes the presence of specific 'homing' receptors expressed on lymphocyte surfaces, capable of targeting re-circulating lymphocytes to mucosal surfaces, together with a characteristic type of immune response, such as IgA, which occurs at a variety of mucosal sites exposed to the environment (Bailey et al., 2005).

Perhaps the most relevant to pig production is the intestinal immune system, as this is the largest mucosal surface in the animal and one that must be highly specialised to cope with the local absorption of nutrients and, consequently, fully able to distinguish between self-, food and commensal bacterial antigens and potentially pathogenic organisms. In addition, opportunistic invasion of host tissues by normally commensal bacteria can have serious consequences, including sepsis and inflammation; therefore, a system is necessary to preserve the homeostatic relationship between the normal, commensal microbiota and the host. This occurs by three main mechanisms: by minimising direct contact between intestinal bacteria and the epithelial cell surface (stratification), by confining penetrant bacteria to their sites of residence in the intestine (compartmentalisation) and by controlling those immune responses which do occur (regulation).

2.1 Structure and barrier function

The intestine provides a partially permeable surface, which must allow the absorption of nutrients from the external environment and prevent unwanted molecules from entering. Changes in the area available for absorption of nutrients dictate the amount that can be absorbed and used by the pig, and the permeability of the intestine influences the types of molecules that can cross the intestinal barrier. Both of these aspects will influence the metabolic and immune systems of the pig, as well as impact the substrates available for use by the intestinal microbiota.

Intestinal villi and crypts, and the epithelial cells which cover them, are integral for increasing the surface area available for absorption of nutrients and, in addition, contain a range of secretory cells, including goblet cells (mucus secreting), Paneth cells (anti-microbial peptides) and enteroendocrine cells (hormone). Note that Paneth cells are confined to the small intestine and are not present in the colon.

Mucus secreted by goblet cells throughout the intestine provides an initial barrier to external molecules and particles. In the colon, there are two layers of mucus: a loose outer layer, which acts as a substrate for bacteria, and a dense inner layer, which prevents the bacteria from reaching the epithelium. In the small intestine, there is only a loose outer layer of mucus, and protection against luminal bacteria is provided by anti-bacterial mediators, such as alpha-defensins produced by Paneth cells, and the mucosal-specific immunoglobulin class, IgA (Johansson and Hansson, 2016). The peristaltic movement of the intestine also facilitates this process.

Beneath this initial mucus barrier, the intestinal epithelium provides a second barrier to molecules through selectively permeable cell membranes and tight junctions between adjacent cells. The integrity of these junctions is crucial for controlling transfer of macromolecules and bacteria between the intestinal lumen and host tissues. Disruption of these tight cell junctions, such as is seen with porcine epidemic diarrhoea virus infection, is detrimental to the health of the pig (Jung et al., 2015).

The lamina propria is a layer of connective tissue below the epithelium, which contains a diverse range of structural cells and leucocytes. In the intestinal lamina propria, the leucocyte population includes plasma cells secreting immunoglobulins (mainly IgA), T-lymphocytes, macrophages and specialised antigen-presenting cells such as dendritic cells. These leucocytes form two distinct types of organisation: highly compartmentalised, discrete tissues like the Peyer's patches and mesenteric lymph nodes; and more diffuse infiltration of the lamina propria with leucocytes along the entire length of the intestine.

Several pathways have been described by which the mucosal immune system samples antigen in the Peyer's patches and the mesenteric lymph nodes. The first involves antigen sampling and transportation by M-cells in the follicle-associated epithelium that overlies the dome region of the Peyer's patch. Following this, specialised dendritic cells take up the antigen and migrate to the T-lymphocyte zones where they present the antigen, resulting in T-lymphocyte activation and migration. In pigs, uptake of antigen into Peyer's patches is believed to be critical for induction of immune responses (Devriendt et al., 2013). Subsequently, there is also signalling to the B-lymphocyte follicles, resulting in induction of responses by B-lymphocytes. There is also evidence that dendritic cells can extend dendrites through the epithelium directly into the lumen by manipulating the tight junctions, allowing them to acquire antigen and, therefore, bypass the need for antigen sampling by M-cells (Rescigno et al., 2001). These dendritic cells then either migrate directly to the mesenteric lymph nodes, where antigen can be presented to the T-lymphocytes, as in the previous pathway, or pass antigen on to migratory dendritic cells for transport to mesenteric lymph nodes (Huang et al., 2000). A third pathway involves intact antigen being absorbed directly across the epithelium, into the lymphatic system and eventually reaching more distant organised lymphoid organs (Telemo et al., 1991). Given the variations in the mechanisms outlined above, there are important considerations when developing, for example, mucosal vaccination strategies, since the most protective immune responses seem to be induced when the local mucosal surface is exposed to the antigen (Sedgmen et al., 2004).

The diffuse immunological architecture of the pig intestine has been well described. This site is usually considered an effector site: that is, immune responses are normally induced in the Peyer's patches and mesenteric lymph nodes as described above, and primed, antigen-specific T- and B-lymphocytes acquire appropriate homing receptors to return to the intestinal lamina propria to express the appropriate function, either tolerance or active immune responses. However, in the pig, the intestinal mucosa is considerably more compartmentalised than many other species. The B-lymphocytes that differentiate into plasma cells secreting immunoglobulins occur around the crypts, but do not appear in the intestinal villi. The villi are dominated by T-lymphocytes and, even here, there is clear spatial separation between CD8+ cytotoxic T-lymphocytes in the epithelium and CD4+ helper T-lymphocytes in the underlying lamina propria. Here, the CD4+ cells interact both with resident dendritic cells and with the capillary endothelial cells: both these cell types appear to be capable of sampling antigens and presenting them to helper T-lymphocytes.

Thus, large components of the mucosal immune system are continuously positioned to recognise and respond to antigens associated with pathogens, commensals and food. For many reasons, it seems unlikely that this architecture is present 'just in case' antigen breaches the normal barrier function of the intestine. Similar architecture is not present underneath the skin, for example, where the barrier to foreign molecules is considerable. As will be discussed later, significant quantities of antigens derived from feed do appear in the blood of normal pigs, demonstrating at least some passage of macromolecules across the intestine. Finally, routine sampling of luminal antigen does seem to occur by mucosal dendritic cells. Thus, the intestinal mucosal immune system seems not to be just a passive structure, waiting for antigen to appear: it is a dynamic architecture in which constantly migrating cells continuously and actively sample luminal antigen, and continuously need to make decisions about the type and level of immune responses to express. This dynamic process has been termed 'controlled inflammation', but is perhaps better viewed as 'active surveillance' in order to regulate the appropriate level inflammation. Regardless, maintaining the level of active immunological commitment within the intestinal mucosa must carry a corresponding level of nutritional cost even in the absence of mucosal inflammation.

2.2 The importance of IgA

Secretory IgA (sIgA) is a major component of the intestinal immune system. It protects the intestinal mucosa from enteric toxins and pathogenic microorganisms, and also helps maintain intestinal homeostasis by regulating the bacterial population. Dimeric IgA is transported across the intestinal epithelium via the polymeric Ig receptor protein and is released into the lumen, along with secretory component. Once in the lumen, sIgA binds to target antigens, coating bacteria, blocking the action of toxins and adhesion molecules, and also acting to confine bacteria within mucin and thus preventing access to intestinal epithelium. IgA antibodies to some components of the normal microbiome have been identified in mouse and are almost certainly present in pigs. However, the mechanisms which dictate different responses to different bacterial taxa are still relatively unknown (Fransen et al., 2015).

While it is often used as a readout for effects of nutritional intervention studies, IgA levels in faeces or in blood can be difficult to interpret as a proxy measure of enteric health. For example, studies carried out by Lewis et al. (2013), which involved feeding a probiotic that had previously been defined in humans and mice, showed an increase in expression of molecules associated with intestinal epithelial barrier function and a decrease in IgA levels in the intestine. In these studies, it seems likely that reduced IgA production was a consequence of reduced challenge with antigen because of improved barrier function: that is, reduced IgA was a proxy for enteric health. However, an increase in IgA levels has also been reported after feeding probiotics to pigs (Scharek et al., 2007) and, in this study, increased IgA was interpreted as an indicator of an effective mucosal immune response. Thus, IgA may offer a useful proxy measure, but care must be taken to consider the underlying mechanisms that may be responsible for the observed changes.

2.3 Immune exclusion and oral tolerance

The actual mechanisms that allow the immune system to discriminate between harmless antigens and harmful pathogens and, specifically, not to respond with damaging responses to food antigens, remain unknown, but several regulatory mechanisms have been identified. One mechanism may be the simple production and secretion of IgA from plasma cells in the intestinal lamina propria into the mucus layer. Antigen-specific IgA can bind to antigen and to mucin, trapping luminal antigens before they gain access to the enterocyte barrier. Alternatively, antigenic macromolecules, which have already been absorbed can be similarly bound by specific IgA in the lamina propria and the complex transported across the epithelium into the lumen, effectively pumping absorbed antigen back out. The combination of these two processes has been termed 'immune exclusion' (Stokes et al., 1975).

However, several groups have shown that weaned piglets do absorb macromolecules from the weaner diet (specifically soya or egg) into blood, despite the process of immune exclusion. Although this may be a consequence of the late appearance of IgA in young piglets. Further, the presence of food antigens in the blood can trigger an active immune response systemically and at the mucosal surface, characterised by serum antibody, inflammation and loss of function, which may be associated with postweaning diarrhoea (Miller et al., 1984). However, once piglets have mounted a primary response to antigens encountered via feed, the response starts to wane, despite continued feeding. If the piglets are then subsequently injected systemically with the same antigen, they mount a much-reduced immune response, or even fail to mount any secondary response. In contrast, responses to unrelated, novel antigens remain normal. The specific response to antigenic molecules in feed appears to be switched off: they have become immunogenically tolerant to the antigen which they have been fed, a mechanism which is termed 'oral tolerance' (Bailey and Haverson, 2006).

Antigen-specific oral tolerance can, therefore, be seen as a mechanism by which a host can prevent potentially damaging immune responses to harmless, non-self-antigens present in food, while still mounting an effective, protective response to potentially pathogenic molecules. Despite this, the actual mechanisms by which the mucosal immune system identifies those antigens to which it needs to make active immune responses (pathogens) and those to which it must make antigen-specific tolerance, rather than active responses, are still unclear. Historically, several mechanisms have been proposed to account for this acquisition of immunological tolerance to food. Including T- and B-cell clonal anergy or deletion (loss of function by food antigen-specific lymphocytes) or induction of antigen-specific regulatory T-lymphocytes (Pabst and Mowat, 2012).

Whether the mucosal immune system also develops antigen-specific tolerance to normal, commensal gut bacteria is less clear. Immune responses against normally commensal bacteria have been implicated in the pathogenesis of inflammatory bowel disease (IBD) in humans and in a number of experimental models of colitis in laboratory rodents. However, it is also clear that some level of specific immune responses against some commensal gut bacteria (antigen-specific T-lymphocytes, IgA) can be present in normal individuals without apparently impairing gut digestive and absorptive functions. Clearly, though, the type and level of response is different between pathogens, commensal bacteria and food, and the mucosal immune system must encompass mechanisms for making those distinctions.

One potential way in which the mucosal immune system may be able to distinguish between commensal and potentially pathogenic microbes is by integrating recognition by the innate immune system and the adaptive immune system. In this model, antigen-specific recognition occurs through T- and B-lymphocyte receptors, whereas non-specific recognition of 'danger' occurs through receptors associated with the innate immune system: only when both are activated are antigen-specific responses triggered. One potential way this may happen is through the use of pattern recognition receptors (PRRs), which recognise and interact with conserved, pathogen-associated molecular patterns (Artis, 2008). There are many types of PRRs, of which the most studied are the toll-like receptor (TLR) family, which may assist with making the distinction between commensal and pathogenic bacteria. Many cell types, including gut epithelial cells, T- and B-lymphocytes and dendritic cells, express a range of different TLRs: for example, TLR-4 recognises microbial lipopolysaccharides, while TLR-5 recognises flagellin. Interestingly, both TLR-2 and TLR-9 appear to be present in newborn animals, suggesting involvement with early colonisation of the intestine with microbes (Tohno et al., 2008). However, it is clear that there are no molecular patterns which are unequivocally associated solely with pathogens or pathobionts and not with commensals, such that simple triggering of one PRR cannot be the sole mechanism for discriminating between 'harmless' and 'dangerous'. If PRRs are the critical factors in discriminating between the two, then the host immune system must be either making the distinction based on integrating information on levels from a range of different PRRs, or on signalling through PRRs in a range of different locations. For example, early work suggested that the expression of TLR-5 on intestinal epithelial cells was confined to the basal and lateral membranes and not the apical membrane, such that triggering would only occur if the tight cell junctions had been breached by luminal bacteria. While conceptually appealing, this has not been supported by more recent work and the involvement of PRRs in discrimination is still unclear.

However, recent interest has identified an alternate pathway which might mediate signalling between innate and adaptive immune systems in the intestinal mucosa, centred around a newly identified set of immunological cells which are now well studied in rodents and humans, but have received little attention in pigs. The innate lymphoid cells (ILCs) do not express T- or B-lymphocyte receptors which would mark them as components of the adaptive immune system, but clearly contribute to responses to intestinal bacteria. The interactions between ILCs and microbiota have not yet been fully explored, but their function is subject to microbial signals, as with T-lymphocytes. Interleukin-22, a cytokine notably produced by proinflammatory Th17 cells and which has a pleiotrophic effect on the mucosal immune system, is also produced by group 3 ILCs, suggesting that the ILCs may communicate with the resident microbiota and may integrate and transmit signals to the adaptive immune system (Belkaid and Hand, 2014).

2.4 Development of the mucosal immune system in young piglets

Probably due to the lack of transport of macromolecules across the placenta, piglets have a poorly developed immune system at birth. The piglet's active immune system (that is, the ability to mount appropriate, active immune responses to novel antigens) is immature, and it relies on passive maternal protection provided through colostrum and milk until weaning. There are several variables that drive the development of the immune system after birth including the external environment, genotype, diet, maternally derived antigen and antibody (Bailey et al., 1994) and the host gut microbiome. However, the time of weaning in most commercial systems is still before the piglet's immune system has reached full maturity, and before it has developed the full ability to distinguish between potential pathogens, commensals and food. Given the large number of novel antigens that piglets are exposed to at weaning, associated with a change in diet, removal of maternal milk antibody, exposure to novel pathogens and commensals, it is hardly surprising that some of the highest levels of morbidity and mortality are seen during the post-weaning period. Historically, there has been widespread use of antibiotics to address infectious diseases during this critical period; however, there is increasing interest in alternative approaches to facilitate earlier maturation of the piglet's immune system.

At birth, both mucosal and systemic tissues contain relatively low numbers of T- and B-lymphocytes, and what B-lymphocytes that are present have a very limited repertoire (that is, the number of novel antigenic structures they can actually recognise is still low). At this stage, they are highly dependent on the repertoire of immunoglobulins they receive via colostrum, from which they absorb IgG, IgM and IgA across the intestinal barrier, and subsequently via milk.

At 1–2 weeks of age, there is evidence of non-specific expansion of B-lymphocytes and Peyer's patches, along with an influx of MHC II+, antigen-presenting cells including dendritic cells. A small number of unconventional, CD2+CD3+CD4-CD8αα+

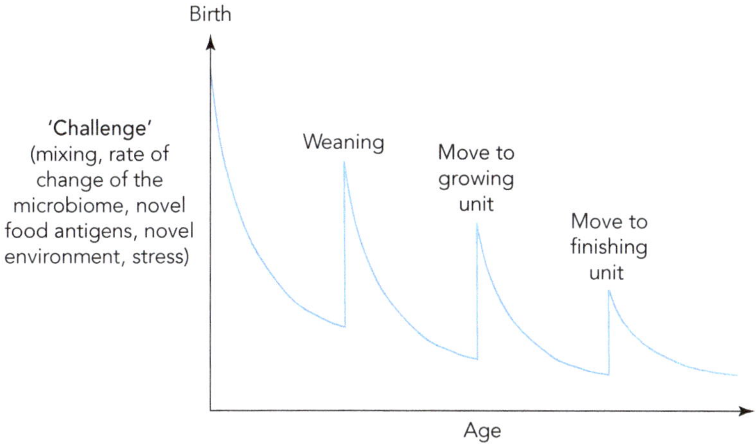

Figure 2 During the lifespan of the pig from birth to slaughter, there are a number of critical points during which the environment changes dramatically – weaning, moving to a new unit, etc. Associated with the potential stress of an environment change, pigs are also subjected to changes in diet and exposure to commensal microbiomes of other pigs, and these result in rapid shifts in the intestinal microbiome and in the challenges which the immune system must cope with.

T-lymphocytes are present during this early stage, before the conventional cell types appear. Conventional T-lymphocytes appear more slowly, with CD4+ T-lymphocytes beginning to appear between 2 and 4 weeks of life. These cells initially express molecules indicating activation, before they settle down into a more resting phenotype. True cytotoxic CD8αβ+ T-lymphocytes do not really appear in numbers until 4 weeks old. Finally, around 6 weeks old, the immunoglobulin repertoire is expanded, contributing to the ability of young piglets to make antigen-specific B-lymphocyte responses (Stokes, 2017). Interestingly, IgA is one of the later characteristics of an effective mucosal immune system to appear in piglets: IgA+ plasma cells do not appear in the intestinal lamina propria until 3–6 weeks of age. Thus, the development of the mucosal immunological architecture in pigs is, again, not consistent with simple inflammatory responses: different cell types are recruited to the intestine in a consistent pattern over many weeks, rather than all cell types appearing together and simply increasing in number.

Superimposed on this staged, gradual development are a number of dramatic changes in husbandry, associated with major changes in exposure to food and microbial antigens, of which the first is weaning. Further 'catastrophic' events, such as movement to growing or finishing facilities, may also result in similar changes in microbiota composition and affect the ability to mount immune responses (Fig. 2).

3 Experimental studies of the links between the immune system and microbiota

It is now well established that there is cross-talk between the gut microbiome and the immune system of the host and that this occurs bi-directionally via direct and indirect mechanisms. Many studies have also confirmed the interplay between the microbiome, the immune system and metabolism in a range of laboratory species, as well as in pigs. While the direction of causality in many of these studies has yet to be elucidated, transplantation of human microbiomes into germ-free mice does replicate the human metabolic and immunological phenotypes, suggesting at least some direct effect of microbiome on host phenotype (Blanton et al., 2016; Crabbe et al., 1968).

The development of a fully functional immune system is reliant on the presence of gut microbiota, as germ-free mice fail to develop normal, mature immune systems (Hooper et al., 2012). Similarly, caesarean-derived, bubble-maintained gnotobiotic piglets fail to develop a normal immune system (Inman et al., 2012). In rodents, colonisation with a single bacterial species may restore some normal immunological function (Atarashi et al., 2011; Round and Mazmanian, 2010). However, these studies are always carried out using laboratory mice in which the normal microbiome has been restricted for nearly a century, which may limit the relevance of these findings to normal mammals. Additional mouse studies have explored the effects of particular bacteria on the development of the immune system. For example, colonisation of gnotobiotic mice with a mixture of 46 different clostridial strains resulted in the expansion of the immunological component of the lamina propria and of systemic regulatory T-lymphocytes (Atarashi et al., 2011). Similarly, colonisation of gnotobiotic piglets with a defined, three-component microbiome recapitulates the early stages of expansion of the mucosal immune system, sequential recruitment of dendritic cells and CD4+ T-lymphocytes, at least. However, care must also be taken in interpreting these studies since, while they provide important information regarding mechanisms, it is difficult to extrapolate to a wild-type intestinal setting, where interactions with other microbes may significantly affect the functions seen with a simple mono- or oligo-colonisation.

In utero, like all other placental mammals, piglets are essentially sterile and acquire their initial 'seed' microbiome during transfer through the vagina at farrowing, and from the skin of the mother during suckling. Further colonisation will occur from the environment over the next few days, but much of this will still be derived from the mother, as a consequence of faecal and dermal contaminations. As piglets begin to play and forage, non-pig-origin components of the microbiome may be acquired. Although some work has been carried out on the development of the microbiome of pigs (Mach et al., 2015), this process has not been studied as well as in human infants. Importantly, it seems from the human studies that the process is not simply a progressive expansion of an initial microbiome as more species are added. Instead, the initial, relatively simple microbiome is almost completely replaced by a new, more complex microbiome, which is then replaced by a third, until a final, relatively stable microbiome is achieved, which may contain none, or relatively few, of the initial strains or species present shortly after birth (Koenig et al., 2011): in human infants, this seems to occur at about three years old (Yatsunenko et al., 2012). This is highly comparable to the core concept of 'succession' described in ecology, where the species structure of an ecosystem changes over time: again, 'pioneering' species rapidly colonise a new environment such as that created by a forest fire and are succeeded by a series of different types of communities, usually of increasing complexity, until a stable 'climax' community is achieved. Current technology has largely focused on determining what microbial species or strains are present in the intestine of individuals or groups, by sequencing and identifying a single gene: usually, the gene encoding the 16S subunit of the ribosome (16S rRNA). However, it is increasingly apparent that we need to think of the intestinal microbiome not as a list of microbial species, but as a dynamic ecosystem in which host, bacteria, viruses and protists interact, and that development of a stable, 'climax' microbiome proceeds through a succession of characteristic early ecosystems. Current techniques used to begin to analyse this complexity include microbial metagenomics, where as many genes as possible are sequenced from a microbial community (Human Microbiome Project Consortium, 2012) and this approach is beginning to be supplemented by microbial transcriptomics, proteomics and metabolomics.

Fundamental to the concept of succession in ecology is that each community determines to a large extent the nature of the succeeding community, rather than successive communities being independent of each other. In order to investigate this phenomenon in piglets, the effect of the early microbiome on long-term carriage was investigated by weaning piglets from several different litters into mixed pens, one piglet from each litter in each pen. The results demonstrated that the microbiomes of piglets continued to be more similar to that of their littermates, rather than to their pen-mates (Stokes, 2017), strongly suggesting that the early, preweaning acquisition of microbiomes was still playing a role in the type of microbial ecosystem establishing after weaning.

A proposed model for this process is shown in Fig. 3. Here, a simplified microbiome 'space' is represented in two dimensions. Soon after birth, piglets acquire an initial microbiome, which subsequently undergoes succession, represented by a shift in microbiome space. However, environmental influences cause the 'trajectories' of piglets under different conditions to gradually diverge, resulting in distinct 'climax' communities. Studies in humans have suggested that, while considerable variation in microbiomes does occur between individuals, the stable climax communities do cluster, roughly, into a small number of so-called 'enterotypes' (Arumugam et al., 2011; Knights et al., 2014). In pigs, two enterotypes have been identified, dominated by Prevotellaceae or Ruminococcaceae, and there seem to be both immunological and performance phenotypes associated with these

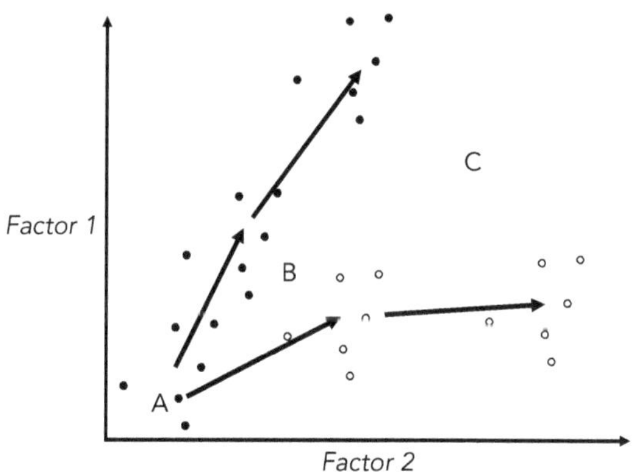

Figure 3 The natural history of microbial ecosystems is to change, becoming more complex (diverse) over time. The complexity of the microbiome is reduced to two dimensions, factor 1 and factor 2, and the microbiome of each pig at a specific time is represented by a single point. Microbiomes are represented at three time points: A, B and C. Where two populations of animals are isolated from each other, differences in environment, together with random effects, will result in gradual divergence of the microbiomes. Because the 'distance' between the two populations becomes greater as the animals age, and because the 'climax' microbiome is more stable, it is likely to be much easier to 'nudge' the microbiome early in life (time point B) than later (time point C).

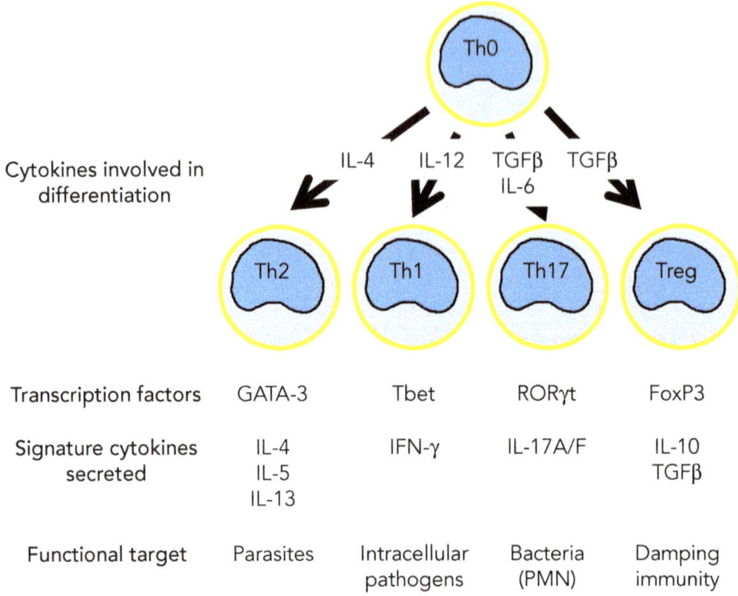

Figure 4 Functional complexity of T-cell subsets. Early studies on the development of the immune system focused on the balance between T-helper 1 (Th1) and T-helper 2 (Th2) T-lymphocytes. However, it is now well established that there are multiple subsets of helper (CD4+) T-lymphocytes, of which the four best-characterised are shown here. Regulatory T-lymphocytes (Tregs) are critically involved in controlling the function of the other subsets and preventing excessive inflammation.

microbial enterotypes (Mach et al., 2015; Ramayo-Caldas et al., 2016), the Prevotellaceae-dominated enterotypes providing performance advantages. In neither humans nor pigs have the pathways or trajectories by which these climax enterotypes establish been determined, but the inference from the model in Fig. 3 is that either they may be vertically transmitted such that the early microbiomes are already different (that is, the enterotypes of piglets are likely to resemble that of their sow) or the early microbiome may be similar but they diverge over time as a consequence of systematic or stochastic differences in their environments. The difference is important for targeting future intervention strategies, since a further implication of the model in Fig. 3 is that deliberate or accidental targeting of the early stages of succession is likely to require less drastic interventions to be effective: targeting 'climax' communities may require catastrophic interventions.

The link between early life experiences, in particular, the type of colonisation by microbiome and the development of inflammatory and allergic disease is one that has been studied in great detail in humans and laboratory rodents, but not in pigs. Epidemiological studies have clearly shown that early microbial colonisation is linked to the likelihood of immune responses to self-antigens (autoimmunity) and harmless environmental antigens (allergy). In particular, there are clear associations between rural and, particularly, farm environments in childhood and protection from allergy and IBD. In pigs, differences in early environment (intensive, extensive, isolator, antibiotics) also affect their microbiome and expression of a range of immunological functions in the intestine (Inman et al., 2010;

Lewis et al., 2012; Mulder et al., 2009, 2011; Schmidt et al., 2011). Consistent with the model proposed in Fig. 3, microbiomes of piglets either left on the sow or moved to a high-containment isolator diverged over time, and this difference was strongly linked to differences in recruitment of dendritic cells and regulatory T-lymphocytes to the intestinal mucosa, along with differences in the amount of interleukin-2 (IL-2) and interleukin-4 (IL-4) produced by systemic and mucosal T-lymphocytes (Inman et al., 2010).

Despite the observations that microbiomes in piglets are directly linked to the type of immune system which develops in the intestinal mucosa, the original idea that microbiota colonisation affected a balance between the more antibody-based immune responses (Th2) and a cellular response (Th1) is now seen as over-simplistic, and involvement of regulatory mechanisms (Tregs) are now seen as critical (Fig. 4). The effects of different microbiomes are now thought to depend on a more complex balancing act between all the T-lymphocyte subsets, in addition to regulating the effects of commensal microbes on the interactions between intestinal enterocytes and dendritic cells (Bailey, 2009).

4 The microbiome and metabolism

4.1 Metabolism and microbiota

It is now well established from studies in mice and humans that there is a strong association between the intestinal microbiome and a range of metabolic conditions, including obesity, metabolic syndrome, environmental enteric dysfunction and diabetes. While none of these are specific health problems in pigs, the level and type of fat deposition are critical for producing the optimal carcase, and the association clearly suggests that the microbiome is an important target for manipulating the metabolism of pigs for optimal performance. Host metabolism can be directly influenced by the intestinal microbiota and, in addition, will affect and be affected by the immune system and its response to the microbiome. Thus, the three components, microbiome, metabolism and immunity, interact strongly with each other and should not be considered in isolation.

4.2 Metabolomic analyses and detection techniques

The intestinal microbiota aids in energy harvest from food intake, fermenting amino acids, and recycling proteins and carbohydrates (Nyangale et al., 2012). Therefore, understanding the role of the intestinal microbiome in the context of host metabolism is a key goal in veterinary and agricultural research. This need is met by the field of metabolomics, a high-throughput technique that involves measurement of low molecular weight metabolites in a cell, tissue or organism under genetic or physiological perturbation. Modern metabolomics research includes two distinct strategies.

Untargeted metabolomics aims to gather information on a complete repertoire of all measurable metabolites within a sample, including potential unknown metabolites, establishing state-specific profiles and characterising global changes. Targeted metabolomics aims to measure a specified and fully characterised subset of known, measurable metabolites. Targeted metabolomics uses existing, increasing knowledge of biochemical pathways to develop a functional understanding of the impact of test conditions on metabolism, while untargeted metabolomics has the potential to identify novel microbial metabolic pathways, although both strategies contribute to further development of the biochemical pathway knowledge base. This is particularly important since, currently, a significant proportion of microbial genes are 'function-unknown', clearly indicating that our understanding of the complexity of microbial metabolic pathways is still relatively limited.

Advancements in technologies mean that high-throughput, next generation methodologies are being increasingly used to analyse both known and unknown metabolites. Even where the pathways in which any specific metabolite features may not yet be clear, these molecules can still act as biomarkers for health and disease when they are consistently found to be correlated with a particular phenotype.

Nuclear magnetic resonance (NMR) spectroscopy and mass spectrometry (MS) are the best-represented tools in metabolomic studies, increasingly augmented with automated high-level computing (as with other meta-omics) (Smirnov et al., 2016). NMR benefits from minimal sample preparation requirements. For example, faecal extracts can be prepared in methanol or water (Jacobs et al., 2008). The method is complicated by the convoluted output caused by surrounding metabolites. For this reason, results libraries are unsuited to NMR studies. Conversely, there are a multitude of MS variations including MS coupled to gas chromatography, liquid chromatography and capillary electrophoresis. Fourier transform infrared spectroscopy offers similar sensitivity to NMR, also enabling high-throughput analyses. Collectively, MS solutions benefit from clear signals, high sensitivity and high selectivity at the expense of lengthy sample extractions and overall speed. Fourier-transform ion cyclotron resonance MS can demonstrate a high dynamic range, overcoming the sensitivity limits of spectroscopic methods alone (D'Anastasio et al., 2009).

Despite the improvements in metabolic profiling, one of the greatest challenges is determining whether the products are a result of microbial or host metabolism. In many cases, they may be host–microbe co-metabolites, meaning the products of bacterial metabolism are absorbed and utilised by the host metabolism or vice versa. The major metabolites produced by the gut microbiota and their effects on the host are discussed in further detail below.

4.3 Specific pathways

Organic acids are metabolites produced by the gut microbiota, which include health biomarkers of interest. This is exemplified by the short-chain fatty acids (SCFAs, discussed separately), lactate, benzoate and hippurate. Benzoate and subsequently hippurate are produced primarily from bacterial metabolism of dietary phenolic compounds. Urinary hippuric acid is considered an obesity biomarker in humans, but is also elevated in healthy humans supplemented with green tea as a possible waste product of chlorogenic and quinic acid metabolism (Mulder et al., 2005). Phenylacetate and phenylpropionate are also organic acid products of gut microbiota, and presence of these organic acids in urine corresponds to poor health in humans (Mulder et al., 2005; Sarkissian et al., 2000). Additionally, hydroxyphenylacetate and phenylacetate are considered biomarkers for colorectal cancer. Lactic acid is a biomarker for healthy vaginal microbiota and is also produced by some human tissues; however, D-lactate can accumulate in the blood and urine of individuals with impaired intestinal carbohydrate adsorption (Bongaerts et al., 2005; Ewaschuk et al., 2005). D-lactic acid accumulation culminates in D-lactic acidosis and intestinal symptoms (Haan et al., 1985; Vella and Farrugia, 1998). In pigs, the dietary inclusion of potassium formate and potassium diformate, organic acid salts, has been reported to improve finishing weight, without altering feed-conversion ratio (Htoo and Molares, 2012). Whether this is a direct effect or mediated through effects on microbiome or host immune system is unclear.

SCFAs have a widely reported strong association with human gut health and are, therefore, among the best-studied metabolites within the context of gut microbiomes. SCFAs are produced when gut bacteria ferment fibres, dietary plant polysaccharides, indigestible oligosaccharides and other otherwise indigestible precursory metabolites after they have passed through the GI tract. Acetate, propionate and butyrate are the main SCFAs produced by the gut microbiota that accumulate in the colon. Cross-conversion of SCFAs within gut microbiomes contributes to overall metabolic flux *in vivo* (den Besten et al., 2013). Evidence exists for cross-conversion from acetate to butyrate and from butyrate to propionate. Several studies report increased butyrate, propionate and/or acetate as health markers. Acetate, the most abundant colonic SCFA, is reported to correlate with feeding high-fat diets in animals, and with predisposition to metabolic syndrome in humans. Acetate is also thought to be a cholesterol precursor and is adsorbed by tissues involved in cholesterol increase. Butyrate is the major source of energy for cells of the colon and is increasingly linked to their function in regulating activation of innate and adaptive immune responses. Microbial butyrate production has been found to be low in humans with ulcerative colitis, a form of IBD, but colitis in at least one rodent model is mediated in part by high-butyrate-producing α-proteobacteria (Ríos-Covián et al., 2016). *Faecalibacterium prausnitzii* is often cited as a commensal gut microbe that produces high amounts of butyrate, and *Faecalibacteria* deficiency may predispose to IBD, but many other organisms produce this metabolite (Barcenilla et al., 2000). Reduced numbers of propionate-producing bacteria in the microbiome have been identified in children at higher risk of asthma (Arrieta et al., 2015). In pigs, butyrate feeding has been used to reduce shedding of *Salmonella* and improve growth rates (Lynch et al., 2017). While the underlying mechanisms are not clear, butyrate feeding does have significant effects on piglet metabolism (Lu et al., 2012).

SCFAs synthesised by gut microbiota can be used for lipid synthesis; however, direct links between lipid metabolism in gut microbiota and subsequent obesity have been investigated in mice and the extent of the correlations is ambiguous (Nicholson et al., 2012; Ghazalpour et al., 2016). One study observed a direct correlation between gut microbiome composition and blood lipid levels (Fu et al., 2015). Specifically, there were correlations between the Christensenellaceae family and Tenericutes phylum and low BMI, lowered triglycerides and higher levels of high-density lipoprotein. A metabolite of gut microbiota-mediated breakdown of the lipid phosphatidylcholine, trimethylamine N-oxide (TMAO) correlates with cardiovascular disease (Wang et al., 2011). Furthermore, Lachnospiraceae, Ruminococcaceae, Prevotella, Treponema, and Bacteroides were correlated with 'fatness' in two porcine breeds (He et al., 2016). While it is clear that the gut microbiome is involved in metabolic pathways of many intermediates of lipid metabolism, including bile acids and SCFAs, the exact biochemical pathways and, importantly, the roles of different components of the ecosystem, which are involved are unknown and the subject of active research.

Bile salts are secreted by hepatocytes into the duodenum, upon lipid consumption, to aid adsorption of dietary lipids and fat-soluble vitamins (A, D, C and K), then most of the remaining bile is re-absorbed into distal ileum. Some bile salts are deconjugated by the gut microbiota, then converted into secondary bile acids or re-absorbed: secondary bile acids are host microbe co-metabolites. Gut microbiota can also modulate bile acid metabolism via downregulation of a tauro-beta-muricholic acid receptor, and affect hepatic bile acid synthesis in mice (Schleiffer and Gairard, 1988). Secondary bile acids produced by Clostridiaceae correlate with predisposition to colon cancer and with cholesterol gallstone disease, which demonstrates the need for organism-specific metabolomic analyses (Bernstein et al., 2005).

Dietary polyphenols are bioactive components found in plant-based foods such as fruits, vegetables and potatoes as well as tea and coffee. Polyphenolic compounds are minimally absorbed in the small intestine and largely absorbed in large intestine, where they are enzymatically modified by gut microbiota (Faria et al., 2014). Phenolic compounds are characterised by hydroxy-phenyl groups and can be structurally classified as flavonoids and non-flavanoids. Flavonoids consist of two benzene rings connected with a pyrone ring, whereas non-flavonoids vary, existing as complex arrangements of connected benzene rings (Ozdal et al., 2016). Several studies report conversion of polyphenols by gut microbiota; however, the reciprocal impact of converted polyphenols on the gut microbiota is poorly understood. For example, caffeic acid is a hydroxycinnamic acid, a major type of non-flavonoid, abundant in fruit. The chemical derivative of caffeic acid, chlorogenic acid, can be converted to caffeine acid by specific bacterial lineages and again to 3-hydroxyphenylacetic acid in the colon of wildtype, but not germ-free mice (Selma et al., 2009). As with SCFAs, there are increasing numbers of studies on polyphenols owing to the reported health benefits in the context of the microbiome (Ozdal et al., 2016). Unfortunately, the bioavailability of polyphenols is not necessarily predictable simply from the levels in feed. Additionally, the effects of supplementation on abundance of polyphenolic metabolites vary within microbiomes. It follows that studies estimating colonic metabolism from the analysis of parent foods *in vitro* must be carefully scrutinised.

Microbial fermentation of proteins and amino acids largely occurs in the distal colon, but has been less well studied than intestinal microbial carbohydrate catabolism (Nyangale et al., 2012). Several protein degradation products are associated with cancer risks, linked to consumption of certain food groups, such as red meat (Hughes et al., 2000). Ammonia from microbial deamination is linked with tumour promotion and amines, typically toxified by the liver, are associated with migraine, hypertension and hepatic coma. In animals fed high-protein diets, these by-products are made in quantity and 'pathogenic' groups of bacteria increase (Richardson et al., 2013). Studies in gnotobiotic mice have demonstrated effects of the gut microbiota on levels of free amino acids in the GI tract.

5 Conclusion

Establishing the causal links between microbiome, metabolism and the immune system has the potential to dramatically change our approach to nutritional and husbandry interventions in pigs. However, rational targeting of interventions to manipulate the microbiome will require a full understanding of the microbiome and the host as an ecosystem, a goal which is several decades away at the moment. The development of high-throughput approaches to sequencing microbial genes and genomes, characterising host immune responses and identifying host and microbial metabolic products will accelerate this process. However, we are already in a position to use these approaches to make empirical associations between components of the microbial ecosystem and performance indicators of growth and health. While empirical associations do not identify mechanisms, they will provide likely targets for intervention, which can be validated in the field, and they also provide valuable indicators for mechanistic studies. Importantly, these empirical associations need to be made in field studies on commercial farms, rather than in experimental husbandry settings. The ability of this field to contribute to the practice, as well as the science, of pig production will depend on the collaboration between academia and industry.

6 Where to look for further information

Short-term research goals should be aimed at identifying empirical associations between composition of the microbial ecosystem (e.g. by 16S rRNA gene sequencing) and measures of productivity in commercial pig production systems. The industry could make significant, rapid advances and identify potential targets for nutritional and management interventions from this kind of empirical association. In the long term, empirical correlations will feed into the more detailed, mechanistic understanding of microbial communities as ecosystems which will allow fully rational design of interventions. A second area of importance will be the timing of interventions: current evidence is suggesting that the earlier an intervention, the higher the probability of achieving a long-term change in the microbiome, metabolism and immune system of pigs. Research needs to identify the optimal time of intervention: should we be working post-weaning, immediately after birth or even in the pregnant sow?

Work in this area is being carried out in a large number of research institutes around the world, a measure of the importance attached to the area by the scientific community. Our own lab, the Mucosal Immunology Group at The University of Bristol, UK, has a long track record of research around the development of the immune system in young piglets and the effects of nutrition and microbial colonisation. A selection of other laboratories includes the Laboratoire de Génétique Animale et Biologie Intégrative (GABI) in the Institut National de la Recherche Agronomique (INRA), France, which has made significant advances in cataloguing swine microbial genomes; the Department of Veterinary and Biomedical Sciences, University of Minnesota, USA, who have studied changes in the microbiome associated with infectious disease; the Department of Agrotechnology and Food Sciences, Wageningen University, the Netherlands, who have developed much of the technology for analysis in humans and domestic livestock; the Department of Animal Science, University of Aarhus; and the Department of Food Science and Technology, University of California Davis.

Much of the important advances in mechanistic understanding of intestinal microbial ecosystems are likely to come from human studies, because of the level of funding. The important resource here is the Human Microbiome Project. In addition, major laboratories include the Gordon lab at Washington University School of Medicine at St Louis, USA; and the Knight lab at the University of California, San Diego, USA.

7 References

Arrieta, M-C., Stiemsma, L. T., Dimitriu, P. A., et al. (2015) Early infancy microbial and metabolic alterations affect risk of childhood asthma. *Science Translational Medicine* 7(307): 307ra152.

Artis, D. (2008) Epithelial-cell recognition of commensal bacteria and maintenance of immune homeostasis in the gut. *Nature Reviews Immunology* 8(6): 411–20.

Arumugam, M., Raes, J., Pelletier, E., et al. (2011) Enterotypes of the human gut microbiome. *Nature* 473(7346): 174–80.

Atarashi, K., Tanoue, T., Shima, T., et al. (2011) Induction of colonic regulatory T cells by indigenous Clostridium species. *Science* 331(6015): 337–41.

Bailey, M. (2009) The mucosal immune system: Recent developments and future directions in the pig. *Developmental and Comparative Immunology* 33(3): 375–83.

Bailey M. (2015) Interaction between immunological system, microflora of the gastrointestinal tract and metabolism. In: *Energy and Protein Metabolism and Nutrition*, pp. 27–33.

Bailey, M. and Haverson, K. (2006) The postnatal development of the mucosal immune system and mucosal tolerance in domestic animals. *Veterinary Research* 37(3): 443–53.

Bailey, M., Haverson, K., Inman, C., et al. (2005) The influence of environment on development of the mucosal immune system. *Veterinary Immunology and Immunopathology* 108(1–2): 189–98.

Bailey, M., Miller, B. G., Telemo, E., et al. (1994) Altered immune response to proteins fed after neonatal exposure of piglets to the antigen. *International Archives of Allergy and Immunology* 103(2): 183–7.

Barcenilla, A., Pryde, S. E., Martin, J. C., et al. (2000) Phylogenetic relationships of butyrate-producing bacteria from the human gut. *Applied and Environmental Microbiology* 66(4): 1654–61.

Belkaid, Y. and Hand, T. W. (2014) Role of the microbiota in immunity and inflammation. *Cell* 157(1): 121–41.

Bernstein, H., Bernstein, C., Payne, C. M., et al. (2005) Bile acids as carcinogens in human gastrointestinal cancers. *Mutation Research* 589(1): 47–65.

Blanton, L. V., Charbonneau, M. R., Salih, T., et al. (2016) Gut bacteria that prevent growth impairments transmitted by microbiota from malnourished children. *Science* 351(6275). Available from: http://www.ncbi.nlm.nih.gov/pubmed/26912898.

Bongaerts, G., Severijnen, R., Skladal, D., et al. (2005) Yeast mediates lactic acidosis suppression after antibiotic cocktail treatment in short small bowel? *Scandinavian Journal of Gastroenterology* 40(10): 1246–50.

Crabbe, P. A., Bazin, H., Eyssen, H., et al. (1968) The normal microbial flora as a major stimulus for proliferation of plasma cells synthesizing IgA in the gut. The germ-free intestinal tract. *International Archives of Allergy and Immunology* 34(4): 362–75.

D'Anastasio, R., Zipfel, B., Moggi-Cecchi, J., et al. (2009) Possible brucellosis in an early hominin skeleton from sterkfontein, *South Africa. PLoS ONE* 4(7): e6439.

den Besten, G., Lange, K., Havinga, R., et al. (2013) Gut-derived short-chain fatty acids are vividly assimilated into host carbohydrates and lipids. *American Journal of Physiology. Gastrointestinal and Liver Physiology* 305(12): G900–10.

Devriendt, B., Goddeeris, B. M. and Cox, E. (2013) The Fcγ receptor expression profile on porcine dendritic cells depends on the nature of the stimulus. *Veterinary Immunology and Immunopathology*, ECMIS 2011 – Escherichia coli and the mucosal immune system: interaction, modulation and vaccination152(1): 43–9.

Ewaschuk, J. B., Naylor, J. M. and Zello, G. A. (2005) D-lactate in human and ruminant metabolism. *The Journal of Nutrition* 135(7): 1619–25.

Faria, A., Fernandes, I., Norberto, S., et al. (2014) Interplay between anthocyanins and gut microbiota. *Journal of Agricultural and Food Chemistry* 62(29): 6898–6902.

Fransen, F., Zagato, E., Mazzini, E., et al. (2015) BALB/c and C57BL/6 Mice differ in polyreactive IgA abundance, which impacts the generation of antigen-specific IgA and microbiota diversity. *Immunity* 43(3): 527–40.

Fu, J., Bonder, M. J., Cenit, M. C., et al. (2015) The Gut Microbiome contributes to a substantial proportion of the variation in blood lipids. *Circulation Research* 117(9): 817–24.

Ghazalpour, A., Cespedes, I., Bennett, B. J., et al. (2016) Expanding role of gut microbiota in lipid metabolism. *Current Opinion in Lipidology* 27(2): 141–7.

Haan, E., Brown, G., Bankier, A., et al. (1985) Severe illness caused by the products of bacterial metabolism in a child with a short gut. *European Journal of Pediatrics* 144(1): 63–5.

He, M., Fang, S., Huang, X., et al. (2016) Evaluating the contribution of gut microbiota to the variation of porcine fatness with the cecum and fecal samples. *Frontiers in Microbiology* 7. Available from: http://www.ncbi.nlm.nih.gov/pmc/articles/PMC5179512/ (accessed 25 May 2017).

Hooper, L. V., Littman, D. R. and Macpherson, A. J. (2012) Interactions between the microbiota and the immune system. *Science* 336(6086): 1268–73.

Htoo, J. K. and Molares, J. (2012) Effects of dietary supplementation with two potassium formate sources on performance of 8- to 22-kg pigs. *Journal of Animal Science* 90 Suppl. 4: 346–9.

Huang, F. P., Platt, N., Wykes, M., et al. (2000) A discrete subpopulation of dendritic cells transports apoptotic intestinal epithelial cells to T cell areas of mesenteric lymph nodes. *Journal of Experimental Medicine* 191(3): 435–44.

Hughes, R., Magee, E. A. and Bingham, S. (2000) Protein degradation in the large intestine: Relevance to colorectal cancer. *Current Issues in Intestinal Microbiology* 1(2): 51–8.

Human Microbiome Project Consortium (2012) Structure, function and diversity of the healthy human microbiome. *Nature* 486(7402): 207–14.

Inman, C. F., Haverson, K., Konstantinov, S. R., et al. (2010) Rearing environment affects development of the immune system in neonates. *Clinical and Experimental Immunology* 160(3): 431–9.

Inman, C. F., Laycock, G. M., Mitchard, L., et al. (2012) Neonatal colonisation expands a specific intestinal antigen-presenting cell subset prior to CD4 T-cell expansion, without altering T-cell repertoire. *PLOS ONE* 7(3): e33707.

Jacobs, D. M., Deltimple, N., van Velzen, E., et al. (2008) (1)H NMR metabolite profiling of feces as a tool to assess the impact of nutrition on the human microbiome. *NMR in Biomedicine* 21(6): 615–26.

Johansson, M. E. and Hansson, G. C. (2016) Immunological aspects of intestinal mucus and mucins. *Nature Reviews Immunology* 16(10): 639–49.

Jung, K., Eyerly, B., Annamalai, T., et al. (2015) Structural alteration of tight and adherens junctions in villous and crypt epithelium of the small and large intestine of conventional nursing piglets infected with porcine epidemic diarrhea virus. *Veterinary Microbiology* 177(3–4): 373–8.

Knights, D., Ward, T. L., McKinlay, C. E., et al. (2014) Rethinking 'Enterotypes'. *Cell Host & Microbe* 16(4): 433–7.

Koenig, J. E., Spor, A., Scalfone, N., et al. (2011) Succession of microbial consortia in the developing infant gut microbiome. *Proceedings of the National Academy of Sciences* 108(Suppl. 1): 4578–85.

Lewis, M. C., Inman, C. F., Patel, D., et al. (2012) Direct experimental evidence that early-life farm environment influences regulation of immune responses. *Pediatric Allergy, Immunology* 23(3): 265–9.

Lewis, M. C., Patel, D. V., Fowler, J., et al. (2013) Dietary supplementation with Bifidobacterium lactis NCC2818 from weaning reduces local immunoglobulin production in lymphoid-associated tissues but increases systemic antibodies in healthy neonates. *British Journal of Nutrition* 110(7): 1243–52.

Lu, H., Su, S. and Ajuwon, K. M. (2012) Butyrate supplementation to gestating sows and piglets induces muscle and adipose tissue oxidative genes and improves growth performance. *Journal of Animal Science* 90 Suppl. 4: 430–2.

Lynch, H., Leonard, F. C., Walia, K., et al. (2017) Investigation of in-feed organic acids as a low cost strategy to combat Salmonella in grower pigs. *Preventive Veterinary Medicine* 139(Pt A): 50–7.

Mach, N., Berri, M., Estellé, J., et al. (2015) Early-life establishment of the swine gut microbiome and impact on host phenotypes. *Environmental Microbiology Reports* 7(3): 554–69.

Miller, B. G., Newby, T. J., Stokes, C. R., et al. (1984) The importance of dietary antigen in the cause of postweaning diarrhea in pigs. *American Journal of Veterinary Research* 45(9): 1730–3.

Mulder, I. E., Schmidt, B., Stokes, C. R., et al. (2009) Environmentally-acquired bacteria influence microbial diversity and natural innate immune responses at gut surfaces. *BMC Biology* 7: 79.

Mulder, I. E., Schmidt, B., Lewis, M., et al. (2011) Restricting microbial exposure in early life negates the immune benefits associated with gut colonization in environments of high microbial diversity. *PLoS ONE* 6(12): e28279.

Mulder, T. P., Rietveld, A. G. and van Amelsvoort, J. M. (2005) Consumption of both black tea and green tea results in an increase in the excretion of hippuric acid into urine. *The American Journal of Clinical Nutrition* 81(1 Suppl.): 256S–260S.

Nicholson, J. K., Holmes, E., Kinross, J., et al. (2012) Host-Gut Microbiota metabolic interactions. *Science* 336(6086): 1262–7.

Nyangale, E. P., Mottram, D. S. and Gibson, G. R. (2012) Gut microbial activity, implications for health and disease: the potential role of metabolite analysis. *Journal of Proteome Research* 11(12): 5573–85.

Ozdal, T., Sela, D. A., Xiao, J., et al. (2016) The reciprocal interactions between polyphenols and gut microbiota and effects on bioaccessibility. *Nutrients* 8(2): 78.

Pabst, O. and Mowat, A. M. (2012) Oral tolerance to food protein. *Mucosal Immunology* 5(3): 232–9.

Ramayo-Caldas, Y., Mach, N., Lepage, P., et al. (2016) Phylogenetic network analysis applied to pig gut microbiota identifies an ecosystem structure linked with growth traits. *The ISME Journal* 10(12): 2973–7.

Rescigno, M., Urbano, M., Valzasina, B., et al. (2001) Dendritic cells express tight junction proteins and penetrate gut epithelial monolayers to sample bacteria. *Nature Immunology* 2(4): 361–7.

Richardson, A. J., McKain, N. and Wallace, R. J. (2013) Ammonia production by human faecal bacteria, and the enumeration, isolation and characterization of bacteria capable of growth on peptides and amino acids. *BMC Microbiology* 13: 6.

Ríos-Covián, D., Ruas-Madiedo, P., Margolles, A., et al. (2016) Intestinal short chain fatty acids and their link with diet and human health. *Frontiers in Microbiology* 7: 185.

Round, J. L. and Mazmanian, S. K. (2010) Inducible Foxp3+ regulatory T-cell development by a commensal bacterium of the intestinal microbiota. *Proceedings of the National Academy of Sciences of the United States of America* 107(27): 12204–9.

Sarkissian, C. N., Scriver, C. R. and Mamer, O. A. (2000) Measurement of phenyllactate, phenylacetate, and phenylpyruvate by negative ion chemical ionization-gas chromatography/mass spectrometry in brain of mouse genetic models of phenylketonuria and non-phenylketonuria hyperphenylalaninemia. *Analytical Biochemistry* 280(2): 242–9.

Scharek, L., Guth, J., Filter, M., et al. (2007) Impact of the probiotic bacteria Enterococcus faecium NCIMB 10415 (SF68) and Bacillus cereus var. toyoi NCIMB 40112 on the development of serum IgG and faecal IgA of sows and their piglets. *Archives of Animal Nutrition* 61(4): 223–34.

Schleiffer, R. and Gairard, A. (1988) Influence of parathyroidectomy on aortic responsiveness to noradrenaline in spontaneously hypertensive rats. *Archives Internationales De Pharmacodynamie Et De Therapie* 292: 189–202.

Schmidt, B., Mulder, I. E., Musk, C. C., et al. (2011) Establishment of normal gut microbiota is compromised under excessive hygiene conditions. *PLoS ONE* 6(12): e28284.

Sedgmen, B. J., Meeusen, E. N. and Lofthouse, S. A. (2004) Alternative routes of mucosal immunization in large animals. *Immunology and Cell Biology* 82(1): 10–6.

Selma, M. V., Espín, J. C. and Tomás-Barberán, F. A. (2009) Interaction between phenolics and gut microbiota: Role in human health. *Journal of Agricultural and Food Chemistry* 57(15): 6485–6501.

Smirnov, K. S., Maier, T. V., Walker, A., et al. (2016) Challenges of metabolomics in human gut microbiota research. *International Journal of Medical Microbiology: IJMM* 306(5): 266–79.

Stokes, C. R. (2017) The development and role of microbial-host interactions in gut mucosal immune development. *Journal of Animal Science and Biotechnology* 8: 12.

Stokes, C. R., Soothill, J. F. and Turner, M. W. (1975) Immune exclusion is a function of IgA. *Nature* 255(5511): 745–6.

Telemo, E., Bailey, M., Miller, B. G., et al. (1991) Dietary antigen handling by mother and offspring. *Scandinavian Journal of Immunology* 34(6): 689–96.

Tohno, M., Shimazu, T., Aso, H., et al. (2008) Molecular cloning and functional characterization of porcine nucleotide-binding oligomerization domain-1 (NOD1) recognizing minimum agonists, meso-diaminopimelic acid and meso-lanthionine. *Molecular Immunology* 45(6): 1807–17.
Vella, A. and Farrugia, G. (1998) D-lactic acidosis: pathologic consequence of saprophytism. *Mayo Clinic Proceedings* 73(5): 451–6.
Wang, Z., Klipfell, E., Bennett, B. J., et al. (2011) Gut flora metabolism of phosphatidylcholine promotes cardiovascular disease. *Nature* 472(7341): 57–63.
Yatsunenko, T., Rey, F. E., Manary, M. J., et al. (2012) Human gut microbiome viewed across age and geography. *Nature* 486(7402): 222–7.

Chapter 4

Disease identification and management on the pig farm

Dominiek Maes, Jeroen Dewulf, Filip Boyen and Freddy Haesebrouck, Ghent University, Belgium

1 Introduction

Despite the importance of animal health, many pig herds worldwide are permanently endemically infected with important pathogens. Examples of such pathogens include porcine reproductive and respiratory syndrome virus (PRRSV), porcine circovirus type 2 (PCV-2), *Mycoplasma hyopneumoniae*, *Actinobacillus pleuropneumoniae*, *Haemophilus parasuis*, pathogenic *Escherichia coli* strains, *Brachyspira* spp. and *Streptococcus* spp. They cause tremendous economic losses to the pig sector. Unfortunately, during the last decades, pig health has not paralleled the improvement levels as observed in pig production and reproductive performance. Prevalence figures of lung lesions in slaughter pigs for instance are comparable to those of 20 years ago (Meyns et al., 2011). This urges for proper control measures to improve the health of pigs.

Endemic diseases are highly prevalent, even in professionally managed pig herds. This is not surprising because many facultative pathogenic bacteria are part of the microbiota and because possible sources of infection and transmission routes are very numerous, for example, sow-to-piglet, pig-to-pig, via semen, aerosol, airborne transmission in pig-dense areas, people, rodents, insects, domestic and feral non-swine animals, birds, fomites,

http://dx.doi.org/10.19103/AS.2017.0013.17

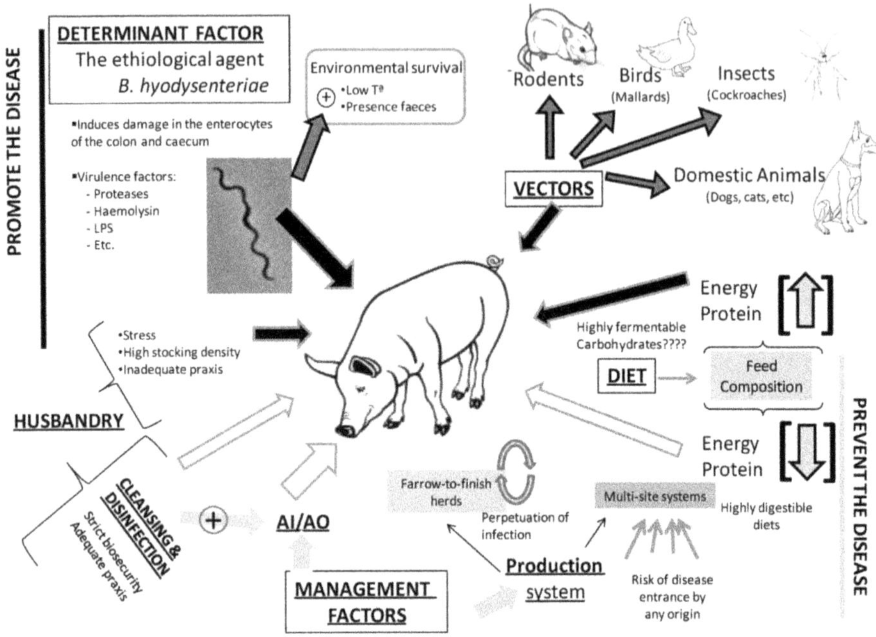

Figure 1 Factors influencing the occurrence and transmission of *Brachyspira hyodysenteriae* (Alvarez-Ordóñez et al., 2013).

carcasses and vehicles (Amass and Baysinger, 2006). Also, often a large number of animals share the same airspace, or are raised on the same location, facilitating the transmission of pathogens. An example of the numerous transmission routes is shown in Fig. 1 for *Brachyspira hyodysenteriae* infections.

In intensive pig production systems, endemic diseases often result from an interaction of different infectious agents, and are influenced by environmental, nutritional and management conditions. Consequently, many pig diseases are multifactorial, with the clinical outcome, the severity of lesions and economic losses being the result of an interplay between pathogens, the animal host defence and the environment. The many interactions between these factors are often not fully known. Therefore, many pig diseases are called 'syndromes' or 'complexes', indicating that the precise aetiology and/or pathogenesis are not fully understood (Table 1).

Endemic pig diseases constitute a major threat for a profitable and sustainable pig production worldwide. They lead to decreased animal performance and carcass quality, more infections with zoonotic agents in humans, increased antimicrobial use, less animal welfare, reduced work satisfaction (and motivation) of the pig producer and finally a poorer image for the pig sector. Not only clinical disease but also subclinical infections, for example, with PRRSV, PCV2, swine influenza, *M. hyopneumoniae* and *A. pleuropneumoniae* do have a measurable negative effect on performance. Decreases in daily weight gain in fattening pigs due to such subclinical infections typically range from 15 to 40 g (Rohrbach et al., 1993; Regula et al., 2000; Maes et al., 2003; Díaz et al., 2012).

Table 1 Examples of pig diseases called 'syndromes' or 'complexes', indicating that the exact aetiology and/or pathogenesis are not fully understood

Porcine reproductive and respiratory syndrome	Postpartum dysgalactia syndrome
Periparturient failing to thrive syndrome	Haemorrhagic bowel syndrome
Post-weaning multisystemic wasting syndrome	Porcine stress syndrome
Porcine dermatitis and nephropathy syndrome	Necrotic ear syndrome
Porcine ulcerative dermatitis syndrome	Porcine respiratory disease complex
Gastric ulceration	

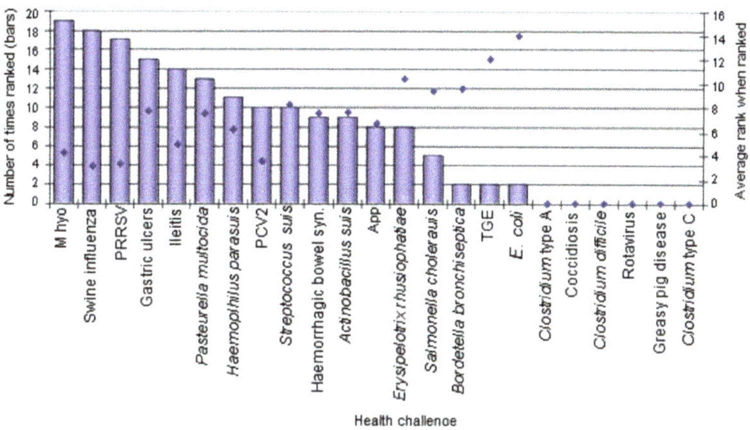

Figure 2 Rank of pathogens in the finishing herd in the United States (The most serious challenge was ranked as 1 and the other challenges were ranked in increasing order. The higher the rank, the less significant the challenge) (Holtkamp et al., 2007).

With clinical disease, production losses are considerably higher (Allan and Ellis, 2000). However, in general, the number of clinically affected herds or animals remains low. Consequently, although decreased performance per pig is lower in subclinically than in clinically infected animals, at a population level, and from an economic perspective, the impact of subclinical infections to the pig sector is likely larger than the effect of clinical disease. Clinically diseased pigs are the tip of the iceberg, and while these animals obviously require appropriate treatment, addressing the potentially widespread subclinical infection 'that lies beneath' is essential (Maes et al., 2012).

Holtkamp et al. (2007) ranked and quantified productivity and economic losses in the United States due to the major swine health challenges based on a survey including 19 large production companies. The veterinarians were asked 1) to rank all significant health challenges in the company over the last year, 2) to provide an estimate of the percentage of animals that were affected annually and 3) to estimate the losses in affected herds for each ranked health challenge. The results for the finishing and nursery units are shown in Fig. 2 and 3, respectively. Estimated losses associated with the four most important health challenges are shown in Table 2.

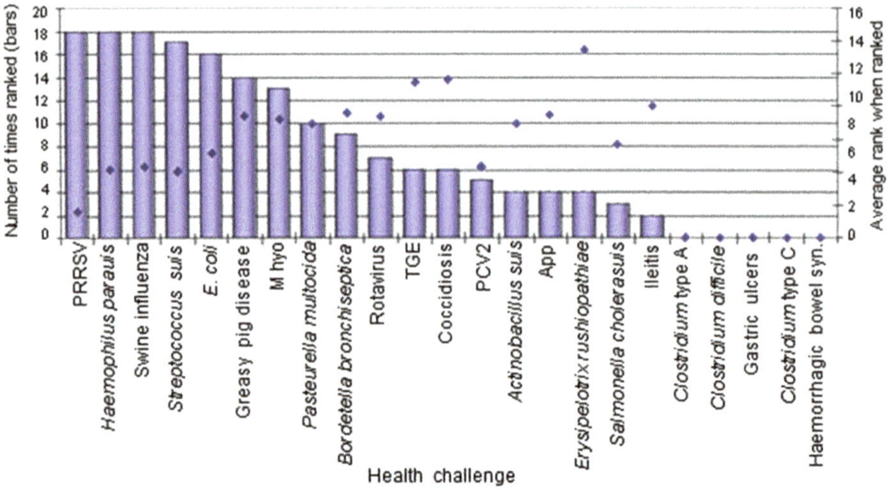

Figure 3 Rank of pathogens in the nursery herd (The most serious challenge was ranked as 1 and the other challenges were ranked in increasing order. The higher the rank, the less significant the challenge) (Holtkamp et al., 2017).

Table 2 Summary of estimated economic losses for top four health challenges in the United States in all stages of production (Holtkamp et al., 2007)

	Losses in affected herds (USD/ pig marketed)			% of animals affected			Average loss for all pigs (USD/pig marketed)			
	Breeding	Nursery	Finishing	Breeding	Nursery	Finishing	Breeding	Nursery	Finishing	Total
PRRSV	7.29	2.86	4.34	41.4	42.8	33.8	4.94	1.23	1.47	7.63
M. hyo-pneumoniae	1.52	1.92	5.84	17.6	10.0	34.3	0.39	0.19	2.00	2.58
Influenza	1.65	1.62	3.37	21.2	26.8	29.9	0.50	0.43	1.00	1.94
PRRS + *M. hyopneu-moniae*			6.69			18.1			1.21	1.21

Niemi et al. (2016) reviewed 130 publications to estimate the cost of production diseases in pigs. Reviewed studies showed great case-by-case variation in the costs of production diseases. Economic losses due to porcine respiratory disease complex were about €6.8 per fattening pig produced by an affected herd. Depending on pathogen and case, these losses ranged from €2 to €19 per fattening pig produced. In the studies analysed, the reduction in returns due to pre-weaning mortality was between €12 and €23 per litter, and due to post-weaning mortality between €2 and €4 per pig. The costs of postpartum dysgalactia syndrome amounted to €95 per affected sow. The costs of lameness in sows ranged from €145 to €180 per lame sow.

This chapter will focus on endemic pig diseases, and discuss the most important issues related to diagnosis and monitoring, and to management and control on the pig farm.

2 Disease identification

2.1 Considerations for a tentative diagnosis

For developing an efficient diagnostic approach, a clear statement of the diagnostic question should be made first (Cannon, 2002). The question might be 'Why did this pig die?' or 'What is causing the clinical signs in this group of pigs?' The diagnostic approaches to address these questions may be different from those addressing questions such as 'When were these pigs exposed to the pathogen?' or 'Are these pigs still free of the pathogen?'. Enteric (e.g. diarrhoea with mortality) and systemic infections (e.g. meningitis, arthritis, with mortality) are the most important disease conditions in suckling and weaned pigs, respiratory diseases (e.g. coughing, sneezing, dyspnoea with mortality) predominate in grow-finishers and reproductive problems and lameness are most common diseases in breeding animals.

The next step in the diagnosis is to obtain a clear description of the clinical problem, including historic information. Relevant questions are: how long are the problems occurring, has the clinical picture remained the same during the last weeks and months, which age groups are affected, which are the morbidity and mortality rates, what is the response to treatment and the current vaccination status and so on. In addition to the clinical picture, information on (changes in) potential risk factors is important, for example, type of farm, purchase policy and health status of possible origin herds, management and biosecurity measures in relation to the disease, air quality, ventilation patterns and nutrition.

The historic and current farm information and the description of the problem can be easily obtained by asking proper questions to the farmer and by carefully examining the pigs and their environment. This information may allow to reach almost a final diagnosis, for example, typical outbreaks of pleuropneumonia, or else, to exclude the role of some pathogens, for example, very high mortality is unlikely to be due to single *M. hyopneumoniae* infections. In most cases, however, further diagnostic work-up is necessary to establish a conclusive diagnosis, to elucidate the pathogens involved and to assess their relative impact in the problem.

2.2 Considerations for a conclusive diagnosis

Necropsies of a representative number of typically affected animals can be very helpful. Commonly found lesions in case of respiratory disease include catarrhal pneumonia (mainly due to infections with *M. hyopneumoniae* and some viral pathogens, e.g. swine influenza), and pleurisy (mainly due to infections with *A. pleuropneumoniae, H. parasuis, S. suis, P. multocida*) (Maes et al., 2001). However, mixed infections are generally the rule and lesions may not appear very typical for one specific pathogen. Performing necropsies is mostly followed by taking samples for further diagnostic examinations, for example, histopathology, immunohistochemistry, pathogen isolation or PCR testing.

In addition to investigating dead or euthanized animals, living animals from the affected age group can be sampled to detect antibodies and/or possible pathogens involved. Different sampling strategies can be used, depending on the age group, the type of sample, the disease and the diagnostic test used (see Section 2.3.2). In case of serology, paired serum samples will often be needed to establish the diagnosis as serum antibodies appear only two to four weeks after infection. In this case, sampling during the acute phase of the disease and a second time several weeks later is required. The type and number of samples and the diagnostic tests to be used depend on the pathogen.

2.3 Monitoring health in pig herds

It may also be interesting to monitor pig herds not suffering from clinical disease in order to assess, for example, freedom of infection, infection levels and variations over time. The results may also serve as a predictor of clinical disease. Detecting disease at an early stage is very important, as treating infections in time may largely prevent or limit adverse effects on the animals' health, welfare and performance (Smart et al., 1989; Opriessnig et al., 2007; Byra et al., 2011). Different parameters can be included in a monitoring programme.

2.3.1 Clinical signs and antibiotic usage

Clinical signs in a herd can be measured on a regular basis using clinical scores, for example, a coughing or diarrhoea score (Pederson et al., 2011; Nathues et al., 2012). Scoring systems are typically based on the number of affected animals and the severity of clinical signs, and allow to describe the clinical signs in more detail and in a quantitative way. Monitoring clinical signs such as coughing can also be automated. This avoids observer bias and allows the farmer to assess the situation in the stables in real time, 24 hours a day (see further).

The registration of antibiotic usage to treat individual or groups of animals is a rather easy tool to monitor the morbidity and disease levels in a farm. Individual Pig Care is an example of a management system for monitoring the health of groups of pigs. It is based on regular, individual observation of pigs so as to permit both early detection of management and health problems and a rapid response (Pineiro et al., 2012).

2.3.2 Animal responses to infection and detection of pathogens

Different sampling strategies can be used either to assess the animal response to infection or to directly detect the pathogen. Sampling of different age groups (cross-sectional sampling and/or serial sampling) allows to investigate a large number of pigs and to assess the infection at group level. Cross-sectional sampling implies the sampling of different age groups during the same herd visit. It has the advantage that the results are obtained quickly and that only one herd visit is needed. Serial sampling implies the sampling of the same animals over time. By sampling the same group (or pigs) over time, more accurate information is obtained on the sequence of infections in these groups. However, different herd visits are needed and results are not readily available. Whenever samples are collected, it is important to use an appropriate sample size to ensure that the conclusions based on the samples are truly accurate. There are ample freeware applications available (e.g. www.winepi.net) allowing easy and straightforward sample size calculation.

To save analysis costs, individual samples may be pooled, and for some pathogens (e.g. PRRS, PCV2), oral fluids obtained from different pigs of a pen can be used (Cuevas-Cordoba and Santiago-Garcıa, 2014). It is evident that samples should be transported in a proper way (e.g. chilled vs. frozen vs. at room temperature). In addition, one should be knowledgeable about the strengths and limitations of the available diagnostic tests and the results should be interpreted properly, taking into account the predictive values of positive and negative test results, and for quantitative tests, the magnitude of the outcome.

Serology is very commonly used to assess infection levels in pig herds. It does not measure the infection itself, but the animal response to infection. Therefore, infection took place some weeks before serum antibodies are detected. Depending on the test, the type of antibodies (IgM vs. IgG) and the pathogen, the time between infection and appearance of antibodies may range from 2 (e.g. acute influenza) to 8 weeks (e.g. *M. hyopneumoniae*) (Sibila et al., 2008). Also, some infected pigs may not seroconvert, especially in case of low infection levels. As a consequence, serology should be interpreted at group level, and not at the individual level. For a proper interpretation of serological data, it is important to know whether pigs have been vaccinated, and which vaccines and vaccination schemes were used.

For many pathogens, ELISA tests are used because of the ease of use, low cost, speed of testing and possible automation. For some pathogens and/or in specific circumstances, other serological tests are used such as the haemagglutination inhibition test, for example, for swine influenza; the immunoperoxidase monolayer assay or serum neutralization test, for example, for PRRSV; and microscopic agglutination test, for example, for leptospirosis. Most serological tests do not allow discrimination between antibodies following infection or following vaccination. For some pathogens, for example, Aujeszky's disease virus, serological tests have been developed that only detect antibodies upon infection with wild virus and not following vaccination with marker vaccines.

In addition to serology, different age groups may be sampled to detect pathogens in blood (e.g. viraemia for PRRSV), tonsil samples (e.g. *A. pleuropneumoniae*), nasal samples, laryngeal or tracheal swabs or BAL fluid (e.g. *M. hyopneumoniae*) and faeces (*Salmonella*, *Lawsonia* (*L.*) *intracellularis*, *B. hyodysenteriae*). The advantage is that the pathogen itself (cultivation, electronmicrosopy), its antigens (ELISA, immunofluorescence, immunohistochemistry, *in situ* hybridization) or its genetic material (PCR) are detected, and that further characterization is possible. Successful cultivation also points to the presence of viable pathogens, and in case of bacterial pathogens, allows to perform antimicrobial sensitivity testing. Bacterial isolation is performed routinely by veterinary diagnostic laboratories and is typically considered the gold standard for unambiguous and specific detection of bacteria involved in morbidity and mortality cases. Isolation of bacterial pathogens is more difficult in case animals have received antimicrobial medication.

In general, PCR testing has a higher sensitivity compared to isolation of the pathogen. For some pathogens such as *M. hyopneumoniae*, different PCR tests are used, for example, conventional PCR, nested PCR or quantitative PCR.

2.3.3 Lesions in slaughter pigs

Respiratory tract lesions can be assessed by means of slaughter checks (Davies et al., 1995). Evaluation of respiratory tract lesions of slaughter pigs is an easy and cheap way to obtain information of the respiratory health of the fattening pigs. If the lesions are monitored from different successive batches of a herd, time evolutions can be assessed. Slaughter checks are also useful to detect subclinical respiratory infections. Most scoring methods are based on a visual subjective estimation of the proportion of lung affected surface and/or volume. Therefore, it can involve possible errors. In addition, most lung lesions are not pathognomonic for a specific pathogen. Presence of severe pleurisy may mask pneumonia lesions as well, and healing of lesions during the fattening period may lead to false-negative results (Noyes et al., 1990). Hence, the subjectivity of the scoring method, regression of lung lesions and the non-specificity of these lesions preclude an aetiologic diagnosis based solely on examination of gross lung lesions at slaughter.

Also other lesions can be monitored in slaughter pigs such as white spots on the liver due to migration of *Ascaris suum* larvae, skin lesions due to sarcoptic mange, joint lesions, gastric ulcers, tail biting lesions, abscesses, and lesions in the intestinal and reproductive tract. These lesions may also help to elucidate possible risk factors on the farm.

2.3.4 Feed and drinking water intake

More and more pig producers monitor feed and drinking water intake, either at barn, room or at pen level. This information is not only interesting to calculate performance, but may also be used as indirect indicators for when animals are stressed (e.g. onset of disease, excessive temperature fluctuations, heat load) or are not performing at an optimal level. The drinking behaviour of diseased pigs has been found to deviate from the consistent patterns seen in healthy pigs (Maselyne et al., 2015). The behavioural changes appear in the subclinical stage of infection, before disease symptoms become visually apparent. If water usage drops more than 30% in a day, or drops for three consecutive days, a potential health challenge may occur.

3 Disease management and control: overview

3.1 Types of disease prevention

Disease prevention may be considered at different levels (Toma et al., 1999) (Table 3). Primary prevention relates to measures that prevent the introduction of pathogens and/or of related determinant factors in a herd. It is the most effective form of prevention. If the pathogen is not present, then disease caused by the pathogen is not possible. Many different strategies to eliminate pathogens from pig herds have been described (Harris and Alexander, 1999; Holst et al., 2015). Eradication programmes for specific diseases, for example, classical swine fever (CSF) and Aujeszky's disease have been successful in the EU. Also endemic infections such as PRRSV, atrophic rhinitis, *B. hyodysenteriae* and *M. hyopneumoniae* have been eliminated successfully from individual herds. Although elimination is the most straightforward measure, there is a permanent risk for re-infections. This is especially the case for herds located in regions with a high pig (herd) density and for pathogens that can spread via airborne transmission, for example, *M. hyopneumoniae*

Table 3 Types of disease prevention in pig herds (adapted from Toma et al., 1999)

	Type of prevention		
	Primary	Secondary	Tertiary
Goal	Absence of pathogen	Absence of disease	Effective treatment of disease
Strategy	Control of risk factors and causes	Control of infection and disease	
Method	Health management programmes (especially external biosecurity)	Health management programmes (external and internal biosecurity, management, nutrition, etc.)	Therapeutic measures

(Maes et al., 2008). To account for these risks, air filtration systems have been installed in modern facilities (see further).

In countries with intensive pig production systems, most herds are endemically infected with several (facultative pathogenic) pathogens. For a number of reasons, it is very difficult and/or not warranted to eliminate all of these pathogens from these herds. Secondary prevention implies that pathogens are still present on the herd, but the infection level is in balance with the herd immunity thanks to optimal management and biosecurity. Finally, clinically diseased pigs should be treated properly as quickly as possible. According to Toma et al. (1999), this is called tertiary prevention. Although this can hardly be considered as prevention, it will remain important in the future. In general terms and especially in farm animals, prevention is better than cure. Therefore, emphasis should be placed on control and prevention of infections.

3.2 Biosecurity measures and management: definition and importance

Biosecurity is the application of management practices that reduce the opportunities for infectious agents to gain access to, or spread within, a food animal production unit (Toma et al., 1999). External biosecurity comprises measures that prevent pathogens from entering the herd, while internal biosecurity relates to preventing the within-herd spread of pathogens. Biosecurity encompasses cleanliness; disinfection; reduction of exposure; management of animals, personnel and goods; and ensuring that animals can be traced. Biosecurity is therefore, essentially, the application of 'informed common sense'. It is an everyday challenge for the farmer. Recent studies have quantified biosecurity on pig herds and the score was expressed as a percentage (Laanen et al., 2013) with 100% being the perfect biosecurity situation and 0% the total absence of any biosecurity measures (www.biocheck.ugent.be). The average external biosecurity score was 65 (range: 45–89) and the average internal score was 52 (range: 18–87). External scores were positively associated with herd size, while internal scores were negatively associated with both 'age of buildings' and 'years of experience of the farmer', indicating that biosecurity is generally better implemented in larger herds, in more modern facilities and by younger farmers. External and internal biosecurity scores were positively associated with performance of fattening pigs. Internal scores were negatively associated with disease treatment incidence, suggesting that improved biosecurity might help in reducing the prophylactic use of antimicrobials. The explained variation in the outcome variables, however, were low, indicating that other factors apart from the biosecurity parameters are also important. Similar results have been obtained in more recent studies (Postma et al., 2016). Biosecurity scoring is particularly interesting for sensitizing pig producers and benchmarking of herds. The most important factors related to external and internal biosecurity are discussed below.

4 External biosecurity

4.1 Purchase policy

Transmission of pathogens occurs very effectively via direct contact between infected and susceptible animals. Tobias et al. (2014) described a ten times higher transmission rate of

A. pleuropneumoniae for direct contact between pigs compared to indirect transmission. Similar findings were obtained for other swine pathogens. Therefore, purchasing animals from other pig herds constitutes a risk for pathogen introduction, especially when it is performed at a regular basis, when replacement rates are high and in case animals originating from different sources. When it is necessary to purchase animals (e.g. to improve genetics), it is important to evaluate the health status of the origin herd and to limit the number of sources. Moreover, if new animals are introduced in a herd, adequate quarantine measures should be taken (Amass and Baysinger, 2006; Maes et al., 2008).

According to Calvar et al. (2014), pig farmers in Bretagne are not fully aware of the risk of purchasing breeding stock in pig farms in Bretagne, except for PRRSV. In an earlier study on quarantine (Calvar et al., 2012), they observed that barely 60% of the breeders know the health status of the delivered animals. On the other hand, the majority of Danish farms have adopted the specific-pathogen-free (SPF) concept and follow a much stricter protocol. To date, approximately 3100 herds in Denmark have a SPF health declaration, while 78% of the sows in Denmark are SPF sows. The SPF system is built on the principle that health status is declared for the connected herds and is taken into account in case of trading between herds. It combines documented high biosecurity at herd level, regular disease testing and publishing of disease status, among other things, according to the certified disease status of the herds (Anonymous, 2015). A previous study on biosecurity in Belgian pig farms showed a good level of implementation of the measures related to purchase policy of animals and semen (Laanen et al., 2013).

Purchasing contaminated semen is also known to be a risk. Pathogen transmission by semen to the sow has been proven for viruses such as CSF virus and PRRSV (Maes et al., 2016). The best way to prevent disease transmission via the semen is to ensure that the boars in artificial insemination centres are free from the pathogen.

4.2 Dead animals and manure

Secreta and excreta from diseased and dead animals are often contaminated with pathogens and pose a risk for direct transmission to susceptible animals. Also indirect transmission of pathogens from cadavers may occur, for example, via fomites, the rendering truck, people (including the veterinarian) and their material, rodents and domestic animals. The risk of transmission via infectious cadavers is well known for foot-and-mouth disease (Scudamore et al., 2002; Hayama et al., 2015). A timely, consequent and careful removal of dead animals is therefore recommended. Many pig pathogens, for example, *Brachyspira* spp. (Hampson, 2012) or *Salmonella enterica* (Carlson et al., 2012) are able to survive for long periods in manure. Therefore, manure or utensils contaminated with manure constitute a risk for introduction of pathogens.

4.3 Feed, water and equipment

Trucks delivering commercial feed to farms enter multiple premises a day often without being (properly) cleaned and disinfected in-between. This causes a risk for pathogen transmission between farms as described by Bottoms et al. (2015). One explanation for the transmission of disease in the recent outbreak of porcine epidemic diarrhoea (PED) virus in North America was the use of contaminated bulk feed containers (Scott et al., 2016). Feed itself should not normally pose a risk of disease introduction since it is generally produced under strict hygienic procedures, although introduction of *Salmonella enterica* via feed has been described (Österberg et al., 2006). Swill feeding, however, causes substantial risks as demonstrated in several CSF outbreaks in the past. Swill feeding is currently banned in the EU (European Commission, 2001). The risk for disease introduction via the feed is much lower for farmers who home mix feed.

Water might also play a role in disease transmission since pathogens may survive in water (Loera-Muro et al., 2013; Szabo and Minamyer, 2014). Duration of survival depends on different factors including traits of both the pathogen and the drinking water. Therefore, a regular (at least once a year) microbiological and chemical evaluation of the water quality is advisable.

4.4 Access check

The farmer, employees, the veterinarian and health professionals, or herd visitors can function as mechanical vectors by spreading pathogens via, for example, their footwear or clothes, but also via their hands. Ribbens et al. (2007) showed that contaminated boots, gloves and coveralls may be involved in the transmission of CSF virus. Amass et al. (2003) showed that showering was able to prevent transmission of *S. suis* from a person to the pigs.

4.5 Vermin and bird control

Rodents and especially rats are well known to spread pathogens (Pearson et al., 2016). Pathogens known to be transmitted by rodents include *L. intracellularis*, *B. hyodysenteriae* and *Salmonella enterica*. The same authors also showed that birds may transmit pathogenic *E. coli* to pigs. Domestic animals such as dogs and cats may also introduce or transmit pathogens. Desrosiers (2011) reported that not only dogs but also insects were able to transmit *B. hyodysenteriae*, and that cats and dogs could transmit *P. multocida*. Saif et al. (2012) reported the shedding of transmissible gastroenteritis as well as PED virus by dogs and cats, and Opriessnig and Wood (2012) described domestic animals as potential reservoirs for *Erysipelothrix rhusiopathiae*.

4.6 Herd location and environment

Some pathogens, for example, *M. hyopneumoniae*, swine influenza virus and PRRSV can be transmitted between herds via airborne transmission (Otake et al., 2010; Desrosiers, 2011). Almost all pathogens, especially respiratory pathogens, can be found in the air in the close proximity of the animals (Tobias et al., 2014). The location of the herd and the density of pig farms in the proximity are thus important risk factors. Maes et al. (2000) reported increased seroprevalences for *M. hyopneumoniae*, swine influenza viruses and Aujeszky's disease virus in herds located in pig-dense areas. Filtration of incoming air, in combination with standard biosecurity procedures, has been demonstrated to prevent transmission of PRRSV into susceptible herds (Alonso et al., 2013). The authors showed that air filtration reduced the risk of introduction of novel PRRSV by approximately 80%, indicating that on large sow farms with good biosecurity in pig-dense regions, approximately four out of five PRRSV outbreaks may be attributable to aerosol transmission. Air filtration may also be helpful to prevent airborne transmission of other pathogens such as swine influenza and *M. hyopneumoniae*.

Feral animals, with an important role of wild boar, are also able to transmit several pig pathogens to domestic pigs.

5 Internal biosecurity

5.1 Production system

All-in, all-out (AIAO) production is an important factor in the control of infectious disease since it can interrupt or decrease the cycle of pathogen transmissions from older to younger pigs. It allows the producer to tailor environmental conditions to a uniform population of pigs and to clean the facilities between groups of pigs. AIAO production also results in better performance and less lung lesions in slaughter pigs (Straw, 1991). Mixing or sorting pigs is a source of stress to the animals and it increases the probabilities of disease transmission. Therefore, an AIAO system in which the same pigs are moved as a group through the different production stages is to be preferred compared to one where pigs are regrouped during transfer from one unit to another.

Early weaning (<3 weeks) can reduce transmission of some pathogens from the sow to the offspring, but it is not allowed to be applied systematically in the EU. In addition, recent studies have shown that early weaning may increase the antimicrobial consumption post-weaning (Postma et al., 2016). Parity segregation has been used in large production systems as a means to control several diseases in the breeding herd. Gilts and their offspring are kept separately from the sows until they reach their second gestation. By that time, they are expected to have acquired the desired immune status and pose no risk for destabilizing the infection level in the herd (Hoy et al., 1986).

5.2 Compartmentalizing, working lines and equipment

Keeping age groups separately and working in a well-defined sequence is recommended, that is, first the youngest animals, followed by the older age groups, thereafter the quarantine and sick bay, and finally the cadaver storage (Neumann, 2012).

Materials and equipment are very important in the transmission of a large number of pathogens. Allaart et al. (2013) described the isolation of *Clostridium perfringens* from bedding material, drinking water, boots, fans and fly strips on farms. Syringes and needles might also serve as transmitters of disease, for example, for PRRSV (Otake et al., 2002) and PCV2 (Patterson et al., 2011).

5.3 Animal stocking density

Decreasing animal density during the different production stages has been shown to reduce the level of disease (Pointon et al., 1985). Crowding may lead to increased transmission of pathogens and stress reactions, making the pigs more susceptible to infectious diseases. Too low stocking densities, however, are financially not justified. Therefore, it is important to find a reasonable compromise between stocking densities that are appropriate for the health of the pigs and those that maximize the returns on the building's cost.

The stocking densities based on EU legislation (Table 4) have been determined in the 1980s, and should not necessarily be considered as optimal values (Dewulf et al., 2007). Studies suggest that for optimal health and production, surface areas per pig in the different age categories may be up to 20% higher (more surface area per pig) than the minimal legal requirements (Hamilton et al., 2003). Further research on the influence of

Table 4 Stocking densities for pigs according to EU legislation (Directive 2008/120/EC)

Average weight (kg)	Minimal required surface area (in m^2) per pig
<10	0.15
10 to 20	0.20
20 to 30	0.30
30 to 50	0.40
50 to 85	0.55
85 to 110	0.65
>110	1.00

different stocking densities on health, welfare and performance is needed, including also economic considerations.

5.4 Management of diseases

Diseased animals may pose a substantial threat to susceptible animals. Correct handling and treatment of diseased animals is therefore of great importance in the reduction of transmission of pathogens.

5.5 Herd size

Study results on the role of herd size as a risk factor for disease are not consistent. Flesja and Solberg (1981) found an increased prevalence of *M. hyopneumoniae* lesions in larger pig herds, whereas in other studies no significant associations could be found between herd size and seroprevalence of *M. hyopneumoniae* (Maes et al., 1999, 2000) or prevalence of *Mycoplasma*-like lung lesions at slaughter (Maes et al., 2001). Biologically plausible reasons for a positive association between herd size and respiratory infections include a greater risk of introduction of the pathogen from outside the herd, greater risk of pathogen transmission within and among herds when the herd is large, and effects of management and environmental factors that are related to herd size. However, owners of large herds might more frequently and more rapidly adopt management and housing practices that mitigate this potential increased risk (Gardner et al., 2002; Laanen et al., 2013).

5.6 Improvement of housing conditions

Housing and/or environmental changes that optimize the climate of the pigs' environment are important in the control of infections. In case of respiratory disease, special attention should be paid to the temperature set points, fan staging, air inlet and curtain settings, sensor placement, heater capacity, drafts and building maintenance (Gonyou et al., 2006). However, making environmental changes for improving the climate in inappropriate or old barns frequently entail extensive remodelling, and therefore, they may be difficult and expensive to implement. Worn slats are more difficult to clean and disinfect, and constitute a risk for enteric pathogens.

5.7 Cleaning and disinfection

Following the correct procedure of cleaning, disinfection and empty period in the compartments/pens will reduce the infection pressure. A lack of cleaning after batches was seen as a risk factor for several zoonotic pathogens such as *Salmonella enterica*, *Campylobacter* spp. and *Listeria monocytogenes* (Fosse et al., 2009; De Busser et al., 2013). Hygiene testing, that is, testing of bacterial contamination of surfaces, or adenosine triphosphate (ATP) analyses of surfaces can be used to check whether the cleaning and disinfection were performed properly.

5.8 Nutrition and drinking water

Nutrition is a key factor for pig health. The feed composition and the physical characteristics of the feed, the feeding level as well as the way of feeding and feeder space per pig are all important. The severity of endemic diseases such as infections with *E. coli* post-weaning, *Brachyspira* spp. and *Salmonella enterica* are largely influenced by nutritional aspects. Feeding coarsely ground meal decreased the survival of *Salmonella enterica* during passage through the stomach (Mikkelsen et al., 2004). A strong reducing effect of fermented liquid feed, including whey, on *Salmonella enterica* shedding and seroprevalence has also been reported (Poljak et al., 2008). The addition of organic acids to feed or drinking water is generally beneficial to reduce *E. coli* and *Salmonella enterica* shedding, but the effects are variable (De Busser et al., 2009; De Ridder et al., 2013).

6 Vaccination and antimicrobial medication

6.1 Vaccination

Improving the immunity status of susceptible animals, for example, by vaccination is a helpful tool to control infectious diseases. Vaccination is generally able to reduce the (risk for) clinical symptoms, lesions and performance losses due to disease. However, most vaccines only provide partial protection, do not prevent infection and are not able to eliminate the pathogens from the herd. The decision whether to implement vaccination is not straightforward. It depends on the pig producer (degree of being 'risk averse'); the type and severity of disease; the available vaccines and expected effects; and other factors such as type of herd, flow of the pigs, infection level, management and housing conditions. The herd veterinarian has a central role for determining which pathogens are playing a key role, and for making a decision in collaboration with the farmer to implement one or more vaccinations and determining the optimal vaccination scheme.

Vaccination schemes are highly variable between herds and also between countries. Some pathogens may be eradicated in some countries and vaccination is prohibited (e.g. Aujeszky's disease virus in many European countries and North America), whereas the virus is still endemic in many other countries and vaccination is used.

Breeding sows are often vaccinated against PRRSV, neonatal *E. coli* infections, parvovirus and erysipelas, and in some herds also against other pathogens such as atrophic rhinitis, swine influenza, PCV-2, *H. parasuis*, *A. pleuropneumoniae*, leptospirosis and so on. In Asian countries, farms may also vaccinate against foot-and-mouth disease, Aujeszky's disease, CSF and Japanese encephalitis. Some vaccinations in sows aim to provide protection to the piglets via maternally derived antibodies, for example, neonatal *E. coli* infections and atrophic rhinitis. Recent studies in high-performing sow herds have shown that colostrum production is quite variable between herds and between sows, and that for many piglets, colostrum intake is insufficient (Decaluwé et al., 2014; Declerck et al., 2015).

Commonly used commercial vaccines in pigs include vaccines against infections with PCV-2, *M. hyopneumoniae*, PRRSV and depending on the country and the herd also vaccines against other pathogens may be applied such as *A. pleuropneumoniae*, swine influenza, *H. parasuis*, post-weaning diarrhoea, oedema disease, Aujeszky's diseases virus and so on.

The costs and benefits of disease control campaigns can be assessed using several methods including gross margin analysis and partial budgeting (Thrusfield, 2007). These are essentially straightforward accounting approaches. Social cost–benefit analysis is the application of a specific technique that allows for the fact that costs and benefits are commonly distributed over time, and sometimes are more than simple financial values.

Maes et al. (2003) conducted a multi-site field study to assess the efficacy and economic profitability of using an inactivated *M. hyopneumoniae* vaccine in 14 pig herds infected by *M. hyopneumoniae* and practising an AIAO production system. Vaccination significantly improved daily weight gain (+22 g), feed conversion ratio (−0.07), medication costs (−0.476 €/pig), prevalence of pneumonia lesions (−14%) and severity of pneumonia lesions (−3%). Mortality rate, severity of coughing and carcass quality were not significantly influenced by vaccination. A partial budget analysis based on significantly improved performance parameters demonstrated that *M. hyopneumoniae* vaccination was economically attractive as it resulted in an increase of the net return to labour with €1.3 per finishing pig sold. The sensitivity of the economic benefit was illustrated towards fluctuations in pig finishing prices.

In general, vaccination is economically justified in clinically affected herds, especially if no major risk factors can be identified or changed in short term. However, vaccination may also be cost-efficient in subclinically infected herds as subclinical infections, for example, with *M. hyopneumoniae*, *A. pleuropneumoniae*, swine influenza, PRRSV and PCV2 also reduce performance (Rohrbach et al., 1993; Regula et al., 2000; Maes et al., 2003). For some important pig pathogens, however, for example, *B. hyodysenteriae* and *S. suis*, no or few commercial vaccines are available. Autogenous vaccines may also be used. These are mostly inactivated vaccines, for example, bacterins based on the pathogen strain(s) isolated from animals affected by the disease. Autogenous vaccines against different pathogens such as *S. suis*, *A. pleuropneumoniae* and *E. coli* have been described (Baums et al., 2010; Geldhof et al., 2012). One of the disadvantages of autogenous vaccines is that often limited data are available on vaccine efficacy and safety (Haesebrouck et al., 2004).

6.2 Antimicrobial medication

Antimicrobials are necessary for treatment of bacterial infections and by doing so to safeguard animal welfare. They should, however, be used judiciously and restrictively to avoid or limit the risk for antimicrobial resistance in both pathogens and the commensal flora. High antimicrobial use and the resulting risk for antimicrobial resistance is an important concern worldwide. Research showed that prophylactic use of antimicrobials is common in many pig herds (Callens et al., 2012). At the same time, variations exist in the average levels of treatment between countries (Sjölund et al., 2016). These variations are beyond the difference in general level of pig health and therefore reflect also habits and attitudes

Table 5 Advantages and disadvantages of vaccination versus antimicrobial medication to control disease in pig herds

Vaccination	Antimicrobial medication
Long-term strategy	Short-term, flexible
More labour-intensive	Less labour-intensive (oral medication)
Against one pathogen (combination vaccines possible)	Against different pathogens (e.g. multiple disease challenges)
Effect until end of fattening period	Effects mainly during medication period
No risk of residues in carcasses	Risk of residues in carcasses
No selection of antibiotic resistance	Selection of antibiotic resistance
Vaccines are available against a limited number of diseases	Against bacterial infections, not against viruses
Partial protection	Partial protection

in relation to antimicrobial use. Repeated prophylactic use without sufficient diagnostic data or using antimicrobials as a substitute for improper management cannot be justified. In addition, it can be expected that the use of critically important antimicrobials such as fluoroquinolones and third- and fourth-generation cephalosporins will be further restricted or (in some countries) even be prohibited for use in veterinary medicine. A comparison between the use of vaccination and antimicrobial medication for the treatment and control of bacterial disease in pigs is presented in Table 5.

7 Future trends in diagnostics and disease monitoring and control

7.1 Precision livestock farming

Precision livestock farming (PLF) has the potential to revolutionize livestock farming and help solve many of the challenges mentioned in the previous sections (Banhazi et al., 2012). It implies the automated (real-time) monitoring or problem detection in livestock. Measurements at the individual level capture the large variation between pigs, and may also be interesting for welfare and health parameters (Berckmans, 2006). Examples include the measurement of feeding and drinking behaviour, body weight, activity, body temperature or other clinical signs such as cough (Cornou and Kristensen, 2013). Maselyne et al. (2015) showed that a radio-frequency identification system for measuring drinking behaviour of fattening pigs could be used as an on-farm detection system for health, welfare and production problems. Infrared temperature measurement equipment is gaining popularity as a diagnostic tool for evaluating human and animal health (Soerensen and Pedersen, 2015).

Systems based on group-level measurements are, however, generally cheaper and easier to implement than systems based on individual measurements. Berckmans et al. (2015) showed that the respiratory distress monitor, a tool that automatically monitors the respiratory health status of a group of pigs, gives earlier warnings compared to a situation where the farmer and veterinarian rely on their own routine observations without the monitor. Sound-based PLF techniques have advantages over other technologies such as cameras or accelerometers because microphones are contactless and relatively cheap, there is no need for a direct line of sight and large groups of animals can be monitored with a single sensor in a room.

Finally, irrespective of the system or the parameter(s) measured, PLF does not aim to replace the farmer but should be considered as a tool to support management decisions.

7.2 Diagnostics

The rapid and accurate establishment of a microbial cause is fundamental to quality porcine health management. Despite dramatic advances in diagnostic technologies, many groups of pigs with suspected infections receive empiric antimicrobial therapy rather than appropriate therapy dictated by the rapid identification of the infectious agent. New tests are needed that can identify one or preferably more specific pathogens, or at a minimum, distinguish between bacterial and viral infections. Also, they should provide information on susceptibility to antimicrobial agents. Tests should be easy to use, at the farm (pen side tests), and provide a rapid, but reliable, result (ideally within an hour).

The expertise of the herd veterinarian to interpret the results will become more important with the advent of newer, more complex tests. The availability of needed tests will lead to improvements in clinical outcomes upon treatment, antimicrobial stewardship, detection and tracking of disease outbreaks, and investigation of unknown pathogens. Emerging technologies could enable the detection and quantification of pathogen burden with new speed, sensitivity and simplicity of use. However, there are significant challenges to the development, regulatory approval, and last-but-not-least, the cost of diagnostic tests based on these new technologies.

7.3 Control measures

Further research into disease prevention and infection control measures such as biosecurity and hygiene procedures are warranted. Many of the current practices are based on empirical expertise, whereas more in-depth research could provide a better insight in and development of preventive measures. In addition, research into the importance of the microbiota for pig health and production is needed.

More and also better vaccines will need to be developed in the future, as for some important pig diseases (e.g. *Brachyspira* spp., *S. suis*) no or very few commercial vaccines are available. Many currently used commercial vaccines only provide partial protection. Therefore, vaccines that confer a better protection are needed. To achieve this, more research is warranted to understand the pathogenesis of the disease, and the protective mechanisms upon infection. It can also be expected that autogenous vaccines, for example, vaccines made to suite a specific herd health situation, may be more frequently used in the future, especially if commercial vaccines are not available or do not provide sufficient protection. Research should also focus on easy administration routes for vaccines, eventually by oral or aerosol administration, and on developing multivalent vaccines. A better understanding of the composition and the role of the pig microbiota is needed, as this may help to better understand the mode of action of probiotics and prebiotics, and may enhance the development of new products to control enteric disease.

7.4 Selection for disease resistance

Many diseases of both farm and companion animals have a variable heritable component. Resistance can be defined as the ability of the host to exert some degree of control over the pathogen life cycle (Bishop, 2012). It encompasses the many ways a host species may be more resistant, for example, less likely to become infected, reduced pathogen proliferation once infected, reduced shedding or transmission of infection. Genetic and genomic studies of disease resistance have been conducted, elucidating the genetic control of between-animal differences in resistance (Bishop and Woolliams, 2014). Notable examples of success include infectious pancreatic necrosis in Atlantic salmon and bovine tuberculosis in dairy cattle. In pigs, Whitworth et al. (2015) edited the gene that makes the CD163 protein which is important during infection with PRRSV. Treated pigs did no longer produce the protein and were protected against clinical disease upon experimental PRRS challenge infection. The results of these studies may be utilized to breed animals for increased resistance. The development of genetic resistance against diseases in pigs has, however, been slow and uneven so far. It has been a long and ultimately unsuccessful battle to develop useful *E. coli*-resistant piglets.

The use of genetic or genomic information must be considered in the broader context of both the breeding goal and the conventional strategies used to control the disease. If the disease is of little importance to the breeder or producer, or is satisfactorily controlled by other means, then there may be little point in attempting to breed for increased resistance. Consequently, most effort should be directed towards costly endemic diseases, for which control by other strategies is proving difficult.

8 Conclusion

Pig health is essential for a sustainable pig production, as infectious diseases may cause substantial economic losses and may affect the quality and safety of pork. As most pathogens in pig herds are present in a subclinical form, monitoring is important to assess the infection levels and to identify the major pathogens involved. Animals in intensive production systems are continuously in a fragile balance. Optimal management, biosecurity and nutrition are needed to maintain health and production targets. Vaccination and medication may be needed as additional control measures. Further research is needed to better understand the host–pathogen interactions and to develop better diagnostics, vaccines and control measures against infectious pig diseases.

9 Where to look for further information

Diseases of Swine. Zimmermann, J., Karriker, L., Ramirez, A., Schwartz, K. and Stevenson, G. (eds), Wiley-Blackwell, Chichester, West Sussex, UK.

Porcine Health Management. The official journal of the European College of Porcine Health Management and the European Association of Porcine Health Management: https://porcinehealthmanagement.biomedcentral.com/.

Journal of Swine Health and Production. The official journal of the American Association of Swine Veterinarians: https://www.aasv.org/shap.html.

Unit of Porcine Health Management, Department of Reproduction, Obstetrics and Herd Health, Faculty of Veterinary Medicine, Ghent University, Belgium: http://www.rohh.ugent.be/v3/research/units/pig_health/.

10 References

Allaart, J., van Asten, A. and Gröne A., 2013. Predisposing factors and prevention of *Clostridium perfringens*-associated enteritis. Comp. *Immunol. Microbiol. Infect. Dis.* 36, 449–64.

Allan, G. and Ellis, J., 2000. Porcine circoviruses: a review. *J. Vet. Diagn. Invest.* 12, 3–14.

Alonso, C., Murtaugh, M., Dee, S. and Davies, P., 2013. Epidemiological study of air filtration systems for preventing PRRSV infection in large sow herds. *Prev. Vet. Med.* 112, 109–17.

Alvarez-Ordóñez, A., Martinez-Lobo, F., Arguello, H., Carvajal, A. and Rubio, P., 2013. Swine dysentery: aetiology, pathogenicity, determinants of transmission and the fight against the disease. *Int. J. Environ. Res. Public Health* 10, 1927–47. https://doi.org/10.3390/ijerph10051927

Amass, S. and Baysinger, A., 2006. Swine disease transmission and prevention. In *Diseases of Swine*, 9th edn, Straw, B., Zimmerman, J., D'Allaire, S. and Taylor, D. (eds), Chapter 68, 1075–98.

Amass, S., Halbur, P., Byrne, B., Schneider, J., Koons, C., Cornick, N. and Ragland, D., 2003. Mechanical transmission of enterotoxigenic *Escherichia coli* to weaned pigs by people, and biosecurity procedures that prevented such transmission. *J. Swine Hlth. Prod.* 11, 61–8.

Anonymous, 2015: SPFSUS. SPF-Sundhedsstyringen – Status June 2015. Available at: http://spfsus.dk/en (accessed 18 November 2016).

Banhazi, T., Lehr, H., Black, J., Crabtree, H., Schofield, P., Tscharke, M. and Berckmans, D., 2012. Precision livestock farming: An international review of scientific and commercial aspects. *IJABE* 5, 1–9.

Baums, C., Brüggemann, C., Kock, C., Beineke, A., Waldmann, K. and Valentin-Weigand, P., 2010. Immunogenicity of an autogenous *Streptococcus suis* bacterin in preparturient sows and their piglets in relation to protection after weaning. *Clin. Vaccine Immunol.* 10, 1589–97.

Berckmans, D. 2006. Automatic on-line monitoring of animals by precision livestock farming. In *Livestock Production and Society*, Geers, R. and Madec, F. (eds), International Society for Animal Hygiene, Wageningen Academic Publishers, Wageningen, The Netherlands, 51–4.

Berckmans, D., Hemeryck, M., Berckmans, D., Vranken, E. and van Waterschoot T., 2015. Animal sound … talks! real-time sound analysis for health monitoring in livestock. *International Symposium on Animal Environment & Welfare*, 23–26 October, 2015, Chongqing, China,·1–8.

Bishop, S., 2012. A consideration of resistance and tolerance for ruminant nematode infections. *Front Genet.* 3, 168. https://doi.org/10.3389/fgene.2012.00168

Bishop, S. and Woolliams, J., 2014. Genomics and disease resistance studies in livestock. *Livest. Sci.* 166, 190–8.

Bottoms, K., Dewey, C., Richardso, K. and Poljak, Z., 2015. Investigation of biosecurity risks associated with the feed delivery: a pilot study. *Can. Vet. J.* 56, 502–8.

Byra, F., Gadbois, P., Cox, W., Gottschalk, M., Farzan, V., Bauer, S. and Wilson J., 2011. Decreased mortality of weaned pigs with *Streptococcus suis* with the use of in-water potassium penicillin G. *Can. Vet. J.* 52, 272–6.

Callens, B., Persoons, D., Maes, D., Laanen, M., Postma, M., Boyen, F., Haesebrouck, F., Butaye, P., Catry, B. and Dewulf, J., 2012. Prophylactic and metaphylactic antimicrobial use in Belgian fattening pig herds. *Prev. Vet. Med.* 106, 53–62.

Calvar, C., Heugebaert, S., Caille, M. E. and Roy, H., 2012. La quarantaine. Des préconisations de techniciens diversifies. Des conduits multiples chez de très bons éleveurs. Rapport d'étude, Chambres d'agriculture de Bretagne.

Calvar, C. and Lemoine, T., 2014. La biosécurité en élevage de production. Rapport d'étude, Chambres d'agriculture de Bretagne.

Cannon, R. 2002. Demonstrating disease freedom-combining confidence levels. *Prev. Vet. Med.* 52, 227–49.

Carlson, S., Barnhill, A. and Griffith, R., 2012. Salmonellosis. In *Diseases of Swine*, Zimmermann, J. J., Karriker, L. A., Ramirez, A., Schwartz, K. J. and Stevenson, G. W. (eds), Wiley-Blackwell, Chichester, West Sussex, UK, 821–33.

Cornou, C. and Kristensen, A., 2013. Use of information from monitoring and decision support systems in pig production: collection, applications and expected benefits. *Livest. Sci.* 157, 552–67.

Cuevas-Cordoba, B. and Santiago-Garcıa, J., 2014. Saliva: a fluid of study for OMICS. *OMICS* 18, 87–97.

Davies, P., Bahnson, P., Grass, J., Marsh, W. and Dial, G., 1995. Comparison of methods for measurement of enzootic pneumonia lesions in pigs. *Am. J. Vet. Res.* 56, 709–14.

De Busser, E., Dewulf, J., Nollet, N., Houf, K., Schwarzer, K., De Sadeleer, L., De Zutter, L. and Maes D., 2009. Effect of organic acids in drinking water during the last 2 weeks prior to slaughter on *Salmonella* shedding by slaughter pigs and contamination of carcasses. *Zoonoses Public Health* 56, 129–36.

Decaluwé, R., Maes, D., Wuyts, B., Cools, A. and Janssens, G., 2014. Piglets' colostrum intake is associates with daily weight gain and survival until weaning. *Livest. Sci.* 162, 185–92.

Declerck, I., Dewulf, J., Piepers, S., Decaluwé, R. and Maes, D., 2015. Sow and litter factors influencing colostrum yield and nutritional composition. *J. Anim. Sci.* 93, 1309–17.

De Ridder, L., Maes, D., Dewulf, J., Pasmans, F., Boyen, F., Méroc, E., Butaye, P. and Van der Stede, Y., 2013. Evaluation of three intervention strategies to reduce the transmission of *Salmonella typhimurium* in pigs. *Vet. J.* 197, 613–8.

Desrosiers, R., 2011. Transmission of swine pathogens: different means, different needs. *Anim. Health Res. Rev.* 12, 1–13.

Dewulf, J., Tuyttens, F., Lauwers, L., Van Huylenbroeck, G. and Maes, D., 2007. De invloed van de hokbezettingsdichtheid bij vleesvarkens op productie, gezondheid en welzijn. *Vl Diergeneesk Tijdschr* 76, 410–6.

Díaz, I., Cortey, M., Darwich, L., Sibila, M., Mateu, E. and Segalés, J., 2012. Subclinical porcine circovirus type 2 infection does not modulate the immune response to an Aujeszky's disease vaccine. *Vet. J.* 194, 84–8. https://doi.org/10.1016/j.tvjl.2012.02.014

European Commission, 2001. Council Directive 2001/89/EC of 23 October 2001 on Community measures for the control of classical swine fever. In European Commission (Ed.), 2001/89/EC. Official Journal of the European Communities, Brussels, Belgium.

Flesja, K. and Solberg, I., 1981. Pathological lesions in swine at slaughter. *Acta Vet. Scand.* 22, 272–82.

Fosse, J., Seegers, H. and Magras, C., 2009. Prevalence and risk factors for bacterial food-borne zoonotic hazards in slaughter pigs: a review. *Zoonoses Public Health* 56, 429–54.

Gardner, A., Willeberg, P. and Mousing, J., 2002. Empirical and theoretical evidence for herd size as a risk factor for swine diseases. *Anim. Health Res. Rev.* 3, 43–55.

Geldhof, M., Vanhee, M., Van Breedam, W., Van Doorsselaere, J., Karniychuk, U. and Nauwynck, H., 2012. Comparison of the efficacy of autogenous inactivated Porcine Reproductive and Respiratory Syndrome Virus (PRRSV) vaccines with that of commercial vaccines against homologous and heterologous challenges. *BMC Vet. Res.* 8, 182.

Gonyou, H., Lemay, S. and Zhang, Y., 2006. Effects of the environment on productivity and disease. In *Diseases of Swine*. 9th edn. Straw, B., Zimmerman, J., D'Allaire, S. and Taylor, D. (eds), Blackwell Publishing Ltd., Oxford, UK, 1027–38.

Haesebrouck, F., Pasmans, F., Chiers, K., Maes, D., Ducatelle, R. and Decostere, A., 2004. Efficacy of vaccines against bacterial diseases in swine: what can we expect? *Vet. Microbiol.* 100, 255–68.

Hamilton, D., Ellis, M., Wolters, B., Schinckel, A. and Wilson, R., 2003. The growth performance of the progeny of two swine sire lines reared under different floor space allowances. *J. Anim. Sci.* 81, 1126–35.

Hampson, D., 2012. Brachyspiral colitis. In *Diseases of Swine*. Zimmermann, J., Karriker, L., Ramirez, A., Schwartz, K. and Stevenson, G. (eds), Wiley-Blackwell, Chichester, West Sussex, UK, 680–96.

Harris, D. and Alexander, T., 1999. Methods of disease control In Straw, B., D'Allaire, S., Mengeling, W. and Taylor, D. (eds), *Diseases of Swine*. Iowa State University Press, Ames, US, 1077–110.

Hayama, Y., Kimura, Y., Yamamoto, T., Kobayashi, S. and Tsutsu, T., 2015. Potential risk associated with animal culling and disposal during the foot-and-mouth disease epidemic in Japan in 2010. *Res. Vet. Sci.* 102, 228–30.

Holtkamp, D., Rotto, H. and Garcia, R., 2007. The economic cost of major health challenges in large US swine production systems. In Proc. AASV, 3–6 March 2007, Orlando, 85–9.

Hoy, S., Mehlhorn, G., Hörügel, K., Dorn, W., Eulenberger, K. and Johannsen, U., 1986. Der Einfluss ausgewählter endogener Faktoren auf das Auftreten entzündlicher Lungenveränderungen bei Schweinen. *Mh. Vet. Med.* 41, 397–400.

Laanen, M., Persoons, D., Ribbens, S., de Jong, E., Callens, B., Maes, D. and Dewulf, J., 2013. Relationship between biosecurity and production/antimicrobial treatment characteristics in pig herds. *Vet. J.* 198, 508–12.

Loera-Muro, V., Jacques, M., Tremblay, Y., Avelar-González, F., Loera Muro, A., Ramírez-López, E., Medina-Figueroa, A., González-Reynaga, H. and Guerrero-Barrera, A., 2013. Detection of *Actinobacillus pleuropneumoniae* in drinking water from pig farms. *Microbiology* 159, 536–44.

Maes, D., 2012. Subclinical porcine circovirus infection: What lies beneath? *Vet. J.* 194, 9.

Maes, D., Deluyker, H., Verdonck, M., Castryck, F., Miry, C., Vrijens, B. and de Kruif, A., 1999. Risk indicators for the seroprevalences of *Mycoplasma hyopneumoniae*, porcine influenza viruses and Aujeszky's disease virus in slaughter pigs from fattening pig herds. *J. Vet. Med. B* 46, 341–52.

Maes, D., Deluyker, H., Verdonck, M., Castryck, F., Miry, C., Vrijens, B., Ducatelle, R. and de Kruif, A., 2001. Non-infectious herd factors associated with macroscopic and microscopic lung lesions in slaughter pigs from farrow-to-finish pig herds. *Vet. Rec.* 148, 41–6.

Maes, D., Segales, J., Meyns, T., Sibila, M., Pieters, M. and Haesebrouck, F., 2008. Control of *Mycoplasma hyopneumoniae* infections in pigs. *Vet. Microbiol.* 126, 297–309.

Maes, D., Van Soom, A., Appeltant, R., Arsenakis, I. and Nauwynck, H., 2016. Porcine semen as a vector for transmission of viral pathogens. *Theriogenology* 85, 27–38.

Maes, D., Verbeke, W., Vicca, J., Verdonck, M. and de Kruif, A., 2003. Benefit to cost of vaccination against *Mycoplasma hyopneumoniae* in pig herds under Belgian market conditions from 1996 to 2000. *Livest. Sci.* 83, 85–93.

Maes, D., Deluyker, H., Verdonck, M., Castryck, F., Miry, C., Vrijens, B. and de Kruif, A., 2000. Herd factors associated with the seroprevalences of four major respiratory pathogens in slaughter pigs from farrow-to-finish pig herds. *Vet. Res.* 31, 313–27.

Maselyne, J., Adriaens, I., Huybrechts, T., De Ketelaere, B., Millet, S., Vangeyte, J., Van Nuffel, A. and Saeys, W., 2015. Measuring the drinking behaviour of individual pigs housed in group using radio frequency identification (RFID). *Animal*, doi:10.1017/S1751731115000774.

Meyns, T., Vansteelant, J., Rolly, E., Dewulf, J., Haesebrouck, F. and Maes, D., 2011. A cross-sectional study of risk factors associated with pulmonary lesions in pigs at slaughter. *Vet. J.* 187, 388–92.

Mikkelsen, L., Naughton, P., Hedemann, M. and Jensen, B., 2004. Effects of physical properties of feed on microbial ecology and survival of *Salmonella enterica* serovar Typhimurium in the pig gastrointestinal tract. *Appl. Environ. Microbiol.* 70, 3485–92.

Nathues, H., Spergser, J., Rosengarten, R., Kreienbrock, L. and Grosse Beilage, E., 2016. Value of the clinical examination in diagnosing enzootic pneumonia in fattening pigs. *Vet. J.* 193, 443–7.

Neumann, E., 2012. Disease transmission and biosecurity. In *Diseases of Swine*. Zimmermann, J., Karriker, L., Ramirez, A., Schwartz, K. and Stevenson, G. (eds), Wiley-Blackwell, Chichester, West Sussex, UK, 552–624.

Niemi, J., Jones, P., Tranter, R. and Heinola, K. 2016. Cost of production diseases to pig farms. In Proc. 24th IPVS congress, Dublin, Ireland, 7–10 June 2016, 302.

Noyes, E., Feeney, D. and Pijoan, C., 1990. Comparison of the effect of pneumonia detected during lifetime with pneumonia detected at slaughter on growth in swine. *J. Am. Vet. Med. Assoc.* 197, 1025–9.

Opriessnig, T., Meng, X-J. and Halbur, P., 2007. Porcine circovirus type 2-associated disease: update on current terminology, clinical manifestations, pathogenesis, and intervention strategies. *J. Vet. Diagn. Invest.* 19, 591–615.

Opriessnig, T. and Wood, R., 2012. Erysipelas. In *Diseases of Swine*. Zimmermann, J., Karriker, L., Ramirez, A., Schwartz, K. and Stevenson, G. (eds), Wiley-Blackwell, Chichester, West Sussex, UK, 2746–79.

Österberg, J., Vågsholm, I., Boqvist, S. and Lewerin, S., 2006. Feed-borne outbreak of Salmonella Cubana in Swedish pig farms: risk factors and factors affecting the restriction period in infected farms. *Acta Vet. Scand.* 47, 13–22.

Otake, S., Dee, S., Corzo, C., Oliveira, S. and Deen, J., 2010. Long-distance airborne transport of infectious PRRSV and *Mycoplasma hyopneumoniae* from a swine population infected with multiple viral variants. *Vet. Microbiol.* 145, 198–208.

Otake, S., Dee, S., Rossow, K., Joo, H., Deen, J., Molitor, T. and Pijoan, C., 2002. Transmission of porcine reproductive and respiratory syndrome virus by needles. *Vet. Rec.* 150, 114–15.

Patterson, A., Ramamoorthy, S., Madson, D., Meng, X., Halbur, P. and Opriessnig, T., 2011. Shedding and infection dynamics of porcine circovirus type 2 (PCV2) after experimental infection. *Vet. Microbiol.* 149, 91–8.

Pearson, H., Toribio, J. A., Lapidge, S. and Hernández-Jover, M., 2016. Evaluating the risk of pathogen transmission from wild animals to domestic pigs in Australia. *Prev. Vet. Med.* 123, 39–51.

Pedersen, K., Stege, H. and Nielsen, J. P., 2011. Evaluation of a microwave method for dry matter determination in faecal samples from weaned pigs with or without clinical diarrhoea. *Prev. Vet. Med.* 100, 163–70.

Pineiro, C., Morales, J., Doncecchi, P., Dereu, A., Macarrilla, J., Bahnolzer, E., Wuyts, N. and Azlor-Marsinach, O., 2012. In Proceedings of 4th European Symposium of Porcine Health Management, Bruges, Belgium, 102.

Pointon, A., Heap, P. and McCloud, P., 1985. Enzootic pneumonia of pigs in South Australia – factors relating to incidence of disease. *Aust. Vet. J.* 62, 98–100.

Poljak, Z., Dewey, C., Friendship, R., Martin, S. and Christensen, J., 2008. Multilevel analysis of risk factors for *Salmonella* shedding in Ontario finishing pigs. *Epidemiol. Infect.* 136, 1388–400. https://doi.org/10.1017/s0950268807009855

Postma, M., Backhans, A., Collineau, L., Loesken, S., Sjölund, M., Belloc, C., Emanuelson, U., Grosse Beilage, E., Stärk, K. and Dewulf, J., 2016. The biosecurity status and its associations with production and management characteristics in farrow-to-finish pig herds. *Animal* 10, 478–89.

Regula, G., Lichtensteiger, C., Mateus-Pinilla, N., Scherba, G., Miller, G. and Weigel, R., 2000. Comparison of serologic testing and slaughter evaluation for assessing the effects of subclinical infection on growth in pigs. *J. Am. Vet. Med. Assoc.* 217, 888–95.

Ribbens, S., Dewulf, J., Koenen, F., Maes, D. and de Kruif, A., 2007. Evidence of indirect transmission of classical swine fever virus through contacts with people. *Vet. Rec.* 160, 687–90.

Rohrbach, B., Hall, R. and Hitchcock, J., 1993. Effect of subclinical infection with *Actinobacillus pleuropneumoniae* in commingled feeder pigs. *J. Am. Vet. Med. Assoc.* 202, 1095–8.

Saif, L., Pensaert, M., Sestak, K., Yeo, S. G. and Jung, K., 2012. Coronaviruses. In *Diseases of Swine*. Zimmermann, J., Karriker, L., Ramirez, A., Schwartz, K. and Stevenson, G. (eds), Wiley-Blackwell, Chichester, West Sussex, UK, 1821–1914.

Scott, A., McCluskey, B., Brown-Reid, M., Grear, D., Pitcher, P., Ramos, G., Spencer, D. and Singrey A., 2016. Porcine epidemic diarrhea virus introduction into the United States: root cause investigation. *Prev. Vet. Med.* 123, 192–201.

Scudamore, J., Trevelyan, G., Tas, M., Varley, E. and Hickman, G. 2002. Carcass disposal: lessons from Great Britain following the foot and mouth disease outbreaks of 2001. *Rev. Sci. Tech. OIE* 21, 775–87.

Sibila, M., Pieters, M., Molitor, T., Maes, D., Haesebrouck, F. and Segales, J., 2009. Current perspectives on the epidemiology of *Mycoplasma hyopneumoniae* infection. *Vet. J.* 181, 221–31.

Sjölund, M., Postma, M., Collineau, L., Lösken, S., Backhans, A., Belloc, C., Emanuelson, U., Groβe Beilage, E., Stärk, K. and Dewulf, J., 2016, Quantitative and qualitative antimicrobial usage patterns in farrow-to-finish pig herds in Belgium, France, Germany and Sweden. *Prev. Vet. Med.*, submitted.

Smart, N., Miniats, O., Rosendal, S. and Friendship, M., 1989. Glasser's disease and prevalence of subclinical infection with *Haemophilus parasuis* in swine in southern Ontario. Can. *Vet. J.* 30, 339–43.

Soerensen, D. and Pedersen, L., 2015. Infrared skin temperature measurements for monitoring health in pigs: a review. *Acta Vet. Scand.* 57, 5. DOI 10.1186/s13028-015-0094-2.

Straw, B., 1991. Performance measured in pigs with pneumonia and housed in different environments. *J. Am. Vet. Med. Assoc.* 198, 627–30.

Szabo, J. and Minamyer, S., 2014. Decontamination of biological agents from drinking water infrastructure: a literature review and summary. *Environ. Int.* 72, 124–8.

Thrusfield, M., 2005. The economics of animal disease (Chapter 50), In *Veterinary Epidemiology*, third edition, Blackwell Science, Oxford, UK, 357–67.

Tobias, T., Bouma, A., van den Broek, J., van Nes, A., Daemen, A., Wagenaar, J., Stegeman, J. and Klinkenberg, D., 2014. Transmission of *Actinobacillus pleuropneumoniae* among weaned piglets on endemically infected farms. *Prev. Vet. Med.* 117, 207–14.

Toma, B., Vaillancourt, J., Dufour, B., Eloit, M., Moutou, F., Marsh, W., Bénet, J. J., Sanaa, M. and Michel, P., 1999. *Dictionary of Veterinary Epidemiology*, Iowa State University, Ames, 199–200.

Whitworth, K., Rowland, R., Ewen, C., Trible, B., Kerrigan, M., Cino-Ozuna, A., Samuel, M., Lightner, J., McLaren, D., Mileham, A., Wells, K. and Prather, R., 2015. Gene-edited pigs are protected from porcine reproductive and respiratory syndrome virus. *Nat. Biotechnol.* 2015; DOI:10.1038/nbt.3434.

Part 2

Welfare issues

Chapter 5

Understanding pig behaviour

Simon P. Turner and Richard B. D'Eath, Scotland's Rural College, UK

1 Introduction

Animal behaviour is central to one of the most popular definitions of welfare that requires that animals should have the opportunity to behave naturally (Kiley-Worthington 1989; Rollin 1993). The opportunity to express natural behaviour is therefore viewed by some as a cornerstone of what it means to be in a good state of welfare. Consequently, the execution of species-typical behaviour is therefore seen as an important welfare assessment tool. According to another definition, good animal welfare is when animals are 'healthy and have what they want' (Dawkins 2008). Highly motivated behaviours may be thwarted when housing or management practices prevent pigs from 'having what they want'. The consequence of thwarting highly motivated behaviours has implications in its own right for the welfare of the individual and its group mates. Behaviour, as well as being an indicator of welfare state, therefore also plays a central role itself in some of the most intractable welfare problems in modern production. Indeed, many welfare problems (e.g. tail biting, belly nosing, fighting, maternal crushing and savaging of neonates, vulva biting, heat stress; Fig. 1 and 2) cannot be understood without knowledge of behaviour of domestic pigs and their wild ancestors. This chapter seeks to provide the basis with which to understand the aetiology of some of the prevalent welfare conditions by referring to the behavioural ecology of pigs and the mismatch between highly motivated systems and the opportunities for their behavioural expression under commercial production. It also considers the evidence for, and causes of, individual variability in the expression of behavioural traits that affect welfare.

http://dx.doi.org/10.19103/AS.2017.0013.18

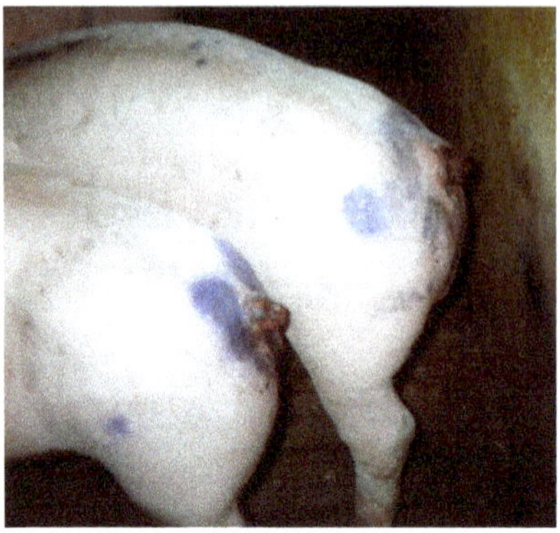

Figure 1 The outcome of severe tail biting in two commercial grower-stage pigs.

Figure 2 Commercial sow wallowing at an outdoor pig unit during warm weather. This form of behavioural thermoregulation is an important method of preventing heat stress in pigs (Image credit: Marianne Farish).

2 Behavioural ecology of pigs

A logical starting point from which to understand the behaviour of modern commercial pigs is to understand the behavioural ecology of their ancestors and to assess to what extent the behavioural repertoire has been adapted by domestication and deliberate artificial selection. Pigs were domesticated between 9 and 11 thousand years ago (Ramos-Onsins et al. 2014; Lega et al. 2016) and, particularly in recent decades, subjected to intense artificial selection for rapid, lean growth and high fecundity. Great morphological and metabolic adaptations to commercial production have occurred as a result of these processes. For example, growth rate has increased several fold from around 0.11 kg/d

Figure 3 Wild boar aggressive interaction targeted towards the head of the recipient (Image credit: Marianne Farish).

Figure 4 Aggressive interaction between unfamiliar commercial domestic pigs (Image credit: Marianne Farish).

for the first year of life of wild boar (Pépin et al. 1987) to around 0.68 kg/d for the first six months of life of domestic pigs (Desire et al. 2015a). Litter size has also shown a similar fold increase from a mean of five in wild boar (Leaper et al. 1999; Nahlik and Sandor 2003) to a mean of 14 in hyper-prolific genetic lines of domestic sows (Rutherford et al. 2013). Despite these gross changes, the behavioural repertoire of commercial pigs has remained remarkably conserved (e.g. Fig. 3 and 4). This can be demonstrated by observational studies of wild pigs and by allowing commercial pigs to live in a naturalistic environment (e.g. Stolba and Wood-Gush 1989; Jensen 1995) or housing wild boar under semi-commercial conditions (Gustafsson et al. 1999; Spinka et al. 2000). What is evident from these studies is that commercial pigs show similar time budgets and behavioural repertoires to wild boar when in a wild environment and that behavioural aberrations induced in commercial pigs by a commercial production environment can be induced by the same environment in wild boar. Furthermore, the behavioural repertoire of commercial pigs is functionally intact as evidenced by the large number of successful populations of feral domestic pigs (e.g. Hone 2002). The behavioural ecology of wild boar and feral pigs and the behaviour of domestic pigs in experimental naturalistic environments therefore remains an entirely relevant source of insight with which to understand the behaviour of modern intensively housed pigs.

Undoubtedly domestication has quantitatively altered the expression of behaviour, for example by reducing the sensitivity to stimuli associated with predators (including humans; Kurz and Marchinton 1972). However, the quality of behavioural expression and the motivational systems that underlie this expression remain. This has major implications for the welfare of individual pigs and their pen mates under commercial production. This section briefly summarises the behaviour of pigs in the wild and draws upon studies of wild boar, feral commercial pigs and commercial pigs released for experimental purposes into a naturalistic environment. Observational studies have reported in great detail on aspects of the undisturbed natural behaviour of pigs (e.g. see Frädrich (1974) and Signoret et al. (1975) for a description of the mating behaviour and the review by D'Eath and Turner (2007) for other aspects of behavioural ecology). Here we focus specifically on those aspects of natural behaviour that have been shown to play a key role in welfare challenges under commercial production. Section 3 then summarises the putative behavioural needs of pigs to be cross-referenced against the ability of farm environments to meet these needs.

2.1 The adaptable pig

Wild boar are omnivorous and exploit a wide variety of habitats, from sea level to mountains and from undisturbed forest to farmland margins. Adaptation to a wide climatic range is achieved by strong reliance upon behavioural means of thermoregulation (shade seeking and wallowing to prevent heat stress (Fig. 2), huddling and nest building to prevent cold stress). Territorial behaviour is shown, but insufficiently to prevent home ranges overlapping. Pigs are diurnally active when undisturbed by humans with a depression in activity around noon (Caley 1997). This basic diurnal pattern is replicated in commercial housing.

2.2 Foraging behaviour

Underground food items such as roots, tubers, bulbs and underground invertebrates are a staple part of the diet (Leaper et al. 1999; Cocca et al. 2007) together with seasonally abundant food (e.g. acorns and beech mast; Groot Bruinderink and Hazebroek 1994; Massei et al. 1996; Herrero et al. 2005). Buried and dispersed food items require a large part (75%) of the activity budget to be dedicated to foraging using the snout as a tool for rooting (Stolba and Wood-Gush 1989; Fig. 5).

2.3 Social behaviour

Wild boar and feral pigs adopt a matrilinear social structure whereby a small number of related sows share a home range and move as a cohesive unit together with their piglets from the most recent breeding season (Gabor et al. 1999; Gonyou 2001). Older juveniles share the home range but move independently from the sow group (boars, Spitz and Janeau 1995; pigs, Stolba and Wood-Gush 1989; feral pigs, Gabor et al. 1999). Sows within a group exhibit a stable, linear dominance hierarchy which is regulated mainly by the avoidance of conflict by subordinates (Mauget 1981; Jensen and Wood-Gush 1984; Gonyou 2001). High social status is conferred by maturity, a large body size and physical strength (Mauget 1981).

Figure 5 A domestic sow on an outdoor commercial farm showing the same rooting behaviour as demonstrated during foraging in the wild (Image credit: Marianne Farish).

Territorial aggression is rare and social groups show mutual avoidance when they come into proximity (Gabor et al. 1999). Aggression between pigs under wild conditions occurs when piglets establish their 'ownership' of teats (the most anterior being most productive and therefore most desirable (Newberry and Wood-Gush 1985)) and between mature boars during the mating season (Mendl 1995). Aggressive interactions can also occur over access to food (Frädrich 1974; Stolba and Wood-Gush 1989; Gabor et al. 1999) but aggression performed for other reasons is typically rare (Mendl 1995). Aggressive contests are ritualised and usually follow several stages of escalation from non-contact parallel walking to damaging fighting with de-escalation and retreat occurring at any point (Gundlach 1968; Frädrich 1974; Barrette 1986). During escalated fighting the head, neck and shoulders are the main target of bites (Kurz and Marchinton 1972; Frädrich 1974; Fig. 3 and 4).

2.4 Mating

The arrival of solitary adult boars into the home range during the mating season triggers the dispersal of sub-adult males and stimulates puberty in females of around 10 months of age who may move to a neighbouring social group or return to their natal group at this time (Frädrich 1974; Gabor et al. 1999). Adult males do not defend a territory but do aggressively compete over access to females (Kurz and Marchinton 1972; Gonyou 2001).

2.5 Parturition, maternal behaviour and early life

Wild boar are seasonally anoestrus and breeding in most years occurs only in the autumn leading to the birth of one litter in the spring (Frädrich 1974; Mauget 1981). In years of food abundance, a second litter may be born (Gundlach 1968; Mauget 1981; Pépin et al. 1987). Feral populations of domestic pigs, whilst being capable of breeding all year, show a reversion to a seasonal peak in fertility (Dzieciolowski et al. 1992).

Parturition ('farrowing') is preceded by the separation of the sow from her social group and the construction of a nest (Gundlach 1968; Spitz 1986; Stolba and Wood-Gush, 1989). The nest is usually in dense undergrowth and comprises a shallow scrape in the earth filled with vegetation to create a pile around 1 m in height (Gundlach 1968; Frädrich 1974). Nest construction is an active period that requires many short journeys to collect materials and rearrangement of these at the site. The main features of next site selection and construction remain intact in domestic sows experimentally released into a naturalistic environment (Jensen 1986; Jensen et al. 1993; Stolba and Wood-Gush 1984, 1989).

The litter sizes of wild boar are highly dependent upon the age of the sow and food availability but a mean of around five piglets is typical (Mauget 1981; Fernández-Llario et al. 1999; Leaper et al. 1999; Nahlik and Sandor 2003). Modern selective breeding has increased the litter size in some genotypes to around 14 born alive but feral domestic populations tend to revert to litter sizes approximating that of wild boar (Kurz and Marchinton 1972; Dzieciolowski et al. 1992). Piglets are born at less than 1% of their adult weight and spend the first few days in the nest (Stangel and Jensen 1991). They are poor at thermoregulation and rely upon huddling with siblings and locating near the udder for warmth.

Sows leave the nest during the first two weeks only for essential maintenance activities and, on re-entering the nest, carefully root the ground before lying down either to arrange nesting material or to check if it is free from piglets (pigs, Petersen et al. 1990). The sow noses the piglets away from where she intends to lie, and will shift position if a piglet becomes trapped beneath her and vocalises (Gundlach 1968; Frädrich 1974). The primary causes of piglet mortality in the wild are predation and the occurrence of late frosts (Seguin et al. 2001; Jedrzejewski et al. 2002). Deaths due to crushing by the sow have not been reported in wild boars or feral domestic sows but are one of the primary causes of neonatal mortality in commercial production. Crushing has been described as a cause of mortality in extensively kept commercial pigs (Newberry and Wood-Gush 1985, 1986), although this may have been due to the intrusion of other adult or sub-adult pigs into the nest area. The risk of crushing in commercial populations is positively correlated with the litter size. Artificial selection for large litter size is, in part, responsible for the discrepancy in crushing incidence in wild compared to commercial populations (Rutherford et al. 2016).

The same pattern of suckling behaviour is seen in wild boar (Gundlach 1968; Horrell 1997) and in free-ranging domestic pigs (Newberry and Wood-Gush 1985; Jensen 1988). This is characterised by continual access to colostrum for the first two hours followed rapidly by a cycle of approximately hourly suckling bouts (Fraser 1980; de Passillé and Rushen 1989). Milk is ejected for only a short period of 15 seconds and competition for access to the most productive teats results in the early life development of a hierarchical 'teat order' in both wild boar and feral and domestic pigs (Fig. 6). Despite this, cross-suckling occurs, whereby a piglet from a different litter sucks from the sow (Delcroix et al. 1985; Eisenberg and Lockhart 1972; Newberry and Wood-Gush 1985).

Figure 6 Early establishment of a stable 'teat order' helps to ensure that piglets are able to feed during the short period of milk ejection (Image credit: Marianne Farish).

Social integration with other litters begins from around one week of age (Gundlach 1968; Jensen and Redbo 1987; Petersen et al. 1989; Stangel and Jensen 1991) and is a gradual process involving little overt aggression. There is a steady rise in play from two days of age until between two and six weeks, followed by a decline that is probably a response to the onset of weaning (Newberry et al. 1988; Stangel and Jensen 1991). Play comprises of locomotor, object and social forms of engagement (Frädrich 1974; Newberry et al. 1988; Newberry and Wood-Gush 1988; Petersen et al. 1989).

Weaning under wild conditions is a gradual process that begins with a rise in the ingestion of solid food at 6–8 weeks and is usually complete by 14–19 weeks (Newberry and Wood-Gush 1985; Jensen 1986; Jensen and Recén 1989). Following weaning pigs associate preferentially with others of the same age but cohabit in the same home range as their mother.

3 Putative behavioural needs of pigs

When pigs' needs in terms of their physical health and nutrition are met, this results in successful growth and reproduction, and meat production for the farmer. However, there may still be a mismatch between behaviours that the pig is motivated to perform and the environment in which it finds itself.

The concept that animals have 'behavioural needs' may be useful here. Behavioural needs arise when internal factors (in the animal's physiological and neurobiological state) lead to high motivation to perform a behaviour irrespective of the external environment (Hughes and Duncan 1988, Duncan, 1998). This simple characterisation has been criticised (Jensen and Toates 1993) because the performance of all behaviours are to some degree affected by *both* internal and external factors. In particular, this is because there is often feedback where the performance of a behaviour either leads or does not lead to a satisfactory end goal (such as construction of a nest, or failure to construct a nest), and the steps in this process may modify what an animal does. However, the basic concept is still useful that animals have needs to perform behaviours which, if prevented from doing so, may lead to signs of suffering.

Through domestication, removal of natural selection pressure may have removed the function of a behaviour to promote fitness but not necessarily its cause. The cause of a behaviour is linked to the reward perceived by the animal from performing the behaviour or, conversely, the frustration felt when it is thwarted (Hughes and Duncan 1988). Preventing the rewards that would normally follow the expression of highly motivated behaviours is at the root of many complex and intractable welfare problems. Therefore, to be confident that an animal has a 'need' to perform a behaviour, we must look towards evidence of how hard it will work to perform the behaviour itself rather than to the fitness consequences that originally explained its evolution.

Domestic pigs housed in a protected indoor environment with ready access to all resources needed for survival are highly motivated to perform behaviours that under these conditions serve no purpose. For example, indoor commercial production conditions provide a warm, dry 'creep' area for the piglets that is free from predation risk and replaces the need for a nest at farrowing. Yet, even when sows have no access to the substrates needed to build a nest and are confined in a farrowing crate, they still show high levels of activity (in the form of changing posture from standing to lying) at the time when they would normally be walking away from the herd to seek isolation from other sows (Haskell and Hutson 1994). This is followed by rooting and pawing movements before giving birth that are reminiscent of the activities required to rearrange material into the form of a nest (Jarvis et al. 2001). These behaviours are seen both in sows giving birth for the first-time (gilts) and to a similar or slightly lesser extent in sows with prior experience of farrowing in these environments. Sows will 'work' to access straw prior to farrowing (Arey 1992) to build a nest in which to farrow. This example shows that the behaviour of pigs when housed in close confinement reflects very closely that of their free-living ancestors and contemporaries, even though the full expression of a behavioural sequence may be prevented by the confined environment (e.g. the sow in the farrowing crate cannot search for, collect and actually rearrange materials). In addition, even when sows are given access to pre-formed nests of straw, they still show nest-building behaviours such as straw carrying and pawing at the ground, and show them to a similar extent as sows given straw that is not already shaped into a nest (Arey et al. 1991). This suggests that the performance of the behaviour itself, rather than the achievement of its end goal, is important to the pig.

This preservation of the behavioural repertoire and underlying motivations suggests that the control systems underlying the causation of behaviour are fully intact even if the behaviours are apparently without purpose in commercial production. Table 1 provides other examples that fit this description whereby pigs show a high motivation to perform a behaviour in a pattern representative of wild pigs even though the behaviour has been divorced from its survival value. Whether these examples constitute true behavioural needs may be contested, and indeed there are likely to be other behaviours for which there is a high motivation that are not included in the table. The examples presented in Table 1 have been selected due to a relatively rich literature describing the motivation pigs show for their performance and the significant welfare consequences that follow under commercial conditions from the thwarting or redirection of this motivation.

Where the outlet of a motivation is redirected, this may be sufficient evidence of the animal having a 'need' to perform such behaviour (Pedersen et al. 2014). An example of this redirection is seen when piglets that normally massage the udder of the sow to stimulate milk let-down redirect this massaging to the belly of other piglets when fed milk by a machine or following early weaning (Rzezniczek et al. 2015). When pigs are housed in close confinement, such as a gestation or farrowing crate, behaviours may become so constrained that it may be difficult to make any judgement about the motivational state of the animal. However, here we can make use of evidence from stress physiology biomarkers and tests that determine how hard an animal is willing to work to access an environment where normal behaviour can be expressed (e.g. Warburton and Mason, 2003; Patterson-Kane et al., 2008; Elmore et al., 2012). These studies show that crated sows show a peak in cortisol levels concurrent with attempts to nest build whilst such a hormonal response is less evident in sows given basic substrates and freedom to turn around (Jarvis et al., 2002; Wischner et al., 2009). Operant conditioning tasks show that a sow's motivation for access to nest building material approaches that of access to food as farrowing approaches (Arey, 1992; Fig. 7). It may also be argued that nest building is not entirely functionless in commercial production. Preventing the execution of this behaviour could contribute to the maternal infanticide ('savaging') seen typically by first-time mothers (gilts) in commercial production but not in the wild (Jarvis et al., 2004; Wischner et al., 2009). The behaviour may be a reflection of a general state of excitability (Ison et al., 2015) and could be exacerbated if the normal sequence of maternal behaviours is prevented.

Table 1 Sources of reduced welfare resulting from a mismatch between the environment in which the ancestors of commercial pigs have evolved and the environment provided in commercial production systems

Situation/ Life stage	Characteristics of the wild environment	Characteristics of the typical indoor farming environment (dependent upon country and production system)	Potential or actual welfare problems resulting from mismatch (*or (in italics) welfare improvements compared to wild pigs*)	References
All ages	Adequate space for the performance of all behaviours	Limited space	Increased aggression, disturbance of resting pigs, exposure to dust, ammonia, increased disease risk	Turner et al., 2000; Hemsworth et al., 2013
All ages	Natural materials abundant (e.g. soil, mud, vegetation)	Full or part-slatted flooring with no access or limited access to materials (e.g. wooden poles, small amount of chopped straw)	Vulva biting in sows, Tail, ear, flank biting in growers and finishers. *Tail docking reduces tail biting risk*	Rizvi et al., 1998; Smulders et al., 2008; Taylor et al., 2010; D'Eath et al., 2014
All ages	Exposure to wide range of climactic conditions; adapt through wallowing and shade-seeking in heat, huddling and nest building in cold	Temperature-controlled environments with heating/ventilation maintained to suit each growth stage. Sprinklers for extreme heat. Reduced need for behavioural thermoregulation	Restricted ability to perform behavioural thermoregulation in case artificial control fails. *Risk of sunburn, heat stroke or chilling reduced*	Fox et al., 2014
All ages	Access to a potential wide range of food types (but can be widely dispersed, sparse or absent) obtained through natural foraging behaviour	Food available at all times, usually uniform type but of high nutritional value. Hunger due to feed restriction in sows can result in abnormal oral behaviour	Lack of food variety and choice. Insufficient outlet for foraging behaviour. Routine hunger in sows. *Sufficient food for healthy growth and reproduction is provided*	Schrøder-Petersen and Simonsen 2001; Jensen et al. 2012; Sapkota et al., 2016
Boars	Live alone, have larger territories than females, compete with other males for natural mating	Boars housed singly; artificial insemination commonly used whereby boars are retained only for oestrus detection	Boar–boar aggression is absent. Possible frustration due to proximity of sows but lack of mating opportunities	Gabor et al., 1999; Gonyou, 2001

(Continued)

Table 1 (*Continued*)

Situation/ Life stage	Characteristics of the wild environment	Characteristics of the typical indoor farming environment (dependent upon country and production system)	Potential or actual welfare problems resulting from mismatch (*or (in italics) welfare improvements compared to wild pigs*)	References
Pregnant sows	Live in sounder groups with stable adult membership	Single housing in gestation crates prevents social interactions; group housing requires regrouping at the beginning of pregnancy. Feed restriction occurs	Single confinement in gestation stalls, or aggression when mixing sow groups leading to skin lesions, lameness, immunosuppression and infection risk. Hunger	Gabor et al., 1999; Gonyou 2001; Jensen et al., 2012; Verdon et al 2015
Farrowing sows	Isolate themselves from group to seek nest site and to nest build and farrow	Moved to farrowing housing where they are single housed but not isolated; often confined in a farrowing crate	Lack of choice of nest site, frustration of nest-building behaviour by confinement in farrowing crates with inadequate space and substrates. Abnormal 'savaging' of piglets by some gilts	Jarvis et al., 2001; 2002; Fernández –Llario, 2004; Chen et al., 2008
Lactating sows	Remain with piglets in nest location or proximity for approximately 10 days post-farrowing	Remain in farrowing accommodation typically for first 3–4 weeks. Confined in farrowing crates	Frustration of filial interactions by crates. Sows cannot reduce contact with their piglets as they age. Chronic stress from prolonged confinement	Petersen et al., 1989; Stangel and Jensen, 1991; Yun et al., 2013
Weaning (sows)	Sow reduces the frequency of suckling and uses behaviour to reduce udder access	Sow has fewer ways to reduce udder access and avoid piglets	Possible frustration of attempts to begin weaning process, and/or to teach piglets about solid food; sudden weaning potentially leads to pain from full udder	
Piglets-Birth	Sows are passive and allow piglets to stand, seek udder and find teat by themselves after birth	Sows are passive, and allow piglets to stand, seek udder and find teat by themselves, although some individual sows may be restless, leading to a risk of piglet crushing	Piglet's experience of this stage is relatively natural. *Artificial shelter/heating reduces piglet deaths due to cold*	Dellmeier and Friend, 1991; Valros et al., 2003

Situation/ Life stage	Characteristics of the wild environment	Characteristics of the typical indoor farming environment (dependent upon country and production system)	Potential or actual welfare problems resulting from mismatch (*or (in italics) welfare improvements compared to wild pigs*)	References
Piglets-Mutilations	Piglets are left intact	Routine artificial management practices of teeth clipping, tail docking (to minimise tail biting injuries) and castration are performed	Acute and possibly chronic pain, potential source of infection	Bonastre et al., 2016; Numberger et al., 2016; Sandercock et al., 2016
Piglets-Suckling	Piglets compete to form a 'teat order'. Piglets vary in size and 'runts' at greater risk of death	Fostering of piglets to equalise litter sizes. Can result in reduced size variation	Disruption of teat order due to fostering increases aggression at the teats. *Weaker piglets may benefit from fostering to a less competitive litter or artificial milk*	Baxter et al., 2013; Devillers et al., 2016
Piglets-Weaning	Weaning is gradual over a period of weeks, being completed between 14 and 19 weeks	Weaning is sudden and relatively early (3–4 weeks). Involves a new environment and regrouping with unfamiliar pigs	Weaning 'growth check', diarrhoea, aggression, stress. Piglets cannot learn from sows about eating solid food	Jensen and Recén, 1989; Berkeveld et al., 2009
Growing pigs	Stable social group composition	Regrouping with unfamiliar pigs to maximise building use and ensure groups of similar weight	Aggression resulting in skin injuries, lameness, immunosuppression and infection risk	Rundgren and Löfquist, 1989; Tan et al., 1991; Morrow-Tesch et al., 1994; D'Eath et al., 2010
Finishing pigs	Juvenile males disperse as puberty approaches	Dispersal impossible	Mounting behaviour leads to lameness, skin injuries and disturbance of rest	Clark and D'Eath, 2013; Hintze et al., 2013

3.1: Case study: tail and ear biting

Tail and ear biting is characterised by the chewing of tail and ear tips most commonly in grower-stage pigs. Ear and tail biting sufficient to cause significant injury has not been reported in either wild boar, feral domestic pigs or domestic pigs released into an experimental natural environment. The behaviour is commonplace in commercial

Figure 7 Given the opportunity, domestic sows show considerable and complex nest building behaviour that mirrors the behaviour of wild boar (Image credit: Marianne Farish).

barren indoor production environments and up to 8% of pigs with undocked tails have tail lesions at slaughter (Hunter et al., 2001). Damage can range from minor lesions to total amputation of ear tips or the whole tail and spinal abscesses. In most major pig producing countries, tail docking of neonatal piglets without anaesthesia is performed routinely to reduce tail biting, although it does not totally prevent it (D'Eath et al. 2016).

Several aetiologies appear to exist for this behaviour (Taylor et al. 2010) but the most common is that the behaviour is a redirected form of the rooting behaviour that accompanies natural foraging. Grower-stage pigs are fed *ad libitum* and can meet their dietary requirements within a very short period from a food source that is energy dense and readily accessible. This contrasts with the high percentage of the active period of a wild pig's day that is devoted to searching for dispersed food that is often hidden and needs to be extracted (Stolba and Wood-Gush 1989). Many factors contribute to the risks of tail and ear biting (see review by D'Eath et al. 2014) but the most important is the absence of high-quality enrichment. Appropriate manipulable material significantly diminishes the likelihood and severity of lesions if it possesses qualities that allow proper feedback from the act of rooting which include being ingestible, destructible, deformable, chewable and odorous (Van de Weerd et al. 2003).

Knowledge of the causation and prevention of tail and ear biting is at the basis of legislation specifying the minimum requirement and quality of enrichment for pigs (e.g. European Union Council Directive 2008/120/EC). Despite legislation and a major research effort over past decades it remains difficult to implement forms of enrichment that are compatible with current commercial practice. As such tail and ear biting remain amongst the most significant, intractable and ubiquitous welfare challenges in the industry and tail docking is performed on the large majority of pigs (e.g. 99% of Irish pigs presented for slaughter with levels of >90% in other countries also; Harley et al. 2012). Consumer concern over the routine use of tail docking as a control strategy for tail biting has led to its ban in a small number of countries (e.g. Finland, Norway, Sweden, Switzerland) and there have been recent voluntary industry commitments to reduce reliance on this method in other regions. It is imperative in coming years that methods are found to adapt management systems to allow more appropriate expression of this rooting motivation, that pigs are adapted through selective breeding such that they do not show tail and ear biting (see Section 4) or that these two strategies are used in parallel.

4 Individual variation in pig behavioural development

So far we have described species-typical behaviour and the continued motivation to perform this behaviour under artificial indoor housing environments. However, behavioural expression is highly individualistic. For example, not all pigs tail or ear bite and not all gilts show infanticidal behaviour towards their piglets. The actual or perceived rewards and punishments resulting from a given course of action are presumably not experienced in the same way for all pigs which affects their motivations and subsequent behavioural expression. This is well seen in the case of aggressive behaviour. The sudden mixing of pigs into new social groups in a confined pen that prevents escape is a routine feature of production, is experienced by most commercial pigs several times in life and results in intense aggression (Desire et al. 2015b). This contrasts with the infrequent and gradual integration of unfamiliar pigs in the wild in which aggression is rare. However, not all commercial pigs fight intensely and understanding the reasons for this individual variation has become an active area of research as indeed it has for each of the behavioural traits that affect pig welfare.

4.1 Case study: to fight or not to fight?

What determines the decision of a pig who is suddenly faced with an unfamiliar animal to initiate an attack against it or to respond to the receipt of an attack with either retaliation or retreat? Pigs encountering another which is apparently matched in resource-holding potential (largely determined by weight in pigs, but influenced by personality; Camerlink et al. 2015) show individual variation in response that is repeatable over time (Mendl 1995; Camerlink et al. 2015; D'Eath et al. 2010; Fig. 8). Variation between individuals in aggressiveness can be linked to many causes. Firstly, the tendency to become involved in aggressive behaviour is heritable in pigs to a similar degree as productivity traits that have been the focus of selective breeding (e.g. the heritability of reciprocated fighting between unfamiliar pigs in two different populations was 0.43 and 0.46; Turner et al. 2008, 2009). No major gene effect has been found for this genetic determination and it is likely that a large number of genomic regions are responsible. Therefore, almost half of the total variance in aggressiveness in a population of the same breed and age can be explained by additive genetic effects.

Both pre- and post-natal factors are likely to be responsible for the remaining between-individual differences in aggressiveness. Jarvis et al. (2006) and Kranendonk et al. (2006) have shown that prenatal stress experienced from subordination of the dam during

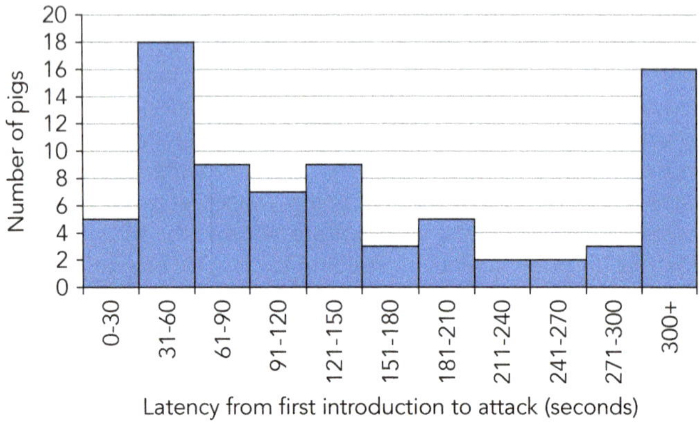

Figure 8 Variation in aggressiveness in a population of commercial grower-stage pigs. The *X*-axis shows the latency from the introduction of a smaller 'intruder' pig into the home pen of a larger 'resident' until the point at which the resident initiated an attack. The test was terminated at 5 minutes after introduction and pigs which had not attacked by this time were given a score of 300+ seconds (Source of data: Irene Camerlink).

pregnancy or exogenous administration of a glucocorticoid to the pregnant dam can modify the aggressive phenotype of the subsequent piglets. In both cases the resulting piglets show more chronic aggression following social regrouping. The neonatal environment is also influential. For example, piglets from large litters or those with a high proportion of females show increased aggressiveness in later life (D'Eath and Lawrence 2004). There may also be differences in the ability of pigs to accurately assess the resource-holding potential of an opponent and compare that against their own (Camerlink et al. 2015), either because of inexperience, differences in cognitive ability or some other mechanism. The dopaminergic and serotonergic systems, amongst others, have a known modulating effect on aggressiveness and the specific balance of these neurochemicals within an individual also determines their aggressive phenotype (D'Eath et al. 2005). Lastly, proximate factors, such as the experience of recent defeat, hunger and poor health can have a profound influence on how willing animals are to engage in aggression (Cloutier et al. 1995; Meerlo et al. 1997; Verdon et al. 2015).

This multitude of determining factors poses a problem when using the quantity of behavioural expression as a window into the emotional state of pigs. How should we view the affective state of a pig which does not attack an opponent compared to a pig which does? Should an unaggressive response be taken as ambivalence to the presence of a new animal, or does this represent a conflict of motivations whereby aggressiveness is ultimately inhibited by fear of pain and defeat? Similarly, it is reasonable to question whether a pig which shows tail biting behaviour has successfully found an outlet for a strong motivation or simply perceived that motivation to be greater than a contemporary who does not tail bite. Answers to such questions require the measurement of affective state of animals differing in behavioural phenotype when exposed to the same external stimulus. Measurement of affect in pigs has thus far focussed on assessing the outcome of imposed treatments at a population level (e.g. Murphy et al. 2014) but its use to assess individual differences in behavioural expression would greatly help us to understand what this variation actually means for the pigs.

5 Future trends

Political and economic drivers are likely to see an increase in interest in maternal behaviour in light of the abolition of farrowing crates in some countries (e.g. Sweden). This will likely focus on the role of maternal behaviour in securing piglet survival in non-crate systems and in understanding individual variation in maternal behaviour (e.g. why sows are consistent over parities in their tendency to crush piglets; Jarvis et al. 2005). This may also encourage greater interest in studying the reasons for individual differences and methods to predict and control savaging of piglets by gilts that has so far received less research interest than perhaps its economic and welfare consequences would justify.

Reduced reliance upon routine tail docking to mitigate the effects of tail biting is likely to enhance multidisciplinary research interest in understanding the consequences of, and solutions to, tail biting. Recent work in this field has included studies on the economic cost-effectiveness of management solutions and ethical analyses of the relative effects of tail biting experienced by some pigs versus tail docking experienced by all (e.g. D'Eath et al. 2016). Work is ongoing, studying the effects of tail biting and tail docking at the mechanistic level (e.g. the neuropathological consequences of tail biting and docking; Sandercock et al. 2016). However, it is unknown what tail biting and docking mean to the emotional lives of pigs and studies on changes in affective state are needed to understand this further.

The costs of observing and recording complex and often unpredictable behaviour present a barrier to both research and commercial management. The development of on-board sensors (e.g. accelerometers, Oczak et al. 2016) or use of automatic visual processing software (e.g. depth image processing, Lao et al. 2016; processing of Euclidian distances, Nasirahmadi, et al. 2016) is a highly active area of research that in future years is likely to routinise the recording of behaviour. At present, software is available that is effective in automatic detection of specific patterns of pig behaviour, but requires excessively regular manual calibration. As routine behavioural recording becomes more feasible this opens the door to on-farm early warning and detection systems to facilitate management of pig behaviour (e.g. alerting farmers to an imminent outbreak of tail biting at a point before this becomes difficult to reverse). Low-cost phenotyping, coupled with the decreasing cost of genotyping and, in the future, DNA sequencing, will also allow breeding for/against behavioural traits that are currently too difficult to include in selection indexes.

To benefit welfare, all of these advances must improve the lives of the animals. Measurement of affective state in livestock is a dynamic and active field. The last decade has seen a substantial expansion of the toolkit of measures that attempt to probe emotional valence in addition to arousal (e.g. see Murphy et al. 2014 for a review). These new measures must be used in addition to those that have been available for some time such as choice tests to examine what it means to pigs when behavioural motivations are thwarted and what it means to recipients of redirected behaviour. The coming years are likely to see measurement of affective state becoming refined, streamlined and more fully incorporated into the mainstream of animal welfare assessment.

6 Conclusion

Understanding pig behaviour firstly requires an appreciation of the behavioural ecology of the species. Onto this we must superimpose the modulating effects of genetic variation, early life development, previous experiences and immediate homeostatic state. Both the quantity and quality of behavioural expression are key tools in assessing welfare state (e.g. Wemelsfelder et al., 2001) whilst behavioural expression in its own right can directly and severely affect animal welfare. Therefore, pig behaviour is studied both as a tool to assess welfare (e.g. monitoring the effects of increased stocking density on pig aggression) but also due to the need to find solutions to the adverse welfare consequences that result from behavioural expression in a suboptimal environment (e.g. injuries from aggression as a result of repeated regrouping).

Much has been learnt about pig behaviour at both a population and individual level. At the grandest of scales, this shows that the relatively short period of domestication and recent efficient selective breeding has done little to modify the behavioural repertoire of pigs compared to the preceding millennia of evolution. We know a great deal more about how behavioural expression, such as the tendency to tail bite, is modified by a complex array of interacting risk factors and homeostatic mechanisms. However, much remains to be understood both at a fundamental level (e.g. the genetic architecture underlying such complex traits as tail biting) and at a practical level (e.g. how to reliably predict an outbreak of tail biting in a manner that can be applied on farms). As yet we still have a poor understanding of what performance of behavioural traits might indicate about affective state, particularly with regard to understanding what individual differences in behavioural expression might tell us about emotional valence.

7 Where to look for further information

Further insight into the behaviour of wild boar is documented in several detailed observational monographs (the richest of which are by Frädrich (1974) and Mauget (1981)). Further studies have detailed the behaviour of domestic pigs released into a wild or semi-wild environment (e.g. Stolba and Wood-Gush 1989). Many welfare assessment approaches have been developed that utilise behaviour as an animal-based indicator of welfare state. The behaviours monitored in these protocols are typically focussed on assessing the outcome of mismatches between behavioural motivation and opportunities for the appetitive and consummatory elements of behavioural expression. Welfare Quality® (http://www.welfarequalitynetwork.net) is the most comprehensive and internationally recognised approach that makes extensive use of behaviour in the assessment of welfare. Welfare Quality includes specific welfare assessment protocols for sows and piglets and for grower and finisher pigs. A further EU-funded consortium ('Farewelldock') is currently studying the effects of tail docking on animal welfare and improved methods of providing appropriate environmental enrichment (http://farewelldock.eu/european-commission-initiative-reduce-tail-docking-improve-enrichment/).

A recognition of the importance of allowing appropriate behavioural expression is implied or explicit in various legislative instruments. As an example, the current directive covering the minimum standards for the protection of farmed pigs in the European Union (Council Directive 2008/120/EC) explicitly states that '*Pigs should benefit from an environment corresponding to their needs for exercise and investigatory behaviour …*' (Paragraph 8). This recognition is then translated into a statutory requirement that '*… pigs must have permanent access to a sufficient quantity of material to enable proper investigation and manipulation activities, such as straw, hay, wood, sawdust, mushroom compost, peat or a mixture of such …*' (Annex 1).

Many sources of information exist to practically support decisions that aim to accommodate pig behavioural requirements in commercially viable production systems. For example, a European Commission Staff Working Document ('*On best practices with a view to the prevention of routine tail-docking and the provision of enrichment materials to pigs*') provides examples of forms of enrichment that are optimal, suboptimal and of marginal value in their ability to meet pigs' desire to root (https://ec.europa.eu/food/sites/food/files/animals/docs/aw-pract-farm-pigs-staff-working-document_en.pdf). A further website (http://www.freefarrowing.org) provides a comprehensive summary of the state of the art in the design of farrowing systems that avoid the need for farrowing crates and allow some opportunities to nest build.

8 References

Arey, D. S. (1992). Straw and food as reinforcers for prepartal sows. *Appl. Anim. Behav. Sci.*, 33: 217–26.

Arey, D. S., Petchey, A. M. and Fowler, V. R. (1991). The preparturient behaviour of sows in enriched pens and the effect of preformed nests. *Appl. Anim. Behav. Sci.*, 31: 61–8.

Barrette, C. (1986). Fighting behavior of wild Sus scrofa. *J. Mammal.*, 67: 177–9.

Baxter, E. M., Rutherford, K. M. D., D'Eath, R. B., Arnott, G., Turner, S. P., Sandøe, P., Moustsen, V. A., Thorup, F., Edwards, S. A. and Lawrence, A. B. (2013). The welfare implications of large litter size in the domestic pig II: management factors. *Anim. Welf.*, 22: 219–38.

Berkeveld, M., Langendijk, P., Soede, N. M., Kemp, B., Taverne, M. A. M., Verheijden, J. H. M., Kuijken, N. and Koets, A. P. (2009). Improving adaptation to weaning: Effect of intermittent suckling regimens on piglet feed intake, growth, and gut characteristics. *J. Anim. Sci.*, 87: 3156–66.

Bonastre, C., Mitjana, O., Tejedor, M. T., Calavia, M., Yuste, A. G., Ubeda, J. L. and Falceto, M. V. (2016). Acute physiological responses to castration-related pain in piglets: the effect of two local anesthetics with or without meloxicam. *Animal*, 10: 1474–81.

Caley, P. (1997). Movements, activity patterns and habitat use of feral pigs (Sus scrofa) in a tropical habitat. *Wildl. Res.*, 24: 77–87.

Camerlink, I., Turner, S. P., Farish, M. and Arnott, G. (2015). Aggressiveness as a component of fighting ability in pigs using a game-theoretical framework. *Anim. Behav.*, 108: 183–91.

Chen, C., Gilbert, C. L., Yang, G., Guo, Y., Segonds-Pichon, A., Ma, J., Evans, G., Brenig, B., Sargent, C., Affara, N. and Huang, L. (2008). Maternal infanticide in sows: Incidence and behavioural comparisons between savaging and non-savaging sows at parturition. *Appl. Anim. Behav. Sci.*, 109: 238–48.

Clark, C. C. A. and D'Eath, R. B. (2013). Age over experience: Consistency of aggression and mounting behaviour in male and female pigs. *Appl. Anim. Behav. Sci.*, 147: 81–93.

Cloutier, S., Beaugrand, J. P. and Laguë, P. C. (1995). The effect of prior victory or defeat in the same site as that of subsequent encounter on the determination of dyadic dominance in the domestic hen. *Behav. Process.*, 34, 293–8.

Cocca, G., Sturaro, E., Dal Compare, L. and Ramanzin, M. (2007). Wild boar (Sus scrofa) damages to mountain grassland. A case study in the Belluno province, eastern Italian Alps. *Ital. J. Anim. Sci.*, 6, 845–7.

Dawkins, M. S. (2008). The science of animal suffering. *Ethology*, 114, 937–45.

D'Eath, R. D. and Lawrence, A. B. (2004). Early life predictors of the development of aggressive behaviour in the domestic pig. *Anim. Behav.*, 67: 501–9.

D`Eath, R. D. and Turner, S. P. (2007). The natural behaviour of the pig. In *The Welfare of Pigs*. Marchant-Forde, J. N. (Ed.). Springer, pp. 13–45.

D'Eath, R. B., Ormandy, E., Lawrence, A. B., Sumner, B. E. H. and Meddle, S. L. (2005). Resident-intruder trait aggression is associated with differences in lysine vasopressin and serotonin

1A (5-HT_{1A}) mRNA expression in the brain of pre-pubertal female domestic pigs (*Sus scrofa*). *J. Neuroendocrinol.*, 17: 679–86.
D'Eath, R. B., Turner, S. P., Kurt, E., Evans, G., Thölking, L., Looft, H., Wimmers, K., Murani, E., Klont, R., Foury, A., Ison, S. H., Lawrence, A. B. and Mormède, P. (2010). Pigs' aggressive temperament affects pre-slaughter mixing aggression, stress and meat quality. *Animal*, 4: 604–16.
D'Eath, R. B., Arnott, G., Turner, S. P. Jensen, T., Lahrmann, H. P., Busch, M. E., Niemi, J. K., Lawrence, A. B. and Sandøe, P., (2014). Injurious tail biting in pigs: how can it be controlled in existing systems without tail docking? *Animal*, 8: 1479–97.
D'Eath, R. B., Niemi, J. K., Vosough Ahmadi, B., Rutherford, K. M. D., Ison, S. H., Turner, S. P., Anker, H. T., Jensen, T., Busch, M. E., Jensen, K. K., Lawrence, A. B. and Sandøe, P. (2016). Why are most EU pigs still tail docked? – economic and ethical assessment of four pig housing and management scenarios in the light of EU directives and animal welfare outcomes. *Animal*, 10: 687–99.
de Passillé, A. M. B. and Rushen, J. (1989). Suckling and teat disputes by neonatal piglets. *Appl. Anim. Behav. Sci.*, 22: 23–38.
Delcroix, I., Signoret, J. P. and Mauget, R. (1985) Communal rearing of the young within the social group of European wild boars. *Ann. Zootech.*, 34: 369.
Dellmeier, G. R. and Friend, T. H. (1991). Behaviour and extensive management of domestic sows (Sus scrofa) and litters. *Appl. Anim. Behav. Sci.*, 29: 327–41.
Desire, S., Turner, S. P., D'Eath, R. B., Doeschl-Wilson, A. B., Lewis, C. R. G. and Roehe, R. (2015a). Genetic associations of short- and long-term aggressiveness identified by skin lesions with growth, feed efficiency and carcass characteristics in growing pigs. *J. Anim. Sci.*, 93: 3303–12.
Desire, S., Turner, S. P., D'Eath, R. B., Doeschl-Wilson, A. B., Lewis, C. R. G. and Roehe, R. (2015b). Analysis of the phenotypic link between behavioural traits at mixing and increased long-term social stability. *Appl. Anim. Behav. Sci.*, 166: 52–62.
Devillers, N., Giraud, D. and Farmer, C. (2016). Neonatal piglets are able to differentiate more productive from less productive teats. *Appl. Anim. Behav. Sci.*, 174: 24–31.
Duncan, I. J. H. (1998). Behavior and behavioral needs. *Poultry Sci.*, 77: 1766–72.
Dzieciolowski, R. M., Clarke, C. M. H. and Frampton, C. M. (1992). Reproductive characteristics of feral pigs in New Zealand. *Acta Theriol.*, 37: 259–70.
Eisenberg, J. F. and Lockhart, M. (1972). *An Ecological Reconnaissance of Wilpattu National Park,* Ceylon, Washington D.C., USA. Smithsonian Institution Press.
Elmore, M. R. P., Garner, J. P., Johnson, A. K., Kirkden, R. D., Patterson-Kane, E. G., Richert, B. T. and Pajor, E. A. (2012). Differing results for motivation tests and measures of resource use: The value of environmental enrichment to gestating sows housed in stalls. *Appl. Anim. Behav. Sci.*, 141: 9–19.
Fernández -Llario, P. (2004). Environmental correlates of nest site selection by wild boar Sus scrofa. *Acta Theriol.*, 49: 383–92.
Fox, J., Widowski, T., Torrey, S. Nannoni, E., Bergeron, R., Gonyou, H. W., Brown, J. A., Crowe, T., Mainau, E. and Faucitano, L. (2014). Water sprinkling market pigs in a stationary trailer. 1. Effects on pig behaviour, gastrointestinal tract temperature and trailer micro-climate. *Livest. Sci.*, 160: 113–23.
Frädrich, H. (1974). A comparison of behaviour in the Suidae. In *The behaviour of ungulates and its relation to management*. Geist, V. and Walther, F. R. (Eds). IUCN Publishers, pp. 133–43.
Fraser, D. (1980). A review of the behavioural mechanism of milk ejection of the domestic pig. *Appl. Anim. Ethol.*, 6: 247–55.
Gabor, T. M., Hellgren, E. C., Van Den Bussche, R. A. and Silvy, N. J. (1999). Demography, sociospatial behaviour and genetics of feral pigs (Sus scrofa) in a semi-arid environment. *J. Zool., Lon.*, 247: 311–22.
Gonyou, H. W. (2001). The social behaviour of pigs. In *Social Behaviour in Farm Animals*. Keeling, L. J. and Gonyou, H. W. (Eds). CABI: Oxon, UK, pp. 147–76.
Groot Bruinderink, G. W. T. A. and Hazebroek, E. (1994) Diet and condition of wild boar Sus scrofa scrofa, without supplementary feeding. *J. Zool.*, 233: 631–48.

Gundlach, H. (1968). Brutfürsorge, Brutpflege, Verhaltensontogenese und Tagesperiodik beim Europäischen Wildschwein (Sus scrofa L.). *Zietschrift fur Tierpsychologie*, 25: 955–95.
Gustafsson, M., Jensen, P., de Jonge, F. H., Illmann, G. and Spinka, M. (1999). Maternal behaviour of domestic sows and crosses between domestic sows and wild boar. *Appl. Anim. Behav. Sci.*, 65: 29–42.
Harley, S., More, S., O'Connell, N., Hanlon, A., Teixeira, D. and Boyle, L. (2012). Evaluating the prevalence of tail biting and carcase condemnations in slaughter pigs in the Republic and Northern Ireland, and the potential of abattoir meat inspection as a welfare surveillance tool. *Vet. Rec.*, 171: 621.
Haskell, M. J. and Hutson, G. D. (1994). Pre-farrowing behaviour of sow and gilts with access to space for locomotion. *Aust. J. Exp. Agr.*, 34: 1099–105.
Hemsworth, P. H., Rice, M., Nash, J., Giri, K., Butler, K. L., Tilbrook, A. J. and Morrison, R. S. (2013). Effects of group size and floor space allowance on grouped sows: Aggression, stress, skin injuries, and reproductive performance. *J. Anim. Sci.*, 91: 4953–64.
Herrero, J., Irizar, I., Laskurain, N. A., Garcia-Serrano, A. and Garcia-Gonzalez, R. 2005. Fruits and roots: wild boar foods during the cold season in the southwestern Pyrenees. *Ital. J. Zool.*, 72: 49–52.
Hintze, S., Scott, D., Turner, S. P., Meddle, S. L., D'Eath, R. B. (2013). Mounting behaviour in finishing pigs: Stable individual differences are not due to dominance or stage of sexual development. *Appl. Anim. Behav. Sci.*, 147: 69–80.
Hone, J. (2002). Feral pigs in Namadgi National Park, Australia: dynamics, impacts and management. *Biol. Conserv.* 105: 231–42.
Horrell, I. (1997). The characterisation of suckling in wild boar. *Appl. Anim. Behav. Sci.*, 53: 271–7.
Hughes, B. O. and Duncan, I. J. H. (1988). The notion of ethological need, models of motivation and animal welfare. *Anim. Behav.*, 36: 1696–707.
Hunter, E. J., Jones, T. A., Guise, H. J., Penny, R. H. C. and Hoste, S. (2001). The relationship between tail biting in pigs, docking procedure and other management practices. *Vet. J.*, 161: 72–9.
Ison, S. H., Wood, C. M. and Baxter, E. M. (2015). The behaviour of pre-pubertal gilts and its relationship to farrowing behaviour in conventional farrowing crates and loose-housed pens. *Appl. Anim. Behav. Sci.*, 170: 26–33.
Jarvis, S., Van der Vegt, B. J., Lawrence, A. B., McLean, K. A., Deans, L. A., Chirnside, J. and Calvert, S. K. (2001). The effect of parity and environmental restriction on behavioural and physiological responses of pre-parturient pigs. *Appl. Anim. Behav. Sci.* 71: 203–16.
Jarvis, S., Calvert, S. K., Stevenson, J., vanLeeuwen, N. and Lawrence, A. B. (2002). Pituitary-adrenal activation in pre-parturient pigs (Sus scrofa) is associated with behavioural restriction due to lack of space rather than nesting substrate. *Anim. Welf.*, 11: 371–84.
Jarvis, S., Reed, B. T., Lawrence, A. B., Calvert, S. K. and Stevenson, J. (2004). Peri-natal environmental effects on maternal behaviour, pituitary and adrenal activation, and the progress of parturition in the primiparous sow. *Anim. Welf.*, 13: 171–81.
Jarvis, S., D'Eath, R. B. and Fujita, K. (2005). Consistency of piglet crushing by sows. *Anim. Welf.*, 14: 43–51.
Jarvis, S., Moinard, C., Robson, S. K., Baxter, E., Ormandy, E., Douglas, A. J., Seckl, J. R., Russell, J. A. and Lawrence, A. B. (2006). Programming the offspring of the pig by prenatal social stress: Neuroendocrine activity and behaviour. *Horm. Behav.*, 49: 68–80.
Jedrzejewski, W., Schmidt, K., Theuerkauf, J., Jedrzejewska, B., Selva, N., Zub, K. and Szymura, L. (2002). Kill rates and predation by wolves on ungulate populations in Bialowieza Primeval Forest (Poland). *Ecology*, 83: 1341–56.
Jensen, M. B., Pedersen, L. J., Theil, P. K., Yde, C. C. and Back Knudsen, K. E. (2012). Feeding motivation and plasma metabolites in pregnant sows fed diets rich in dietary fiber either once or twice daily. *J. Anim. Sci.*, 90: 1910–19.
Jensen, P. (1986) Observations on the maternal behaviour of free-ranging domestic pigs. *Appl. Anim. Behav. Sci.*, 16: 131–42.
Jensen, P. (1988) Maternal behaviour and mother- young interactions during lactation in free-ranging domestic pigs. *Appl. Anim. Behav. Sci.*, 20: 297–308.

Jensen, P. (1995). The weaning process of free-ranging domestic pigs-within- litter and between-litter variations. *Ethology* 100: 14–25.

Jensen, P. and Redbo, I. (1987). Behavior during nest leaving in free-ranging domestic pigs. *Appl. Anim. Behav. Sci.*, 18: 355–62.

Jensen, P. and Recen, B. (1989). When to wean-observations from free-ranging domestic pigs. *Appl. Anim. Behav. Sci.*, 23: 49–60.

Jensen, P. and Toates, F. M. (1993). Who needs 'behavioural needs'? Motivational aspects of the needs of animals. *Appl. Anim. Behav. Sci.*, 37: 161–81.

Jensen, P. and Wood-Gush, D. G. M. (1984). Social interactions in a group of free-ranging sows. *Appl. Anim. Behav. Sci.*, 12: 327–37.

Jensen, P., Vestergaard, K. and Algers, B. (1993). Nestbuilding in free-ranging domestic sows. *Appl. Anim. Behav. Sci.*, 38: 245–55.

Kiley-Worthington, M. (1989) Ecological, ethological, and ethically sound environments for animals: toward symbiosis. *J. Agr. Ethics*, 2: 323–47.

Kranendonk, G., Hopster, H., Fillerup, M., Ekkel, E. D., Mulder, E. J. H. and Taverne, M. A. M. (2006). Cortisol administration to pregnant sows affects novelty-induced locomotion, aggressive behaviour, and blunts gender differences in offspring. *Horm. Behav.*, 49: 663–72.

Kurz, J. C. and Marchinton, R. L. (1972). Radiotelemetry studies of feral hogs in South Carolina. *J. Wildl. Manag.*, 36: 1240–8.

Lao, F. Brown-Brandl, T., Stinn, J. P., Liu, K., Teng, G. and Xin, H. (2016). Automatic recognition of lactating sow behaviors through depth image processing. *Comput. Electron. Agr.* 125: 56–62.

Leaper, R., Massei, G., Gorman, M. L. and Aspinall, R. (1999). The feasibility of reintroducing Wild Boar (Sus scrofa) to Scotland. *Mamm. Rev.*, 29: 239–59.

Lega, C., Raia, P., Rook, L. and Fulgione, D. (2016). Size matters: A comparative analysis of pig domestication. *Holocene*, 26: 327–32.

Massei, G., Genov, P. V. and Staines, B. W. (1996). Diet, food availability and reproduction of wild boar in a Mediterranean coastal area. *Acta Theriol.*, 41: 307–20.

Mauget, R. (1981). Behavioural and reproductive strategies in wild forms of Sus scrofa (European wild boar and feral pigs). In *The Welfare of Pigs*, Volume 11. Sybesma, W. (Ed.). The Hague: Martinus Nijhoff, pp. 3–13.

Meerlo, P., Overkamp, G. J. F. and Koolhaas, J. M. (1997). Behavioural and physiological consequences of a single social defeat in Roman high- and low-avoidance rats. Psychoneuroendocrinology 22, 155–68.

Mendl, M. (1995). The social behaviour of non-lactating sows and its implications for managing sow aggression. *Pig J.*, 34: 9–20.

Morrow-Tesch, J. L., McGlone, J. J. and Salak-Johnson, J. L. (1994). Heat and social stress effects on pig immune measures. *J. Anim. Sci.*, 72: 2599–609.

Murphy, E., Nordquist, R. E. and van der Staay, F. J. (2014). A review of behavioural methods to study emotion and mood in pigs, Sus scrofa. *Appl. Anim. Behav. Sci.*, 159: 9–28.

Nahlik, A. and Sandor, G. (2003) Birth rate and offspring survival in a free-ranging wild boar Sus scrofa population. *Wildl. Biol.*, 9: 37–42.

Nasirahmadi, A., Hensel, O., Edwards, S. A. and Sturm, B. (2016). Automatic detection of mounting behaviours among pigs using image analysis. *Comput. Electron. Agr.* 124: 295–302.

Newberry, R. C. and Wood-Gush, D. G. M. (1985). The suckling behaviour of domestic pigs in a semi-natural environment. *Behaviour*, 95: 11–25.

Newberry, R. C. and Wood-Gush, D. G. M. (1986). Social relationships of piglets in a semi-natural environment. *Anim. Behav.*, 34: 1311–18.

Newberry, R. C. and Wood-Gush, D. G. M. (1988). Development of some behaviour patterns in piglets under semi-natural conditions. *Anim. Prod.*, 46: 103–9.

Newberry, R. C., Wood-Gush, D. G. M. and Hall, J. W. (1988). Playful behaviour of piglets. *Behav. Processes*, 17: 205–16.

Numberger, J., Ritzmann, M., Ubel, N., Eddicks, M., Reese, S. and Zols, S. (2016). Ear tagging in piglets: the cortisol response with and without analgesia in comparison with castration and tail docking. *Animal*, 10: 1864–70.
Oczak, M., Maschat, K., Berckmans, D., Vranken, E. and Baumgartner, J. (2016). Can an automated labelling method based on accelerometer data replace a human labeller? – Postural profile of farrowing sows. *Comput. Electron. Agr.*, 127: 168–75.
Patterson-Kane, E. G., Pittman, M. and Pajor, E. A. (2008). Operant animal welfare: productive approaches and persistent difficulties. *Anim. Welf.*, 17: 139–48.
Pedersen, L. J., Herskin, M. S., Forkman, B., Halekoh, U., Kristensen, K. M. and Jensen, M. B. (2014). How much is enough? The amount of straw necessary to satisfy pigs' need to perform exploratory behaviour. *Appl. Anim. Behav. Sci.*, 160: 46–55.
Pépin, D., Spitz, F., Janeau, G. and Valet, G. (1987). Dynamics of reproduction and development of weight in the Wild boar (Sus scrofa) in South-west France. *Int. J. Mammal.Biol.*, 52: 21–30.
Petersen, H. V., Vestergaard, K. and Jensen, P. (1989). Integration of piglets into social groups of free-ranging domestic pigs. *Appl. Anim. Behav. Sci.*, 23: 223–36.
Petersen, V., Recen, B. and Vestergaard, K. (1990). Behaviour of sows and piglets during farrowing under free-range conditions. *Appl. Anim. Behav. Sci.*, 26: 169–79.
Ramos-Onsins, S. E., Burgos-Paz, W., Manunza, A. and Amills, M. (2014). Mining the pig genome to investigate the domestication process. *Heredity* 113: 471–84.
Rizvi, S., Nicol, C. J. and Green, L. E. (1998). Risk factors for vulva biting in breeding sows in south-west England. *Vet. Rec.*, 143: 654–8.
Rollin, B. E. (1993) Animal welfare, science, and value. *J. Agr. Environ. Ethics*, 6(Suppl. 2): 44–50.
Rundgren, M., Löfquist, I. (1989). Effects on performance and behaviour of mixing 20kg pigs fed individually. *Anim. Prod.*, 49: 311–15.
Rutherford, K. M. D., Baxter, E. M., D'Eath, R. B., Turner, S. P., Arnott, G., Roehe, R., Ask, B., Sandøe, P., Moustsen, V. A., Thorup, F., Edwards, S. A., Berg, P. and Lawrence, A. B. (2013). The welfare implications of large litter size in the domestic pig I: biological factors. *Anim. Welf.*, 22: 199–218.
Rzezniczek, M., Gygax, L., Wechsler, B. and Weber, R. (2015). Comparison of the behaviour of piglets raised in an artificial rearing system or reared by the sow. *Appl. Anim. Behav. Sci.*, 165: 57–65.
Sandercock, D. A., Smith, S. H., Di Giminiani, P. and Edwards, S. A. (2016). Histopathological characterization of tail injury and traumatic neuroma development after tail docking in piglets. *J. Comp. Pathol.*, 155: 40–9.
Sapkota, A., Marchant-Forde, J. N., Richert, B. T. and Lay, D. C. (2016). Including dietary fiber and resistant starch to increase satiety and reduce aggression in gestating sows. *J. Anim. Sci.*, 94: 2117–27.
Schrøder-Petersen, D. L. and Simonsen, H. B. (2001). Tail biting in pigs. *Vet. J.*, 162: 196–210.
Seguin, J. F., Thibault, J. C., Torre, J., Bayle, P. and Vigne, J. D. (2001). The diet of young Golden Eagles Aquila chrysaetos in Corsica: Foraging in a manmade mammal fauna. Ardea, 89: 527–35.
Signoret, J. P., Baldwin, B. A., Fraser, S. and Hafez, E. S. E. (1975). The behaviour of swine. In *The Behaviour of Domestic Animals*. Hafez, E. S. E. (Ed.). London: Bailliere, Tindall and Cox, pp. 295–329.
Smulders, D., Hautekiet, V., Verbeke, G., Geers, R. (2008). Tail and ear biting lesions in pigs: an epidemiological study. *Anim. Welf.*, 17: 61–9.
Spinka, M., Illmann, G., de Jonge, F., Andersson, M., Schuurman, T. and Jensen, P. (2000) Dimensions of maternal behaviour characteristics in domestic and wild x domestic crossbred sows. *Appl. Anim. Behav. Sci.* 70, 99–114.
Spitz, F. (1986). Current state of knowlege of wild boar biology. *Pig News Info.*, 7: 171–5.
Spitz, F. and Janeau, G. (1995). Daily selection of habitat in wild boar (Sus scrofa). *J. Zool.*, 237: 423–34.
Stangel, G. and Jensen, P. (1991). Behaviour of semi-naturally kept sows and piglets (except suckling) during 10 days post-partum. *Appl. Anim. Behav. Sci.*, 31: 211–27.
Stolba, A. and Wood-Gush, D. G. M. (1984). The identification of behavioural key features and their incorporation into a housing design for pigs. *Annales Recherche Veterinaire*, 15: 287–98.

Stolba, A. and Wood-Gush, D. G. M. (1989). The behaviour of pigs in a semi-natural environment. *Anim. Prod.*, 48, 419–25.

Tan, S. S. L., Shackleton, D. M., Beames, R. M. (1991). The effect of mixing unfamiliar individuals on the growth and production of finishing pigs. *Anim. Prod.*, 52: 201–6.

Taylor, N. R., Main, D. C. J., Mendl, M. and Edwards, S. A. (2010). Tail-biting: a new perspective. *Vet. J.*, 186: 137–47.

Turner, S. P., Ewen, M., Rooke, J. A., Edwards, S. A. (2000). The effect of space allowance on performance, aggression and immune competence of growing pigs housed on straw deep-litter at different group sizes. *Livest. Prod. Sci.*, 66: 47–55.

Turner, S. P., Roehe, R., Mekkawy, W., Farnworth, M. J., Knap, P. W. and Lawrence, A. B. (2008). Bayesian analysis of genetic associations of skin lesions and behavioural traits to identify genetic components of individual aggressiveness in pigs. *Behav. Genet.*, 38: 67–75.

Turner, S. P. Roehe, R., D'Eath, R. B. Ison, S. H., Farish, M., Jack, M. C., Lundeheim, N., Rydhmer, L. and Lawrence, A. B. (2009). Genetic validation of post-mixing skin injuries in pigs as an indicator of aggressiveness and the relationship with injuries under more stable social conditions. *J. Anim. Sci.*, 87: 3076–82.

Valros, A., Rundgren, M., Pinka, M., Saloniema, H. and Algers, B. (2003). Sow activity level, frequency of standing-to-lying posture changes and anti-crushing behaviour – within sow repeatability and interactions with nursing behaviour and piglet performance. *Appl. Anim. Behav. Sci.*, 83: 29–40.

Van de Weerd, H. A., Docking, C. M., Day, J. E. L., Avery, P. J. and Edwards, S. A., (2003). A systematic approach towards developing environmental enrichment for pigs. *Appl. Anim. Behav. Sci.*, 84, 101–18.

Verdon, M., Hansen, C. F., Rault, J.-L., Jongman, E., Hansen, L. U., Plush, K. and Hemsworth, P. H. (2015). Effects of group housing on sow welfare: A review. *J. Anim. Sci.*, 93: 1999–2017.

Warburton, H. and Mason, G. (2003). Is out of sight out of mind? The effects of resource cues on motivation in mink, Mustela vison. *Anim. Behav.*, 65: 755–62.

Wemelsfelder, F., Hunter, T. E. A., Mendl, M. T. and Lawrence, A. B. (2001). Assessing the 'whole animal': a free choice profiling approach. *Anim. Behav.*, 62, 209–20.

Wischner, D., Kemper, N. and Kriter, J. (2009). Nest building behaviour in sows and consequences for pig husbandry. *Livest. Sci.*, 124: 1–8.

Yun, J., Swan, K. M., Vienola, K., Farmer, C ., Oliviero, C., Peltoniemi, O. and Valros, A. (2013). Nest-building in sows: Effects of farrowing housing on hormonal modulation of maternal characteristics. *Appl. Anim. Behav. Sci.*, 148: 77–84 .

Chapter 6

Defining and ensuring animal welfare in pig production: an overview

Paul H. Hemsworth, University of Melbourne, Australia

1 Introduction

The publication of Ruth Harrison's book *Animal Machines* in 1964 led to a marked increase in public concern about the welfare of farm animals in confinement systems. In response, the British Government established the Brambell Committee (Brambell et al., 1965) with its terms of reference 'To examine the conditions in which livestock are kept under systems of intensive husbandry and to advise whether standards ought to be set in the interests of their welfare, and if so what should they be?' Amongst its numerous recommendations, the Committee called for research into veterinary medicine, stress physiology, animal science and, in particular, animal behaviour. The Brambell Committee thus provided an agenda for animal welfare science and, together with the public concern about the welfare of farm animals in confinement systems, animal welfare science emerged as an interdisciplinary field of research in the 1970s (Fraser, 2008).

It is widely recognized that most of the animals that society uses can suffer. As argued by Mellor et al. (2009), in using these animals for our purposes we exercise varying degrees of control over the quality and duration of their lives, which provides us therefore with the opportunity to manage them humanely. Moreover, using them for our own purposes,

http://dx.doi.org/10.19103/AS.2017.0013.19

not theirs, requires us to do so. Therefore, we have an ethical 'duty of care' towards the animals in our control and this translates into a practical obligation to keep their welfare at acceptable levels.

2 Animal welfare and its assessment

The following is a scientific perspective of animal welfare and its assessment that is now widely accepted in the published literature (e.g. see reviews by Fraser, 2008; Mellor et al., 2009; Fraser and Nicol, 2011; Hemsworth et al., 2015a).

2.1 Animal welfare

Animal welfare is a state and it is generally agreed that animal welfare relates to experienced sensations, that is, how the animal feels. These experiences arise as the integrated outcomes of sensory and other neural inputs from within the animal and from its environment. These inputs are processed and interpreted by the animal's brain according to its species-specific and individual nature, and its past experience. The integrated outcomes of this processing represent the animal's current experience (its welfare status), and this changes as the balance and nature of the inputs change. These experiences are all subjective, varying in their affective or emotional contents and, based on human experience, are likely to include negative affective experiences such as thirst, hunger, nausea, pain and fear, and positive affective experiences such as satiety, contentment, companionship, curiosity and playfulness.

Animal welfare is more than the mere absence of negative experiences since animal welfare, as a state experienced by an animal, can vary on a continuum from very bad to very good. Most research during at least the last 40 years concentrated on preventing and ameliorating negative states. However, there is increasing societal interest in providing domesticated animals with the opportunity for positive affective experiences, and consequently, there is an increasing research focus on positive welfare states in these animals.

2.2 Conceptual frameworks for assessment of animal welfare

There are basically three conceptual frameworks that are used to assess animal welfare, namely, biological functioning, affective state and natural living.

Biological functioning

The rationale for this conceptual framework is that difficult or inadequate adaptation will generate welfare problems for animals. This conceptual framework emphasizes that animals use a range of behavioural and physiological responses to assist them to cope with challenges, and while biological regulation in response to challenges occurs continuously, successful adaptation is not always possible. Marked challenges may overwhelm an individual's capacity to adapt and lead to its death. However, less severe challenges can still have significant biological costs, leading to growth, reproductive, health and other impairments, which may reflect and/or result in welfare problems for the animal. Thus, animal welfare is at risk in environments in which adaptation is difficult given the animals' genetic background.

Conceptualized in these terms, it is the biological cost of stress that is the key to understanding the associated welfare implications. How well an animal is coping with the challenges it faces will be reflected in the normality of its biological functioning and fitness, with severe risks to welfare associated with the most extreme coping attempts. The behavioural responses include stereotypies, redirected behaviours and displacement activities and the physiological stress responses to challenges, while also varied and complex, nearly always include activation of the sympatho-adrenal-medullary (SAM) system and the hypothalamo-pituitary-adrenal (HPA) axis, with the consequent increase in synthesis of catecholamines and glucocorticoids, respectively. The extent to which these coping attempts are or are not succeeding is reflected in the biological costs to the animal, such as deterioration in growth efficiency, reproduction and health (injury or disease).

This conceptual framework is useful in assessing risks to animal welfare, that is, negative welfare states. Indeed, the majority of studies on pig welfare have employed the biological functioning framework to infer compromised pig welfare on the basis that suboptimal biological functioning accompanies negative affective states, such as hunger, pain, fear, helplessness, frustration and anger. The measures used have included behavioural variables, such as aggression; physiological variables, such as circulating concentrations of cortisol, neutrophil: lymphocyte ratio and immunoglobulin A; and fitness variables, such as lameness, skin lesions, liveweight change and reproductive performance.

Affective state

The second conceptual framework emphasizes that the welfare of an animal derives from its capacity for affective experiences. Thus, the welfare state is likely to be negative when the predominant affects experienced are unpleasant, and vice versa.

It is well recognized that affective experiences are generated both by sensory inputs that reflect the animal's internal functional state and by other sensory inputs that reflect the animal's perception of its external circumstances. Thus, preference research, in which the strength of the preference for a chosen environmental option or the motivation to perform a type of behaviour is measured, has been used by some scientists to make inferences about animal welfare. The rationale for these inferences is that preferences are influenced by the animal's affective experiences, which have evolved to motivate behaviour in order to avoid harm and facilitate survival, growth and reproduction. Furthermore, others have suggested that animals make choices that are in their best interest: it is likely that animals will avoid aversive stimuli and choose rewarding stimuli.

Other approaches utilized in assessing affective experiences, particularly negative affective experiences, include measures of behaviour, such as fear, pain and illness behaviours; cognitive bias, such as deviation in judgement; and physiology, such as activation of the SAM system and the HPA axis, as well as employing the intuitive perception of human observers using an approach known as qualitative behavioural assessment. While there are inconsistencies in the literature, behaviours such as play, affiliative behaviours and some vocalizations appear to be the most promising indicators for currently assessing positive affective experiences in animals.

Natural living

This conceptual framework, albeit not often well enunciated in the literature, is predicated on the view that the welfare of animals is improved when they can express their normal behaviour. For some people this also implies that the animal should be raised in a 'natural' environment and allowed to behave in 'natural' ways. However, the concept of natural is usually too poorly defined to provide a sound basis for animal welfare assessment, and thus when applied uncritically it may lead to poorer welfare instead of an improvement. There is a need to define natural behaviours that are desirable or undesirable in terms of animal welfare and to clarify the rationale for their inclusion or exclusion. Indeed, there is an increasing focus on highly-motivated behaviours.

3 Community, animal welfare and public education

Most people accept that humans have a moral obligation towards farm, companion and laboratory animals. In addition to the undeniable benefits that these animals provide to humans, the domestication of these animals has increased their dependence on humans and thus necessitates this obligation. How individuals make decisions on the acceptability, or otherwise, of a specific animal use is not entirely clear, but these decisions appear to be affected by a number of considerations including scientific information on the harms and benefits to the animal caused by the animal use (Hemsworth and Coleman, 2011). However, what is at question for most people is the extent of this obligation, particularly in relation to the standards of welfare that society should provide to these animals.

Thus, ultimately the question of the acceptability of a specific animal use is an ethical one in which science can provide factual evidence that needs to be utilized to provide objectivity to the answer (Matthews and Hemsworth, 2012). However, the development of a clear consensus on an ethically and scientifically defensible philosophy on animal welfare is obviously difficult. A society's attitudes to the use of and obligations to farm, companion and laboratory animals are extremely disparate, influenced by demographic factors, religion and culture, and vary over time with economic and ideological changes (Teutsch, 1987; Palmer and Sandoe, 2011).

The public is often a key driver of animal welfare change since public views affect decision makers at the political, regulatory, retail and industry levels. Furthermore, failure to meet the expectations of the public can lead to increased litigation, increased regulations and increasing consumer demands, all of which hamper the success of the livestock industry (Arnot, 2009). While societal concerns in general dictate the need for animal welfare standards and animal welfare legislation, consumers and stakeholders in the meat supply value chain affect not only the supply and market shares of animal food products but also the animal welfare standards, some of which can often exceed legislative requirements (Matthews and Hemsworth, 2012; Vanhonacker and Verbeke, 2014). Public attitudes about animal welfare are often based on limited knowledge, and the public's beliefs are largely acquired from the mass media, perhaps filtered by opinion leaders (Coleman et al., 2015). Coleman (2010) concluded that to address both the mismatch between the public's perceived and actual knowledge of livestock practices and public welfare concerns, there needs to be accurate and reciprocal communication between the livestock industries and the community. In addition to transparency to the public in farming techniques, Coleman et al. (2016) proposed a clear articulation of the implications of these farming practices for both food quality on the one hand and animal welfare on the other. They concluded that purely economic arguments are unlikely to be sufficient if the public cannot be reassured about welfare concerns. Furthermore, the fostering of a culture shift amongst the livestock industries is desirable to sensitize them to changing cultural values about animal quality of life. Industry responses need to have a balance between listening to community requirements and a preparedness to defend a practice if, on balance, it is considered the best in terms of healthy food (which is very prominent in consumer thinking), economics and welfare (Coleman, 2004; Martin and Shepheard, 2011). Thus accurate and reciprocal communication between the livestock industries and the community should facilitate a well-informed community able to make rational choices and industries responding appropriately to community expectations.

4 Common welfare concerns with pig production: an overview

4.1 Introduction

Animal production evokes both public concerns and policy debates on the conditions that guarantee food security, public health, environmental quality and animal welfare (Vanhonacker et al., 2012). While individuals differ considerably, this attention on animal welfare appears to centre on concerns about the negative effects of intensification of animal production on animal welfare, such as a general lack of social contact, an inability to exercise and a restricted choice of stimuli (such as other conspecifics and additional features of the physical environment) to interact with (Barnett et al., 2001; Te Velde et al., 2002; Vanhonacker et al., 2009). The Eurobarometer surveys (e.g. European Commission, 2007), which provide a snapshot of attitudes to animal welfare, indicate that the European community largely considers farm animal welfare to be important, with the welfare of laying hens identified to be at the greatest risk, followed by pigs. Coleman (2008) concluded that, although there is variability, most countries throughout the Western world show similar patterns of attitudes to farm animal welfare.

There is a general agreement on the more controversial animal welfare issues in pig production (see reviews by Borell et al., 1997; Barnett et al., 2001; Faucitano and Schaefer, 2008; Marchant-Forde, 2009). Some of these issues have been shown to result in adverse consequences and thus reduced pig welfare, whereas others are unresolved or may prove to be perceptions. The issues that appear to be most frequently raised and some reasons for raising them are briefly considered here.

The issues that appear to be most frequently raised include close confinement of breeding sows during gestation, farrowing crates, barren environments, restricted feeding of breeding sows, indoor production, surgical husbandry procedures and sow lameness. These issues will be considered in this section and the next three sections. Other issues, such as transport, genetics and high productive lines, infectious diseases, peri-parturient disorders, early weaning of piglets, flooring and bedding, dusty environments, and high concentrations of ammonia and other irritating gases, are well covered elsewhere (refer to Borell et al., 1997; Marchant-Forde, 2009a).

4.2 Indoor and outdoor production systems

One of the reasons why housing of farm animals changed markedly post-Second World War was that consumers and governments in Western societies wanted cheap and safe food (Hodges, 2000). Through research and industry development, better housing, management, health and animal genetics have increased productivity, improved food quality and lowered the cost of food. While these changes in animal production methods have reduced or eliminated a number of welfare problems – such as predation, thermal stress, some infectious diseases and nutritional stress – these changes in production methods have exacerbated or have created other welfare problems such as restricted space and social contact and barren environments (Cronin et al., 2014).

The modern indoor intensive production system, particularly for pigs and poultry, is intensive and thus considered today by some sectors of the community to be inherently 'bad' because of lack of space, 'barrenness' of the environment and the reliance on technology (Barnett et al., 2001; Mellor et al., 2009). In contrast, outdoor housing is typically extensive and so considered by some to be inherently 'good' because it provides a more 'natural' environment and choice for the animal in performing a number of behaviours over a relatively large area, and the lower technological inputs provide for fewer equipment breakdowns that may adversely affect welfare.

While extensive livestock production systems are generally not considered to involve 'housing', extensive systems do impose restrictions on animals, albeit with considerable freedom and there are different welfare risks including frequency of inspections, climatic conditions and natural disasters. Nevertheless, the main focus of welfare concerns has been on intensive production systems. While the major responses to these concerns have varied, a typical response has been the development of alternative systems, sometimes based on previous, more traditional farming practices, such as free-range pig and poultry systems. Often these systems have merely replaced one set of welfare problems with another. One reason for this substitution of welfare problems when a new housing system is developed is that there is insufficient understanding of the animals' requirements, including those for space, social contact and environmental enrichment, and without this understanding it is difficult to design appropriate housing systems.

The most contentious welfare issues involve indoor housing, particularly of breeding sows, and while these are considered in this chapter, there are potential welfare issues in outdoor pig production. These issues include the implications of nose-ringing, shelter, litter desertion, wallow design and management, thermoregulation in winter and any implications of lower back fat, and the suitability of different genotypes (Barnett et al., 2001; Marchant-Forde, 2009a).

5 Common welfare concerns with pig production: confinement, floor space and group size

5.1 Confinement of gestating, parturient and lactating sows

Individual housing in stalls for gestating sows and farrowing crates for parturient and lactating sows has been a controversial issue for many decades. As a consequence of legislative, consumer and retailer pressure on the housing of sows in gestation stalls (Matthews and Hemsworth, 2012), many countries have moved to, or are moving to, group-housing systems for gestating sows. Recent research has therefore focussed on the design and management aspects of group housing, such as floor space allowance, group size, method of mixing sows and nutrition, hunger and feeding method for group-housed pregnant sows and some of these factors are considered later.

The majority of sows in the EU and the United States are confined in farrowing crates during parturition and lactation (Baxter et al., 2012). Farrowing crates save space and labour and facilitate ease of inspection and intervention, but they were also designed to reduce piglet mortality (Barnett et al., 2001; Johnson and Marchant-Ford, 2009). Criticisms of the farrowing crate have generally focussed on sow welfare, since they limit the movement of sows and the opportunity for sows to perform natural behaviours, such as nest building and interaction with the piglets (Barnett et al., 2001).

There is evidence that primiparous sows in loose farrowing and lactation housing systems have improved maternal behaviour, based on increased interactions with their piglets and increased responsiveness to piglet vocalizations (Cronin and Smith, 1992; Cronin et al., 1996; Thodberg et al., 2002). These improvements in maternal behaviour may have welfare benefits for piglets since nest-building activity, response to piglet distress calls, nose contact with piglets during posture changes and restlessness when piglets are removed are reported to negatively correlate with risk of piglet crushing (Andersen et al., 2005). There is also limited evidence that piglets reared in farrowing crates may be deprived of some benefits relating to social development. For example, piglets reared in loose farrowing and lactation housing systems show more play behaviour and less injurious behaviour, such as nibbling, sucking or chewing another piglet (Oostindjer et al., 2011; Singh et al., 2017) and piglets reared in multi-litter group lactation systems are less aggressive post-weaning (Li and Wang, 2011; Verdon et al., 2016).

While there are concerns about the variability in pre-weaning piglet mortality in loose farrowing and lactation housing systems (Baxter et al., 2012; Mousten et al., 2013), a contentious point with these loose housing systems is that live born piglet mortality is usually higher than with farrowing crates (Cronin et al., 2014). However, some recent reviews of the literature conclude that comparable levels of total piglet survival, one of the main economic and welfare parameters for evaluating alternatives to farrowing crates, have been achieved in some non-crate farrowing systems (Baxter et al., 2012; Morrison et al., 2012). For example, collation of a large database of literature by Baxter et al. (2012) indicates that pen systems designed to meet the animals' biological needs such as separate dunging and lying areas and rails or sloped walls to assist sow posture changes and protect piglets, had a similar total piglet mortality (16.6%) to conventional farrowing crates (18.3%) and modified farrowing crates, such as turnaround/ellipsoid (16.3%) and hinged/swingside crates (17.4%). In a large comparative study in Switzerland (655 farms comprising 63 661 litters), Weber et al. (2007) found no difference in total piglet mortality (stillbirths and live born deaths) between loose farrowing pens and farrowing crates (17.2% and 18.1% total mortality, respectively), but crushing was higher in loose pens and mortality due to other causes was higher in farrowing crates. Similarly, in a cohort study of 112 breeding farms in England (2143 litters, with 6.5% of piglets stillborn and 12% live born pre-weaning mortality), Kilbride et al. (2012) found no difference in pre-weaning mortality or numbers reared between non-crated and crated systems. As in the study by Weber et al. (2007), more piglets were crushed when reared in non-crated systems and more piglets died of other causes in crated systems. While such comparisons of systems in industry are difficult because of variation within systems in design features and variation both within and between systems in management, these findings indicate that total piglet mortality in non-crated systems can be similar to that in conventional crated systems.

Since the majority of pre-weaning piglet mortalities occur within the first 2–3 days post-partum and are mainly caused by crushing, Johnson and Marchant-Forde (2009) concluded that farrowing crates can safeguard piglet survival and welfare during nest occupation in the farrowing phase, especially limiting early pre-weaning mortality. Consequently, there is interest in brief confinement of sows only during parturition and early lactation, when piglet mortality is most at risk. Danish research has shown that confinement of sows for the first 4 days post-partum is effective in maintaining early piglet survival (Moustsen et al., 2013; Hales et al., 2015; Hales Pedersen, 2015). While concern for sow welfare has been raised with confinement because of restricting nest building (Lawrence et al., 1994; Jarvis et al., 2002), there is a lack of comparative data on multiparous sows. There is evidence based on cortisol concentrations that gilts experience acute stress responses at entry to either farrowing crates or pens (Cronin et al., 1991). However, gilts in crates showed increased stress during the 24 h pre-partum period and parturition (Lawrence et al., 1994; Jarvis et al., 1998), but not the day following parturition (Cronin et al., 1991; Lawrence et al., 1994; Jarvis et al., 1998). Gilts in crates had similar cortisol concentrations to those in pens during the first three weeks post-partum (Cronin et al., 1991), while multiparous sows in crates had similar cortisol concentrations prior to and during parturition and higher cortisol concentrations from days two to five post-partum (Oliviero et al., 2008).

Although gilts in pens with straw spend considerable time in straw-directed behaviour, and gilts in crates spend considerable time in floor- and fixture-directed behaviour (Lawrence et al., 1994), sows are generally very inactive for several days after parturition in loose or confined housing (Hales Pedersen, 2015). Clearly, more comprehensive research on sow welfare post-partum is required. One of the difficulties in reviewing research on loose housing is the large variation in design features, such as space and bedding, which may markedly affect sow welfare.

Therefore, there will be continuing development of the farrowing and lactation housing systems during parturition and early in lactation, as well as the remainder of lactation to safeguard sow and piglet welfare. However, any farrowing and lactation housing system will need to be economical for large-scale commercial production (Johnson and Marchant-Forde, 2009a; Vosough Ahmadi et al., 2011). A common conclusion in most comparative studies and reviews on piglet mortality is that the maternal characteristics of sows and the quality of stockpersonship will be integral to the success of loose farrowing and lactation systems.

5.2 Space and group size for gestating sows

Research indicates that floor space affects sow aggression, injuries and stress, particularly early after mixing. Reducing floor space available for gilts and sows within the range of 1.0–3.0 m^2/animal increases aggressive behaviour and stress, as assessed by plasma cortisol concentrations, and reduces both immune competence (in some studies), as reflected in cell-mediated responses, and reproductive performance based on farrowing rate (Barnett 1997; Barnett et al., 1992; Hemsworth et al., 1986, 2013, 2015b; Weng et al., 1998; Remience et al., 2008). These results indicate that a space allowance for gilts and sows of 1.4 m^2 per animal is likely too small and a recent experiment showed a consistent linear effect of floor space on aggression and stress early after mixing, with aggression and cortisol concentrations reducing with increasing space from 1.45 m^2 to 2.9 m^2/sow (Hemsworth et al., 2015b). Furthermore, the effects of space on aggression and stress are most pronounced soon after mixing (e.g. Hemsworth et al., 2013, 2015b), indicating that sows may adapt to reduced floor space, albeit with risks to reproductive performance. The effects of space on the prevalence of skin injuries in gilts and sows are contradictory (Barnett et al., 1992; Barnett 1997; Weng et al., 1998; Salak-Johnson et al., 2007; Remience et al., 2008; Hemsworth et al., 2013).

These results on the effects of space highlight the need to reduce aggression and stress in the immediate post-mixing period. The strategy of staged-gestation penning, with more space immediately after insemination and less space later in gestation, appears to address both animal welfare and economic considerations but requires further examination.

While it might be expected that sow aggression would increase in large groups because of the greater number of pairs required to establish dominance relationships, there is no evidence that large group sizes of 30 or 80 sows affect stress (Hemsworth et al., 2013). However, reports on effects of group size on aggression and injuries are contradictory. The duration of aggression, but not the prevalence of injuries, increased in groups of 40 than smaller groups in one large study (Taylor et al., 1997), whereas injuries, but not frequency of aggressive interactions and stress, increased in larger groups in another study (Hemsworth et al., 2013).

Thus, other factors such as floor space and competition for feed or feeding space (discussed later) may have a greater impact on aggression, injuries and stress than group size. However, floor space may interact with other design features of the pen. For example, providing barriers to allow subordinates to avoid dominant sows reduces aggression in pigs (Marchant et al., 1995; Marchant-Forde and Marchant-Forde, 2005). Furthermore, the incorporation of feeding stalls in pens reduces aggression and stress in gilts (see review by Verdon et al., 2015).

Floor space for fattening pigs is less contentious than for breeding sows, probably because the effects of inadequate space allowance on pig productivity are both well studied and easier to demonstrate (O'Connell, 2009). Spatial restriction of fattening pigs leading to reduced access to feeders, increased stress and consequently reduced growth performance will also have adverse welfare implications.

6 Common welfare concerns with pig production: stereotypies and injuries

6.1 Restricted feeding/hunger in gestating sows

The feed intake of breeding sows is commonly restricted during gestation for productivity and lameness reasons (Meunier-Salaün et al., 2001). While sufficient for maintenance, and for some growth and foetal development, operant conditioning studies indicate that conventional restricted feeding during gestation results in pigs being hungry for a considerable period of the day (Hutson 1991; Lawrence and Terlouw 1993).

Lawrence and Terlouw (1993) proposed that feed restriction and the inability to express foraging behaviour are a major cause of the development of oral stereotypies in sows, such as repetitive chain and bar manipulations, vacuum chewing and excessive drinking. Stereotypies have been shown to be reduced when feed allowance of concentrate is increased (Appleby and Lawrence, 1987; Terlouw and Lawrence, 1993; Bergeron et al., 2000). Furthermore, increasing digestible energy has been shown to reduce vacuum chewing (Bergeron and Gonyou, 1997).

High-fibre diets, similar in dietary energy and major nutrient levels, fed to sows markedly increased feeding time (Robert et al., 1993; Brouns et al., 1994; Ramonet et al., 1999), and using multiple regression, Robert et al. (1997) showed that this increased feeding time accounted for much of the differences in level of stereotypies between diets. Thus, as concluded by Bergeron et al. (2006), these results support the view that expressing foraging behaviour can reduce stereotypies, regardless of nutrient intake.

6.2 Barren environments

Indoor production systems are considered by some to provide barren environments for animals (Barnett et al., 2001) and barren environments have been implicated in the development of stereotypies in captive animals. Stereotypies are relatively invariant motor acts, repeated frequently that have no apparent function or goal and they can occur in prolonged (days or weeks) conflict or thwarting situations in which there is no chance of escape (Dawkins, 1980; Mason, 1991). Stereotypies may originate from redirected behaviours (and other abnormal behaviours including displacement behaviours, vacuum behaviours, etc.) if the conflict or thwarting persists. Once developed, stereotypies can become ritualized to the extent that they become part of the behavioural repertoire and persist even in the absence of the original eliciting, stimuli/conditions (Mason, 1991). In pigs, stereotypies are essentially oral activities and include vacuum chewing ('sham-chewing'), head waving, chewing of bars, and licking, chewing or nosing of various available objects (Fraser, 1975; Stolba et al., 1983; Terlouw et al., 1991). Stereotypies have been reported in a range of housing systems including tethered, stall-housed and group-housed sows (Barnett et al., 2001; Schouten and Rushen, 1992).

There has been, and still is, ongoing discussion and research on the welfare significance of stereotypies. But stereotypies in captive animals have been generally viewed either as an adaptive coping response to the captive environment or as the inappropriate output in a conflict or thwarting situation (Mason and Latham, 2004).

As discussed earlier, feed restriction and the inability to express foraging behaviour have been proposed as a major cause of the development of oral stereotypies in sows (Lawrence and Terlouw, 1993). Furthermore, as considered above, stereotypies may develop from other prolonged conflict or thwarting situations and there is evidence that stereotypies can develop in response to barren or restricted environments (Mason, 1991; Würbel et al., 1998).

Stress can obviously compromise the welfare of animals. Research to date on strategies to reduce animal stress in captive animals has focussed on modifying the animal's environment through the identification and elimination of the main stressors and improved stress resilience through provision of environmental enrichment or genetic selection (Galindo et al., 2011; Hocking et al., 2011). Environmental enrichment can be defined as an increase in the biological relevance of captive environments by appropriate modifications, resulting in an improvement of the biological functioning of captive animals (Newberry, 1995). Evidence of improved biological functioning may include increased lifetime reproductive success, increased inclusive fitness or a correlate of these such as improved health. In general, enrichment, therefore, should present objects or situations that have functional relevance to the animal and act with a foreseeable rewarding outcome (Newberry, 1995). Enriched environments may mitigate deleterious stress effects on neurobiological systems and endocrine profiles, and promote stress adaptability in both humans (Southwick et al., 2005) and rodents (Lehmann and Herkenham, 2011). However, a limitation of many enrichment programmes for captive animals is that the provision of stimuli such as toys, music, chains or rubber, have little functional relevance to the animals (Newberry, 1995), and thus provides them with little to no reward.

Thus, lack of stimulation in the pig's environment may lead to boredom and stereotypies, but the welfare implications of intensive, indoor and non-bedded systems in current pig production systems are poorly understood. The European Union Commission directive 2001/93/EC states 'pigs must have permanent access to a sufficient quantity of material to enable proper investigation and manipulation activities, such as straw, hay, wood, sawdust, mushroom compost or a mixture of such that does not compromise the health of animals.' However, the extent to which this provides functional enrichment as defined by Newberry (1995) has been questioned (Marchant-Forde, 2009b), and clearly further research to identify effective enrichment for pigs is required (Van de Weerd and Day, 2009).

6.3 Foot and leg injuries/lameness

Lameness is common in the pig industry, and is considered a major welfare concern that has the potential to induce pain and discomfort for extended periods of time (Main et al., 2000; Barnett et al., 2001). The primary cause of lameness is pain (Cockram and Hughes, 2011) and thus these injuries are therefore considered to be associated with some level of suffering in the animal (Dawkins, 1998). Lameness, which is loosely defined as impaired movement or deviation from normal gait (Cockram and Hughes, 2011), is common in commercial sows (Chapinal et al., 2010). The causes of lameness range from inflammation and pain (Cockram and Hughes, 2011) and infected skin and claw lesions (Verlade, 2007), to broken bones (Marchant-Forde, 2009c). The USDA has ranked lameness as the third most common reason for culling sows on farm (15% of sows; USDA, 2007), and Stalder et al. (2004) reported that leg soundness is one of the most commonly identified reasons for the involuntary culling of sows.

Quality of flooring is widely considered to be essential to pig welfare as it is likely to have a direct effect on foot health and the culling rate from lameness (Barnett et al., 2001). This aspect has been reviewed by Borell et al. (1997) and, for sows, focusses on the common use of partially slatted pens. Important factors include the space between the slats, the roughness of the surface and the edge design (Boon and Wray, 1989). Smooth surfaces and deep litter systems may cause excessive hoof growth, leading to lameness (Geyer, 1979), but some abrasion is necessary for foot health (Webb and Nilsson, 1983), while foraging substrates can reduce injury caused by slipping (Heinonen et al., 2013). In a study of 15 herds, Gjein and Larssen (1995) showed an overall lameness rate of 13.1% in sows with increased risks associated with concrete versus plastic slats and poor floor hygiene. The quality of flooring interacts with the thermal requirements and pen design on lameness in adult pigs. For outdoor sows, site and soil type are significant factors affecting lameness (Thornton, 1990).

7 Common welfare concerns with pig production: surgical husbandry procedures

There is no doubt that sectors of the general community are increasingly questioning the welfare impact of and the need for some husbandry procedures, particularly surgical interventions that clearly cause pain (Coleman, 2008). A common definition of pain is 'an unpleasant sensory and emotional experience, associated with actual or potential tissue damage, or described in terms of such damage' (International Association for the Study of Pain, 1979). There is a range of behavioural and physiological indicators of pain (Benson, 2004; Hemsworth et al., 2009), but the difficulty is that these indicators are not specific to pain *per se*; other conditions in animals may have similar behavioural and physiological indicators, for example, sickness, fear and chronic stress. Therefore, while there is debate about the usefulness of behavioural and physiological measures in specifically measuring the level of pain experienced by animals, issues of animal welfare are broader than pain. Animals show a range of behavioural and physiological responses to aversive stimulation, including those situations involving an element of pain. These responses are widely accepted in science as indicators of reduced welfare.

While the most serious risks to animal welfare are generally those situations involving long-term exposure or frequent exposure to a stressor, the implications of an acute stressor, such as surgical husbandry procedures which elicits a range of behavioural and physiological responses, require attention. These behavioural (defensive behaviours including avoidance) and physiological (activation of the SAM system and the HPA axis) responses are adaptive and the magnitude of these acute responses to husbandry procedures can be useful to assess the intensity of the stressor. This view is predicated on the idea that the magnitude and duration of an acute stress response reflects the intensity of the stressor (Harbuz and Lightman, 1992), and thus can be useful in differentiating between the relative aversive or noxious components of stressors under comparison.

7.1 Tail docking

Tail docking is a common prophylactic measure routinely used worldwide to reduce tail-biting in pigs. Although the incidence of tail biting appears to be highly variable, a review of the published literature suggests that 30–70% of farms experience the problem, with estimates of the prevalence of lesioned tails on farm varying widely but of the order of 1–5% (ESFA, 2007). The problems with tail biting today are at least as severe as they were 50 years ago (Schroder-Petersen and Simonsen, 2001). While tail biting is generally associated with intensification of pig production (Smith and Penny, 1986), it occurs in outdoor systems (Walker and Bilkei, 2005).

Tail biting is a potentially serious welfare problem; light chewing on the tail can lead to wound and bleeding and more severe consequences such as infection, spinal abscess, paralysis and, in extreme cases, death (Fraser, 1987; Schroder-Petersen and Simonsen, 2001). Van Putten (1969) lists the possible consequences of tail biting as restlessness, poor growth, possible paralysis and mortality due to infections, and condemnation of the carcass.

Tail biting is considered to be a multi-factorial syndrome. Factors influencing tail-biting appear to include external factors such as manipulable objects or substrates that can be chewed and manipulated, and particularly straw, space, stocking density and group size; indoor and outdoor climate; crowding, flooring, and food and feeding system; and internal factors such as genetics, gender, age and weight, gastrointestinal discomfort and health status (see reviews of Schroder-Petersen and Simonsen, 2001; ESFA, 2007; Taylor et al., 2010).

Understanding the causation of tail biting is difficult because of its sporadic and unpredictable occurrence and the fact that many factors have been associated with tail biting. In a review of the literature, Taylor et al. (2010) proposed that tail biting may not be a single homogenous behavioural category but rather comprise at least three different forms of behaviour, termed 'two-stage' (pre- and damaging stages), 'sudden forceful' and 'obsessive' tail biting, that each has differing underlying motivations. It is useful to summarize the proposal of Taylor et al. (2010) to highlight the difficulty in understanding the aetiology of tail biting and appreciating both future research direction and possible solutions. The 'pre-damage stage', in which the tail is lightly mouthed and manipulated causing no visible damage or distress to the recipient, is considered to be a normal extension of the pigs' natural foraging and exploratory behaviour. However, at some point, this manipulation may break the skin and once bleeding occurs, the problem can rapidly escalate as other pigs are attracted to the tail, leading to harmful injuries, the 'damaging stage'. Thus, preventing pigs from developing a manipulatory interest in tails by reducing the level of foraging/exploratory behaviour, or providing more appropriate substrates for foraging/exploratory behaviour, may be the key to preventing this form of tail biting. Sudden forceful tail biting involves a pig's tail being seized and forcefully pulled or bitten, generally without a prior period of gentle manipulation. Sudden forceful tail biting is more commonly seen when pigs are unable to access a desired resource, such as feeders. Thus, preventing this type of tail biting may require identification of situations which lead to competitive/frustrating interactions between pigs, such as competition for resources such as space, food, water and preferred lying areas. Obsessive tail biting is similar to the previous form of tail biting in that tails are forcefully pulled and bitten; however, this appears to be performed by one or a few individual animals in the group that appear to be focussed or fixated on biting tails, persistently looking for tails to be bitten. While either of the other two forms of tail biting may proceed it, obsessive tail biters may find the act of tail biting more rewarding than accessing the resource, thus resulting in tail biting itself becoming a consummatory, rather than appetitive, behaviour. It is also possible since this behaviour is extensively performed and its occurrence is similar in frequency and intensity to that of stereotypies, that a degree of pathological change has taken place, for example with a dietary or health problem.

Removal of the distal half of the tail is a common industry practice: the remaining section of the tail is sufficiently sensitive to chewing to cause the recipient pig to respond (Fraser and Broom, 1997). Tail docking is usually performed by removing at least half of the tail using either scalpels, scissors/wire cutters or by cautery with a hot iron.

Tail docking of pigs increased squeals with higher peak vocal frequencies during the treatment (Morrison et al., 2013; Marchant-Forde et al., 2009; Noonan et al., 1994), escape attempts (Morrison et al., 2013), standing with head lowered (Morrison et al., 2013) and posterior 'scooting' (Sutherland et al., 2008). While Prunier et al. (2005) did not find any differences in plasma profiles of cortisol and ACTH during the first 3 hours following docking in 1-day-old piglets, tail docking elevated cortisol concentrations 15 and 30 minutes post-treatment, with lower responses with the use of cauterization rather than side-cutter pliers (Morrison et al., 2013; Sutherland et al., 2008). Marchant-Forde et al. (2009) found no difference in cortisol concentration 45 minutes post-treatment using side-cutter pliers or cauterization.

Whilst there is evidence that tail docking causes acute, short-term behavioural and physiological responses, there is concern that tail docking may cause chronic, long-term pain in pigs since neuromas were observed in the tail stumps of docked pigs (Simonsen et al., 1991; Done et al., 2003; Herskin et al., 2015). While Sandercock et al. (2011) found no difference in the behavioural responses to mechanical or noxious stimuli of pigs with docked or intact tails, the implications of neuromas which develop in the damaged nerves on longer-term pain clearly require further study, since the presence of neuromas might lead to altered nociceptive thresholds.

7.2 Castration

Male pigs are routinely castrated to eliminate boar taint in pork. As in other livestock industries, castration also reduces the management problems associated with aggressive and sexual behaviour and handling (Carroll et al., 2006). The main method of castration of pigs is the surgical removal of the testicles with a blade (e.g. scalpel) or pulling until the cord tears.

Research indicates, in general, that castration in young pigs results in a short-term moderate acute stress response and that castration is likely to be painful for piglets at any age. Castrated piglets vocalized more frequently during treatment than sham-castrated piglets (Taylor et al., 2001). Furthermore, it has been shown that castration reduces sucking and standing and increases lying for six hours (McGlone et al., 1993), increases standing or sitting and reduces lying for two hours (Taylor et al., 2001) and reduces sucking and massaging the udder for 2.5 hours, increases inactivity while awake for 2.5 hours and increases scratching and tail wagging for 24 hours (Hay et al., 2003). Prunier et al. (2005) found that castration increased plasma ACTH and cortisol concentrations from 5 to 30 minutes and from 15 to 90 minutes, respectively, while Carroll et al. (2006) found that castration increased plasma cortisol concentrations for 24 hours. In contrast, Carroll et al. (2006) found no effects of castration on nursing, lying, standing or sitting during the first 2 h after treatment, and Hay et al. (2003) found no effects of castration on urinary corticosteroids and catecholamines concentrations from 1 to 4 days following castration. Age at castration within the range of 1 to 20 days appears to have little influence on behaviour (McGlone et al., 1993; Taylor et al., 2001) or cortisol concentrations (Carroll et al., 2006).

The results of studies on the effects of castration on growth generally indicate little or no effects. McGlone et al. (1993), Hay et al. (2003) and Carroll et al. (2006) found no effect of castration on weight gain, but Kielly et al. (1999) found that castration at 3 days of age but not 10 days temporarily reduced weight gain in piglets. It is of interest that the stress associated with handling appeared to increase with age in the study by Carroll et al. (2006).

Castration is usually performed without anaesthesia or post-operative analgesia. In 2010, the European Union passed regulations to ensure that all piglets are castrated using analgesia/anaesthesia. Lidocaine is the most common local anaesthetic tested but has no pain-relieving effects post-surgery (Haga and Ranheim, 2005). Non-steroidal anti-inflammatory drugs (NSAIDs) are being increasingly studied and being licenced for use in food-producing animals. Several researchers have suggested that lidocaine in combination with the NSAID meloxicam may provide the most comprehensive pain relief for surgically castrated piglets. Indeed, some recent research has shown promising results. Meloxicam reduced plasma cortisol concentrations, isolation behaviour and cranial skin temperature in castrated piglets, findings that are consistent with effective pain relief (Tenbergen et al., 2014; Bates et al., 2014). A practical limitation with anaesthetics and analgesics is the additional handling that may be required to deliver them and the consequences of the additional handling. It has been shown with piglet castration and tail docking that the fear and stress arising from handling *per se* may be as stressful as the pain component (Marchant-Forde et al., 2009). However, in the study by Bates et al. (2014), meloxicam was successfully transferred from the sow to piglets through milk, which provides a less invasive and less time-consuming method of administration. The use of anaesthesia or post-operative analgesia to reduce chronic or long-lasting pain in commercial settings requires investigation.

Vaccination to control boar taint appears to be a practical and effective alternative to surgical castration. In addition, vaccination to control boar taint provides distinct welfare advantages over surgical castration. Immunization against gonadotropin-releasing factor (GnRF) late in the finishing stage of production prevents testicular growth and activity while retaining most of the growth and efficiency benefits of intact boars (Dunshea et al., 2001; Jaros et al., 2005). These effects of vaccination to control boar taint occur without some of the welfare impacts, such as acute stress and pain associated with surgical castration in the young male pig or the subsequent high levels of aggression and mounting behaviour that occur in the intact male at 19 to 22 weeks of age (Cronin et al., 2003; Rydhmer, et al ., 2006; Brewster and Nevel, 2013). It is of interest that Allison et al. (2009) in a meta-analysis of data from 15 field studies (total of 4540 piglets) found a slight but significant increase in the percentage of piglets that died or were withdrawn prior to weaning in surgically castrated piglets than in piglets immunized against GnRF (5.0 vs 3.4%). These results suggest that, in addition to pain and acute stress effects, surgical castration may affect mortality of piglets.

8 Safeguarding animal welfare

8.1 Science and education

Science provides the means to understand the impact of animal use on the animal, and thus science has an important role in underpinning societal decisions on animal use and the acceptability or otherwise of attendant conditions and compromises. Science, therefore, should continue to have a prominent role in underpinning our decisions on animal use and the attendant conditions and compromises. Exclusion of science can result in emotive or self-interested arguments from sectional groups dominating community debate. This is not to say that such arguments should be ruled out, quite the reverse, as they reflect, in part, current community values. However, they should contribute to, not pre-empt, the debate (Hemsworth et al., 2015). For example, the publicly engaging concepts of 'free range' and 'capacity to express natural behaviour' among domesticated animals can lead to compromised welfare when implemented in circumstances which, on the face of it, suggest that welfare would be improved. For example, in a study of 1,486 UK flocks, Weeks et al. (2012) found that mortality of hens over a 52-week laying period was 9.5% for free-range hens compared to 5.4% for hens in cages.

Obviously, societal interests in animal welfare also include consideration of wider issues such as human health, economic and social implications, and environmental impacts (Fisher and Mellor, 2008; Mellor and Bayvel, 2008). Decisions on acceptable animal use can therefore involve difficult and complex choices and consequently they may remain controversial.

In terms of education, there needs to be transparency to the public in farming techniques and in particular a clear articulation of the implications both for food quality on the one hand and for animal welfare on the other. Furthermore, the fostering of a culture shift among the livestock industries is desirable to sensitize them to changing cultural values about animal quality of life. Industry responses need to have a balance between listening to community requirements on the one hand and a preparedness to defend a practice on the other if, on balance, it is considered the best in terms of healthy food, economics and welfare. This is essential for a well-informed community able to make rational choices and if industry is to respond appropriately to community expectations (Coleman et al., 2016).

While science and community values underpin the establishment, amendment or validation of industry welfare standards and practices, it is critical to deliver industry education, through staff training. As discussed in the next section, training of farm managers and stockpeople is essential to safeguard animal welfare.

8.2 Management

The principle that management, including supervising and managing animals, affects farm animal welfare is widely recognized within the livestock industries. However, the manner in which management affects animal welfare, both directly and indirectly, is probably not fully appreciated (Hemsworth and Coleman, 2009, 2011).

Management at the farm level

At the level of farm management, human resource management practices, including employee selection and training, animal management practices, such as best practice in housing and husbandry, and implementation of welfare protocols and audits, all impact on farm animal welfare. Implementation of welfare protocols and audits will be considered under assurance schemes.

Management at the stockperson level

At the stockperson level, together with the opportunity to perform tasks well, stockpeople require a range of well-developed husbandry skills and knowledge to effectively care for farm animals. Appreciating the factors that affect work performance, as well as where deficiencies exist at the levels of both the stockperson and the farm, is the first step in developing a strategic programme to ensure that stockpeople have well-developed husbandry skills and knowledge, as well as access to the appropriate facilities and opportunity to use these skills and knowledge to effectively care for and manage farm animals (Hemsworth and Coleman, 2009, 2011).

Knowing and being skilled at the techniques that must be used to accomplish a task are clearly prerequisites to being able to perform that task and thus these job-related characteristics will be limiting factors on job performance in situations where specific technical skills and knowledge are required to perform the tasks. Most stockperson training programmes target husbandry competencies rather than application of these competencies.

There has been an ever-increasing body of evidence accumulating since the 1980s of the effects of human interactions on farm animal fear and stress responses. This body of research has been recently reviewed by Hemsworth and Coleman (2011) and Coleman and Hemsworth (2014) but is briefly summarized here.

Field studies on farm animals, including pigs, revealed variation in fear of humans by farm animals that could not be explained by farm characteristics or animal genetics. The variation in behavioural fear responses was strongly related to increased stress responses and reduced productivity. This led to the hypothesis that this variation in fear was caused by human factors, giving rise to investigations directed towards identifying the human characteristics responsible for these effects. Laboratory studies and correlational and intervention studies in commercial settings on a number of farm animal species provided consistent evidence of causal relationships between stockperson attitudes (based on beliefs), stockperson behaviour, animal fear responses, animal stress physiology and animal productivity, and provided evidence of causal relationships between these variables (see Hemsworth and Coleman, 2011). A key advance in understanding opportunities to manipulate human–animal relationships in the field is that the antecedents of stockperson behaviour are their attitudes and, in particular, their beliefs about their behaviour, their animals' behaviour, fear and stress, and the effectiveness and appropriateness of specific handling behaviours.

The efficacy of cognitive–behavioural interventions to improve human–farm animal relationships was demonstrated by the finding among dairy and pig stockpersons that cognitive–behavioural training can improve attitudes and human behaviour towards animals, with consequent reductions in animal fear and improvements in productivity (Coleman et al., 2000; Hemsworth et al., 1994, 2002). Cognitive–behavioural techniques basically involve changing a person's behaviour by first targeting both the beliefs that underlie the behaviour (attitude) and the behaviour in question, and second, maintaining these changed beliefs and behaviours. This process of inducing behavioural change is a comprehensive procedure in which all of the personal and external factors that are relevant to the behavioural situation are explicitly targeted. This includes addressing common perceived barriers to change, addressing defensiveness about previous behaviour, changing habits, providing follow-ups to reinforce changes as well as changing the relevant attitudes and behaviour. This approach to training was also shown to be practical and effective among a wide range of stockpersons working in a variety of situations, providing strong evidence for introducing this type of training into the livestock industries.

The training programme used as an experimental tool during research in the pig industry has been commercialized for on-farm use, and is called 'ProHand' (a contraction of the 'Professional Handling of Pigs Program', Animal Welfare Science Centre, 2011). Furthermore, as part of the European Union Sixth Framework programme, ProHand principles guided the development of the Welfare Quality training packages for stockpeople working in the European pig industry (Ruis et al., 2010). An important characteristic of this programme is that it is based on scientific research and its effectiveness in improving welfare has been demonstrated by properly designed intervention studies.

Technical skills and knowledge are important attributes of the work performance of stockpeople, and clearly, training targeting these attributes is important in improving animal welfare. Furthermore, training programmes that target the attitude and behaviour of stockpeople offer considerable opportunity to improve pig welfare. Improved human–animal interactions may also enhance job-related characteristics, such as job satisfaction, motivation and commitment, thereby potentially improving the stockperson's job performance and career prospects (Coleman and Hemsworth, 2014). Thus, including training targeting the attitudes and behaviours of stockpeople towards pigs in conjunction with the technical skills and knowledge of stockpeople is likely to not only reduce the stress associated with handling and husbandry procedures involving humans, but also improve the motivation in stockpeople to learn new technical skills and knowledge and to apply these competencies to the management of the animals under their care.

8.3 Design of the housing system

While there is a focus in intensive animal production on housing systems, research indicates that the design of the housing system is probably more important for animal welfare than is generally recognized (Barnett et al., 2001; Rushen and de Passillé, 1992). For example, the previous discussion on the design features of group housing of sows such as the amount and quality of space and protection of sows at feeding, indicate that it is clear that variations in design may produce substantial variation in risks to welfare.

The focus on different housing systems ignores many of the important factors that can affect animal welfare such as the design of the system *per se* and the quality of management, particularly stockpersonship. Research on animal welfare in new and modified housing systems, as well as current but contentious systems, needs to be attentive to the design contributions of these systems to animal welfare.

8.4 Welfare monitoring in the field

The assessment of welfare in a livestock enterprise can be used to demonstrate compliance with policy, law, and regulatory standards, and to assure both consumers and non-consumers that certain welfare standards are being met. Welfare assessment in a livestock enterprise can also be used to assist owners and managers to monitor and improve the welfare of their livestock. Welfare assessment in the field requires practical and effective welfare measures and two types of measures are used: environmental and management components (input measures); and validated, repeatable and feasible animal-based welfare indices (outcome or output measures).

In recent years, the assessment and monitoring of animal welfare has shifted from the conventional approach of evaluating the environment and resources required to ensure good welfare, to focussing on animal-based measures of welfare. An example of this is the animal welfare assessment at a farm or on-site level in the European Union Welfare Quality® project (Botreau et al., 2009)).

Animal-based measures can provide a direct assessment of the animal's welfare state, and while environmental parameters will offer information regarding potential or current welfare risks, they fail to directly reflect the welfare state of the animal (Colditz et al., 2014). Nevertheless, Main et al. (2014) suggest that outcome measures are unlikely to replace all environment measures, particularly where welfare science has shown that the resources provided contribute to genuine welfare benefits. Furthermore, it is well recognized that there are some challenges in assessing some animal-based measures, particularly behaviour measures, in a reliable, consistent and time-efficient way (Main et al., 2014).

Butterworth et al. (2011), Colditz et al. (2014) and Main et al. (2014) argue that compliance and market assurance schemes might measure compliance with certain aspects of the welfare of animals but they do not provide clear directions on how to improve animal welfare. Indeed, these authors propose a continuous improvement approach utilizing both regular monitoring of pre-defined welfare criteria (input- (environmental and management) and outcome-based), benchmarking performance to identify targets for improvement and a management system to ensure preventive and corrective action to maximize levels of these criteria. Further, the authors suggest that such an animal welfare risk assessment and management scheme lends itself to providing evidence for compliance and assurance schemes. In addition, these authors propose utilizing so-called production variables, many of which farmers already collect such as growth, reproduction, health, culling statistics and environmental measures. Thus, this multi-pronged approach provides opportunities to benchmarking livestock welfare, both within farms and within the livestock industry for use by individual farmers, industry and stakeholders, to provide compliance evidence and market assurance and, probably most importantly, to assess animal welfare and risk in order to continuously improve animal welfare.

9 Conclusion

Improvements in pig welfare are likely to occur in the future through research and adoption in the areas of animal management and housing design. The focus of discussion on different housing systems tends to overlook the importance of the design of the housing system on pig welfare. Research clearly indicates that the design of the housing system is probably more important for animal welfare than is generally recognized, and clearly, research on animal welfare in new and modified housing systems, as well as current but contentious systems, needs to be cognizant of the design contributions of these systems to pig welfare.

Irrespective of the housing system, the skills, knowledge and motivation of stockpeople to effectively care for and manage their animals are integral to the standard of welfare experienced by the pigs. Attitudes influence not only the manner in which stockpeople handle pigs, but also their motivation to care for pigs. Thus, training targeting technical skills and knowledge as well as the attitudes and behaviours of stockpeople should be a primary component of the human resource management practices at a farm.

Welfare monitoring in the field is an important tool to continuously improve animal welfare. It allows benchmarking, both within farms and within the industry, for use by individual farmers, industry and stakeholders, provides compliance evidence and market assurance and, probably most importantly, offers the opportunity to incorporate animal welfare risk assessment and management in the routine management programme of the farm.

The public is also a key driver of animal welfare change and thus there needs to be transparency to the public in farming techniques and a clear articulation of the implications for both food quality on the one hand and animal welfare on the other. However, industry responses need to have a balance between listening to community requirements and a preparedness to defend a practice if, on balance, it is considered the best in terms of healthy food, economics and welfare.

10 Where to look for further information

There are a number of reviews on animal welfare and its assessment (Fraser, 2008; Mellor et al., 2009; Mellor, 2012; Hemsworth et al., 2015) and the welfare of commercial pigs has been reviewed by Borell et al. (1997), Barnett et al. (2001), Faucitano and Schaefer (2008), Marchant-Forde (2009) and Verdon et al. (2015). The topic of human–animal interactions in the livestock industries has been reviewed by Hemsworth and Coleman (2011) and Coleman and Hemsworth (2014) and details of the commercialized training programme targeting attitudinal and behavioural change in stockpeople ('ProHand') are available at http://www.animalwelfare.net.au/article/training-programs.

11 References

Allison, J., Helmholdt, E., Pearce, M., Brock, F. and Crane, J. (2009), 'Comparison of mortality (animal withdrawal) rates in male fattening pigs reared using either physical castration or vaccination with Improvac® as the method to reduce boar taint'. In: *Eur. Symp. Porcine Health Man.*, 28–29 August 2009, Copenhagen, Denmark.

Andersen, I. L., Berg, S. and Bøe, K. E. (2005), 'Crushing of piglets by the mother sow (*Sus scrofa*) — purely accidental or a poor mother?' *Appl. Anim. Behav. Sci.*, 93, 229–43.
Animal Welfare Science Centre (2011), *Training Programs: Targeting Human-animal Interactions to improve Animal Welfare and Productivity*, Available: http://www.animalwelfare.net.au/article/training-programs
Appleby, M. C. and Lawrence, A. B. (1987), 'Food restriction as a cause of stereotypic behaviour in tethered gilts'. *Anim. Prod.*, 45, 103–10.
Arnot, C. (2009), 'Protecting our freedom to operate: Earning and maintaining public trust and our social license'. In: *Proc. The Southwest Nutrition and Management Conference*, Arizona, USA.
Barnett, J. L. 1997. Modifying the design of group pens with individual feeding places affects the welfare of pigs. In: *Proc. 5th Int. Symp. American Soc. Agri. Engineers.*, (eds), Bottcher, R. W. and Hoff, S. J., St. Joseph, MI. pp. 613–18.
Barnett, J. L., Hemsworth, P. H., Cronin, G. M., Newman, E. A., McCallum, T. H. and Chilton, D. (1992), 'Effects of pen size, partial stalls and method of feeding on welfare-related behavioural and physiological responses of group-housed pigs'. *Appl. Anim. Behav. Sci.*, 34, 207–20.
Barnett, J. L., Hemsworth, P. H., Cronin, G. M., Jongman, E. C. and Hutson, G. D. (2001), 'A review of the welfare issues for sows and piglets in relation to housing'. *Aust. J. Agric. Res.*, 52, 1–28.
Bates, J. L., Karriker, L. A., Stock, M. L., Pertzborn, K. M., Baldwin, L. G. (2014), 'Impact of transmammary delivered meloxicam on biomarkers of pain and distress in piglets after castration and tail docking'. *PLoS ONE*, 9(12), e113678.
Baxter, E. M., Lawrence, A. B. and Edwards, S. A. (2012), 'Alternative farrowing accommodation: welfare and economic aspects of existing farrowing and lactation systems for pigs'. *Animal*, 6, 96–117.
Benson, G. J. (2004), 'Pain in farm animals: nature, recognition, and management'. In: *The Well-Being of Farm Animals: Challenges and Solutions*, (eds), Benson, G. J. and Rollin, B. E., Blackwell Publishing, Iowa, USA, pp. 61–84.
Bergeron, R. and Gonyou. H. W. (1997), 'Effects of increasing energy intake and foraging behaviours on the development of stereotypies in pregnant sows'. *Appl. Anim. Behav. Sci.*, 53, 259–70.
Bergeron, R., Bolduc, J., Ramonet, Y., Meunier-Salaün, M. C. and Robert, S. (2000), 'Feeding motivation and stereotypies in pregnant sows fed increasing levels of fiber and/or food'. *Appl. Anim. Behav. Sci.*, 70, 27–40.
Bergeron, R., Badnell-Waters, A. J., Lambton, S. and Mason, G. (2006), 'Stereotypic oral behaviour in captive ungulates: foraging, diet and gastrointestinal function'. In: *Stereotypic Animal Behaviour. Fundamentals and Applications to Welfare*, (eds), Mason, G. and Rushen, J., 2nd Edition, CABI, Wallingford, Oxfordshire, UK, pp. 19–27.
Boon, C. R. and Wray, C. (1989), 'Building design in relation to the control of diseases of intensively housed livestock'. *J. Agric. Eng. Res.*, 43, 149–61.
Borell, E. von, B. D. M., Csermely, D., Dijkhuizen, A. A., Edwards, S. A., Jensen, P., Madec, F. and Stamataris, C. (1997), 'The welfare of intensively kept pigs'. In: *Report of the Scientific Veterinary Committee*, European Union, Brussels, Belgium.
Botreau, R., Veissier, I. and Perny, P. (2009), 'Overall assessment of animal welfare: strategy adopted in Welfare Quality®'. *Animal Welfare*, 18, 363–70.
Brambell, F. W. R., Barbour, D. S., Barnett, M. B., Ewer, T. K., Hobson, A., Pitchforth, H., Smith, W. R., Thorpe, W. H. and Winship, F. J. W. (1965), Report of the Technical Committee to Enquire into the Welfare of Animals Kept Under Intensive Husbandry Systems. Her Majesty's Stationery Office, London, United Kingdom.
Brewster, V. and Nevel, A. (2013), 'Immunocastration with Improvac™ reduces aggressive and sexual behaviours in male pigs'. *Appl. Anim. Behav. Sci.*, 145, 32–6.
Brouns, F., Edwards, S. A. and English, P. R. (1994), 'Effect of dietary fibre and feeding system on activity and oral behaviour of group housed gilts. *Appl. Anim. Behav. Sci.*, 39, 215–23.
Butterworth, A., Mench, J. A. and Wielebnowski, N. (2011), 'Practical strategies to assess and improve welfare'. In: *Animal Welfare*, (eds), Appleby, M. C., Hughes, B. and Mench, J. CABI, Wallingford, Oxfordshire, UK, pp. 200–14.

Carroll, J. A., Berg, E. L., Straunch, T. A., Roberts, M. P. and Kattesh, H. G. (2006), 'Hormonal profiles, behavioral responses and short-term growth performance after castration of pigs at three, nine, or twelve days of age'. *J. Anim. Sci.*, 84, 1271–8.

Chapinal, N., Ruiz de la Torre, J. L., Cerisuelo, A., Gasa, J., Baucells, M. D., Coma, J., Vidal, A. and Manteca, X. (2010), 'Evaluation of welfare and productivity in pregnant sows kept in stalls or in 2 different group housing systems'. *J. Vet. Behav.*, 5, 82–93.

Cockram and Hughes, (2011), 'Health and Disease'. In: *Animal Welfare*, (eds), Appleby, M. C., Hughes, B. and Mench, J., CABI, Wallingford, Oxfordshire, UK, pp. 120–37.

Colditz, I., Ferguson, D., Collins, T., Matthews, L., and Hemsworth, P. (2014), 'A prototype tool to enable farmers to measure and improve the welfare performance of the farm animal enterprise: The Unified Field Index'. *Animals*, 4, 446–62.

Coleman, G. (2004), 'Public attitudes to animal research'. In *ANZCCART Conference*, Christchurch NZ, 18–19 August 2003, pp. 78–86.

Coleman, G. J. (2008), 'Public perceptions of animal pain and animal welfare'. *OIE Technical Series (World Organisation for Animal Health)* 10, 26–37.

Coleman, G. J. (2010), 'Educating the public: information or persuasion?'. *J. Vet. Med. Ed.*, 37, 74–82.

Coleman, G. J. (2015), 'Public attitudes relevant to farm animal welfare policy'. *Farm Policy J.*, 12(3), 45–57.

Coleman, G. J. and Hemsworth, P. H. (2014), 'Training to improve stockperson beliefs and behaviour towards livestock enhances) welfare and productivity'. *Scientific and Technical Review of the Office International des Epizooties (Paris)* 33(1), 131–7.

Coleman, G. J., Hemsworth, P. H., Hay, M. and Cox, M. (2000), 'Modifying stockperson attitudes and behaviour towards pigs at a large commercial farm'. *Appl. Anim. Behav. Sci.*, 66, 11–20.

Coleman, G. J., Jongman, E. C., Greenfield, L. and Hemsworth, P. H. (2016), 'Farmer and public attitudes toward lamb finishing systems'. *J. Appl. Anim. Welfare Sci.*, DOI:10.1080/10888705.2015.1127766.

Cronin, G. M. and Smith, J. A. (1992), 'Suckling behaviour of sows in farrowing crates and straw-bedded pens'. *Appl. Anim. Behav. Sci.*, 33, 175–89.

Cronin, G. M., Simpson, G. J. and Hemsworth, P. H. (1996), 'The effects of the gestation and farrowing environments on sow and piglet behaviour and piglet survival and growth in early lactation'. *Appl. Anim. Behav. Sci.*, 46, 175–92.

Cronin, G. M., Barnett, J. L., Hodge, F. M., Smith, J. A. and McCallum, T. H. (1991), 'The welfare of pigs in two farrowing/lactation environments: cortisol responses of sows. *Appl. Anim. Behav. Sci.*, 32, 117–27.

Cronin, G. M., Dunshea, F. R., Butler, K. L., McCauley, I., Barnett, J. L. and Hemsworth, P. H. (2003), 'The effects of immuno-castration and surgical-castration on the behaviour and consequently growth of group-housed, male finisher pigs'. *Appl. Anim. Behav. Sci.*, 81, 111–26.

Cronin, G.M., Rault, J-L. and Glatz, P.C. (2014), 'Lessons learned from past experience with intensive livestock management systems'. *Rev. sci. tech. Off. int. Epiz.*, 33 (1), 139-151.

Dawkins, D. S. (1980), *Animal Suffering: the Science of Animal Welfare*, Chapman and Hall, London, UK.

Dawkins, M. S. (1998), 'Evolution and animal welfare'. *Quarterly Review of Biology* 73, 305–28.

Done S. H., Guise J., and Chennels, D. (2003), 'Tail biting and tail docking in pigs. *Pig J.*, 51, 136–54.

Dunshea, F. R., Colantoni, C., Howard, K. (2001), 'Vaccination of boars with a GnRH vaccine (Improvac) eliminates boar taint and increases growth performance'. *J. Anim. Sci.*, 79, 2524–35.

EFSA, (2007), 'Scientific report on the risks associated with tail biting in pigs and possible means to reduce the need for tail docking considering the different housing and husbandry systems'. *Annex EFSA J.*, 611, 1–13 (Question no. EFSA-Q-2006–013, 1–98).

European Commission (2007), 'Attitudes of EU citizens towards Animal Welfare'. *Special Eurobarometer 270/Wave* 66.1.

European Union (2010), *European Declaration on Alternatives to Surgical Castration of Pigs*. Available: http://ec.europa.eu/food/animal/welfare/farm/docs/castration_pigs_declaration_en.pdf.

Faucintano, L. and Schaefer, A. L. (eds) (2008). *Welfare of Pigs from Birth to Slaughter*. ed. Faucintano, L. and Schaefer, A. L., Wageningen Academic Publishers, Wageningen, The Netherlands.

Fisher, M. W. and Mellor, D. J. (2008), 'Developing a systematic strategy incorporating ethical, animal welfare and practical principles to guide the genetic improvement of dairy cattle'. *New Zealand Vet. J.*, 65, 100–6.

Fraser, D. (1975). 'The effect of straw on the behaviour of sows in tether stalls'. *Anim. Prod.*, 21, 59–68.

Fraser, D. (1987), 'Attraction to blood as a factor in tail biting by pigs'. *Appl. Anim. Behav. Sci.*, 17, 61–8.

Fraser, D. (2008), *Understanding Animal Welfare: The Science in its Cultural Context*, Wiley-Blackwell, West Sussex, United Kingdom.

Fraser, A. F. and Broom, D. M. (1997), *Farm Animal Behaviour and Welfare*, Third Edition, CABI, Wallingford, Oxfordshire, UK.

Fraser, D. and Nicol, C. J. (2011), 'Preference and motivation research'. In: *Animal Welfare*, (eds). Appleby, M. C., Hughes, B. and Mench, J., CABI, Wallingford, Oxfordshire, UK, pp. 183–99.

Galindo, F., Newberry, R. C., and Mendl, M. (2011), 'Social Conditions'. In: *Animal Welfare*, (eds), Appleby, M. C., Hughes, B. and Mench, J., CABI, Wallingford, Oxfordshire, UK, pp. 228–45.

Geyer, H. (1979), 'Morphologie und wachstum der schweineklaue. Grundlagen für stallbodengestaltung und klauenpathologie'. *Schweizer Archiv für Tierheilkunde* 121, 275–93.

Gjein, H. and Larssen, R. B. (1995), 'The effect of claw lesions and claw infections on lameness in loose housing of pregnant sows'. *Acta Veterinaria Scandinavica* 36, 451–9.

Haga, H. A. and Ranheim, B. (2005), 'Castration of piglets: the analgesic effects of intratesticular and intrafunicular lidocaine injection'. *Vet. Anaesth. Analg.*, 32, 1–9.

Hales, P. J. (2015), Loose housing or temporary confinement of sows in designed farrowing pens. *PhD Thesis*. University of Copenhagen, Copenhagen, Denmark.

Hales, J., Moustsen, V. A., Devreese, A. M., Nielsen, M. B. F. and Hansen, C. F. (2015), 'Comparable farrowing progress in confined and loose housed hyper-prolific sows'. *Livestock Sci.*, 171, 64–72.

Harbuz, M. S. and Lightman, S. L. (1992), 'Stress and the hypothalamo-pituitary-adrenal axis: acute, chronic and immunological activation'. *J. Endocrin.*, 134, 327–39.

Hay, M., Vulin, A., Genin, S., Sales, P. and Prunier, A. (2003), 'Assessment of pain induced by castration in piglets: behavioral and physiological responses over the subsequent 5 days'. *Appl. Anim. Behav. Sci.*, 82, 201–18.

Heinonen, M., Peltoniemi, O. and Valros, A. (2013), 'Impact of lameness and claw lesions in sows on welfare, health and production'. *Livestock Sci.*, 156, 2–9.

Hemsworth, P. H. and Coleman, G. J. (2009), 'Animal welfare and management'. In: *Food Safety Assurance and Veterinary Public Health Volume 5. Welfare Productions Animals: Assessment and Management Risks*, (eds), Smulders, F. J. M. and Algers, B., Wageningen Academic Publishers, the Netherlands, pp. 133–47.

Hemsworth, P. H. and Coleman, G. J. (2011), 'Human-Livestock Interactions: The Stockperson and the Productivity and Welfare of Farmed Animals', 2nd Edition CAB International, Oxon United Kingdom.

Hemsworth, P. H., Coleman, G. J. and Barnett, J. L. (1994), 'Improving the attitude and behaviour of stockpeople towards pigs and the consequences on the behaviour and reproductive performance of commercial pigs'. *Appl. Anim. Behav. Sci.*, 39, 349–62.

Hemsworth, P. H., Barnett, J. L., Karlen, G. M., Fisher, A. D., Butler, K. L. and Arnold, N. A. (2009), 'Effects of mulesing and alternative procedures to mulesing on the behaviour and physiology of lambs'. *App. Anim. Behav. Sci.*, 117, 20–7.

Hemsworth, P. H., Barnett, J. L., Hansen, C. and Winfield, C. G. (1986), 'Effects of social environment on welfare status and sexual behaviour of female pigs. II. Effects of space allowance'. *Appl. Anim. Behav. Sci.*, 16, 259–67.

Hemsworth, P. H., Coleman, G. J., Barnett, J. L., Borg, S. and Dowling, S. (2002), 'The effects of cognitive behavioral intervention on the attitude and behavior of stockpersons and the behavior and productivity of commercial dairy cows'. *J. Anim. Sci.*, 80, 68–78.

Hemsworth, P. H., Mellor, D. J., Cronin, G. M. and Tilbrook, A. J. (2015a), 'Scientific assessment of animal welfare', *New Zealand Vet. J.*, 63, 24–30.

Hemsworth, P. H., Rice, M., Nash, J., Giri, K., Butler, K. and Tilbrook, A. J. and Morrison, R. S. (2013), 'Effects of group size and floor space allowance on grouped sows: Aggression, stress, skin injuries and reproductive performance'. *J. Anim. Sci.*, 91, 4953–64.
Hemsworth, P. H., Rice, M., Morrison, R. S., Butler, K. and Tilbrook, A. J. (2015b), 'Effects of floor space allowance on aggression and stress in grouped sows'. *Proc. 49th Congr. Intern. Soc. Appl. Ethol.*, (eds), Yasue, T., Ninomiya, S., Uetake, K. and Morita, S., Sapporo, Japan, pg. 65.
Herskin, M. S., Thodberg, K. and Jensen, H. E. (2015), 'Effects of tail docking and docking length on neuroanatomical changes and healed tail tips of pigs'. *Animal*, 9, 677–81.
Hocking, P. M., D"Eath, R. B. and Kjaer, J. B. (2011), 'Genetic Selection'. In: *Animal Welfare*, (eds), Appleby, M. C., Hughes, B. and Mench, J., CABI, Wallingford, Oxfordshire, UK, pp 263–78.
Hodges, J. (2000), 'Why livestock, ethics and quality of life?'. In: *Livestock, Ethics and Quality of Life*, (eds), Hodges, J. and Han, K., CABI, Wallingford, Oxfordshire, UK, pp. 1–26.
Hutson, G. D. (1991), 'A note on hunger in the pig: sows on restricted rations will sustain an energy deficit to gain additional food'. *Animal Production* 52, 233–5.
International Association for the Study of Pain. (1979), 'Pain terms: a list with definitions and notes'. *Pain* 6, 249–52.
Jaros, P., Burgi, P., Stark, K. D. C., Claus, R., Hennessy, D. and Thun, R. (2005), 'Effect of active immunization against GnRH on androstenone concentration, growth performance and carcass quality in intact male pigs'. *Livestock Prod. Sci.*, 92, 31–8.
Jarvis, S., Lawrence, A. B., McLean, K. A., Chirnside, J., Deans, L. A., Calvert, S. K. (1998), 'The effect of environment on plasma cortisol and beta-endorphin in the parturient pig and the involvement of endogenous opioids'. *Anim. Reprod. Sci.*, 52, 139–51.
Jarvis, S., Calvert, S. K., Stevenson, J., van Leeuwen, N. and Lawrence, A. B. (2002), 'Pituitary–adrenal activation in pre-parturient pigs (*Sus scrofa*) is associated with behavioural restriction due to lack of space rather than nesting substrate'. *Anim Welfare*, 11, 371–84.
Johnson, A. K. and Marchant-Forde, J. N. (2009), 'Welfare of pigs in the farrowing environment'. In: *The Welfare of Pigs*, (ed.), Marchant-Forde, J. N., Springer Science and Business Media, New York City, USA, pp. 141–88.
Kielly, J., Dewey, C. E. and Cochran, M. (1999), 'Castration at 3 days temporarily slows growth of pigs'. *Swine Health Prod.*, 7, 151–3.
KilBride, A. L., Mendl, M., Statham, P., Held S., Harris, M., Cooper, S. and Green, L. E. (2012), 'A cohort study of preweaning piglet mortality and farrowing accommodation on 112 commercial pig farms in England'. *Preventive Vet. Med.*, 104, 281–91.
Kirkden, R. D., Broom, D. M. and Andersen, I. L. (2013), 'Piglet mortality: Management solutions'. *J. Anim. Sci.*, 91, 3361–89.
Lawrence, A. B. and Terlouw, E. M. C. (1993), 'A review of behavioral factors involved in the development and continued performance of stereotypic behaviors in pigs'. *J. Anim. Sci.*, 71, 2815–25.
Lawrence, A. B., Petherick, J. C., McLean, K. A., Deans, L. A., Chirnside, J., Vaughan, A., Clutton, E. and Terlouw, E. M. C. (1994), 'The effect of environment on behaviour, plasma cortisol and prolactin in parturient sows'. *Appl. Anim. Behav. Sci.*, 39, 313–30.
Lehmann, M. L. and Herkenham, M. J. (2011), 'Environmental enrichment confers stress resiliency to social defeat through an infralimbic cortex-dependent neuroanatomical pathway'. *Neurosci.*, 31, 6159–73.
Li, Y., and Wang, L. (2011), 'Effects of previous housing system on agonistic behaviours of growing pigs at mixing'. *App. Anim. Behav. Sci.*, 132, 20–6.
Main, D. C. J., Clegg, J., Spatz, A. and Green, L. E. (2000), 'Repeatability of a lameness scoring system for finishing pigs'. *Vet. Record*, 147, 574–6.
Main, D., Mullan, S., Atkinson, C., Cooper, M., Wrathall, J. and Blokhuis, H. (2014), 'Best practice framework for animal welfare certification schemes'. *Trends Food Sci. Tech.*, 37, 127–36.
Marchant-Forde, J. N. (ed.) (2009a), *The Welfare of Pigs*. ed. Marchant-Forde, J. N., Springer Science and Business Media, New York City, USA.

Marchant-Forde, J. N. (2009b), 'Future perspectives of the welfare of pigs'. In: *The Welfare of Pigs*, (ed.), Marchant-Forde, J. N., Springer Science and Business Media, New York City, USA, pp. 331–42.

Marchant-Forde, J. N. (2009c), 'Welfare of dry sows'. In: *The Welfare of Pigs*, (ed.), Marchant-Forde, J. N., Springer Science and Business Media, New York City, USA, pp. 95–139.

Marchant-Forde, J. N. and Marchant-Forde, R. M. (2005), 'Minimizing inter-pig aggression during mixing'. *Pig News Info.* 26, 63N–71N.

Marchant-Forde, J. N., Lay, D. C., McMunn, K. A., Cheng, H. W., Pajor, E. A. and Marchant-Forde, R. M. (2009), 'Postnatal piglet husbandry practices and well-being: the effects of alternative techniques delivered separately'. *J. Anim. Sci.*, 87, 1479–92.

Marchant, J. N., Mendl, M. T., Rudd, A. R. and Broom, D. M. (1995), 'The effect of agonistic interactions on the heart rate of group-housed sows'. *Appl. Anim. Behav. Sci.*, 46, 49–56.

Martin, P. and Shepheard, M. (2011), 'What is meant by the social license?'. In: *Defending the Social License of Farming: Issues, Challenges and New Directions for Agriculture*, (eds), Williams, J. and Martin, P., CSIRO Publishing, Collingwood Victoria, Australia.

Mason, G. J. (1991), 'Stereotypies: a critical review'. *Anim. Behav.*, 41, 1015–37.

Mason, G. J. and Latham, N. R. (2004), 'Can't stop, won't stop: is stereotypy a reliable animal welfare indicator?'. *Animal Welfare* 13(Supplement), S57–69.

Matthews, L. R. and Hemsworth, P. H. (2012), 'Drivers of change: law, international markets, and policy'. *Anim. Front.*, 2, 40–5.

McGlone, J. J., Nicholson, R. I., Hellman, J. M. and Herzog, D. N. (1993), 'The development of pain in young pigs associated with castration and attempts to prevent castration-induced behavioral change'. *J. Anim. Sci.*, 71, 1441–6.

Mellor, D. J. and Bayvel, A. C. D. (2008), 'New Zealand's inclusive science-based system for setting animal welfare standards. *Appl. Anim. Behav. Sci.*, 113, 313–29.

Mellor, D. J., Patterson-Kane, E. and Stafford, K. J. (2009), 'The Sciences of Animal Welfare', *Wiley-Blackwell Publishing*, Oxford, United Kingdom.

Meunier-Salaün, M. C., Edwards, S. A. and Robert, S. (2001), 'Effect of dietary fiber on the behaviour and health of the restricted sow. *Anim. Feed Sci. Technol.*, 90, 53–69.

Morrison, R. S., Sawyer, K. S. B., Kells, N. J., Johnson, C. B. and Hemsworth, P. H. (2013), 'Stress responses of two-day old piglets to tail docking'. In: *Manipulating Pig Production XIV. Proc. 14th Biennial Conf. Aust. Pig Sci. Assoc.*, (eds), Pluske, J. R. and Pluske, J. M., pp.128.

Moustsen, V. A., Hales, J., Lahrmann, H. P., Weber, P. M. and Hansen, C. F. (2013), 'Confinement of lactating sows in crates for 4 days after farrowing reduces piglet mortalities'. *Animal* 7, 648–54.

Newberry, R. C. (1995), 'Environmental enrichment: increasing the biological relevance of captive environments'. *Appl. Anim. Behav. Sci.*, 44, 229–43.

Noonan, G. J., Rand, J. S., Priest, J., Ainscow, J. and Blackshaw, J. K. (1994), 'Behavioural observations of piglets undergoing tail docking, teeth clipping and ear notching'. *Appl. Anim. Behav. Sci.*, 39, 203–13.

O'Connell, N. E., (2009), 'Housing the fattening pig'. In: *The Welfare of Pigs*, (ed.), Marchant-Forde, J. N., Springer Science and Business Media, New York City, USA, pp. 189–210.

Oliviero, C., Heinonen, M., Valros, A., Halli, O. and Peltoniemi, O. A. T. (2008). 'Effect of the environment on the physiology of the sow during late pregnancy, farrowing and early lactation'. *Anim. Reprod. Sci.*, 105, 365–77.

Oostindjer, M., van den Brand, H., Kemp, B. and Bolhuis, J. (2011), 'Effects of environmental enrichment and loose housing of lactating sows on piglet behaviour before and after weaning'. *Appl. Anim. Behav. Sci.*, 134, 31–41.

Palmer, C. and Sandoe, P., (2011), 'Animal Ethics'. In 'Animal Welfare', In: *Animal Welfare*, (eds), Appleby, M. C., Hughes, B. and Mench, J., CABI, Wallingford, Oxfordshire, UK, pp. 1–12.

Prunier, A., Mounier, A. M., and Hay, M. (2005), 'Effects of castration, tooth resection or tail docking on plasma metabolites and stress hormones in young pigs'. *J. Anim. Sci.*, 83, 216–22.

Putten van, G. (1969), 'An investigation into tail biting among fattening pigs. *Brit. Vet. J.*, 125, 511–7.

Ramonet, Y., Meunier-Salaun, M. C. and Dourmad, J. Y. (1999), 'High-fiber diets in pregnant sows: digestive utilization and effects on the behavior of the animals'. *J. Anim. Sci.*, 77, 591–9
Remience, V., Wavreille, J., Canart, B. V., Meunier-Salaün, M. C., Prunier, A., Bartiaux-Thill, N., Nicks, B. and Vandenheede, M. (2008), Effects of space allowance on the welfare of dry sows kept in dynamic groups and fed with electronic sow feeder'. *Appl. Anim. Behav. Sci.*, 112, 284–96.
Robert, S., Matte, J. J., Farmer, C., Girard, C. L. and Martineau, G. P. (1993), 'High-fibre diets for sows: effects on stereotypies and adjunctive drinking'. *Appl. Anim. Behav. Sci.*, 37, 297–309.
Robert, S., Rushen, J. and Farmer, C. (1997), 'Both energy content and bulk of feed affect stereotypic behaviour, heart rate and feeding motivation of female pigs'. *Appl. Anim. Behav. Sci.*, 54, 161–71.
Ruis M., Coleman G. J., Waiblinger, S., Windschnurer, I. and Boivin, X. (2010), 'A multimedia-based cognitive-behavioural intervention program improves attitudes and handling behaviours of stockpeople in livestock farming'. *Adv. Anim. Biosci.*, 1, 175.
Rushen, J. and de Passillé, A. M. B. (1992), 'The scientific assessment of the impact of housing on animal welfare: A critical review'. *Can. J. Anim. Sci.*, 72, 721–43.
Rydhmer, L., Zamaratskaia, G., Andersson, H. K., Algers, B., Guillemet, R., and Lundstrom, K. (2006), 'Aggressive and sexual behaviour of growing and finishing pigs reared in groups, without castration'. *Acta Agriculturae Scandinavica Section A*, 56, 109–19.
Salak-Johnson, J. L., Niekamp, S. R., Rodriguez-Zas, S. L., Ellis, M. and Curtis, S. E. (2007), 'Space allowance for dry, pregnant sows in pens: Body condition, skin lesions and performance'. *J. Anim. Sci.*, 85, 1758–69.
Sandercock, D. A., Gibson, I. F., Rutherford, K. M. D., Donald, R. D., Lawrence, A. B., Brash, H. M., Scott, E. M. and Nolan, A. M. (2011), 'The impact of prenatal stress on basal nociception and evoked responses to tail docking and inflammatory challenge in juvenile pigs'. *Physiol. Behav.*, 104, 728–37.
Schouten, W. and Rushen, J. (1992), 'Effects of naloxone on stereotypic and normal behaviour of tethered and loose-housed sows'. *Appl. Anim. Behav. Sci.*, 33, 17–26.
Schroder-Petersen, D. L. and Simonsen, H. B. (2001), 'Tail biting in pigs'. *Vet. J.*, 162, 196–210.
Simonsen, H. B., Klinken, L. and Bindseil, E. (1991), 'Histopathology of intact and docked pig tails'. *Brit. Vet. J.*, 147, 407–12.
Singh, C., Verdon, M., Cronin, G.M. and Hemsworth, P.H. (2017). The behaviour and welfare of sows and piglets in farrowing crates or lactation pens. Animal, Accepted; Accepted 1 November 2016. DOI:10.1017/S1751731116002573.
Smith, W. J. and Penny, R. H. C. (1986), 'Behavioural problems including vices and cannibalism'. In: *Diseases of Swine*, (eds), Leman, A. D., Straw, B., R. D. Glace, Mengeling, W. L., Penny, R. H. C. and Scholl, E., E. 6th edition, Iowa State University Press, Ames, USA, pp. 761–22.
Southwick, S. M., Vythilingam, M. and Charney, D. S. (2005), 'The psychobiology of depression and resilience to stress: implications for prevention and treatment'. *Annu. Rev. Clin. Psychol*, 1, 255–91.
Stalder, K. J., Knauer, M., Baas, T. J., Rothschild, M. F. and Mabry, J. W. (2004), 'Sow Longevity'. *Pig News and Information* 25(2), 53N–74N.
Stolba, A., Baker, N. and Wood-Gush, D. G. M. (1983), 'The characterisation of stereotyped behaviour in stalled sows by informational redundancy'. *Behaviour* 87, 157–81.
Sutherland, M. A. Bryer, P. J. Krebs, N. and McGlone, J. J. (2008), 'Tail docking in pigs: acute physiological and behavioural responses'. *Animal*, 292–7.
Schroder-Peterson, D. L. and Simonsen, H. B. (2001), 'Tail biting in pigs'. *Vet. J.*, 162:196–210.
Taylor, A. A., Weary, D. M., Lessard, M. and Braithwaite, L. (2001), 'Behavioural responses of piglets to castration: the effect of piglet age'. *Appl. Anim. Behav. Sci.*, 73, 35–43.
Taylor, N. R., Main, D. C. J., Mendl, M. and Edwards, S. A. (2010), 'Tail-biting: a new perspective'. *Vet. J.*, 186, 137–47.
Taylor, I. A., Barnett, J. L. and Cronin, G. M. (1997), 'Optimum group size for pigs. In: *Proc. 5th. Int. Symp. Am. Soc. Agri. Eng.*, (eds), Bottcher, R. W. and Hoff, S. J., St. Joseph, Michigan, USA, pp. 965–71.

Tenbergen, R., Friendship, R., Cassar, G., Amezcua, M. R. and Haley, D. (2014), 'Investigation of the use of meloxicam for reducing pain associated with castration and tail docking and improving performance in piglets'. *J. Swine Health Prod.*, 22, 64–70.

Terlouw, E. M. C. and Lawrence, A. B. (1993), 'Long-term effects of food allowance and housing on development of stereotypies in pigs'. *Appl. Anim. Behav. Sci.*, 38, 103–26.

Terlouw, E. M. C., Lawrence, A. B., Ladewig, J., De Passille, A. M., Rushen, J. and Schoutem, W. G. P. (1991), 'Relationship between plasma cortisol and stereotypic activities in pigs'. *Behav Process* 25, 133–53.

Teutsch, G. M. (1987), 'Intensive farm animal management seen from an ethical standpoint'. In: *Ethical, Ethological and Legal Aspects of Intensive Farm Animal Management*, (eds), von Loeper, E., Martin, G., Muller, J., Nabholz, A. and van Putten, G., Tierhaltung, 18, pp. 9–40.

Thodberg, K., Jensen, K. H. and Herskin, M. S. (2002), 'Nursing behaviour, postpartum activity and reactivity in sows: effects of farrowing environment, previous experience and temperament'. *Appl. Anim. Behav. Sci.*, 77, 53–76.

Thornton, K. (1990), *Outdoor Pig Production*, Farming Press Ltd., Ipswich, UK.

USDA. (2007), 'Swine 2006', *Reference of Swine Health and Management Practices in the United States*, Part 1, 2006, TJSDA: APHIS:VS, CEAH, Fort Collins, Colorado, Number N475.1007.

Van de Weerd, H. A. and Day, J. E. L. (2009), 'A review of environmental enrichment for pigs housed in intensive housing systems. *Appl. Anim. Behav. Sci.*, 116, 1–20.

Verdon, M., Hansen, C. F., Rault, J.-L., Jongman, E., Hansen, L. U., Plush, K. and Hemsworth, P. H. (2015), 'Effects of group-housing on sow welfare: A review'. *J. Anim. Sci.*, 93, 1999–2017.

Verdon, M., Morrison, R. S. and Hemsworth, P. H., (2016), 'Rearing piglets in multi-litter group lactation systems: effects on piglet aggression and injuries post weaning.' *Appl. Anim. Behav. Sci.*, 183, 35–41.

Verlarde, A. (2007), 'Agonistic behaviour'. In *On Farm Monitoring of Pig Welfare*, (eds), Verlarde, A., and Geers, R., Wageningen Academic Press, Wageningen, The Netherlands, pp. 53–6.

Vosough, A. B., Stott, A. W., Baxter, E. M., Lawrence, A. B., and Edwards, S. A. (2011), 'Animal welfare and economic optimisation of farrowing systems'. *Animal Welfare*, 20, 57–67.

Walker, P. K. and Bilkei, G. (2005), 'Tail-biting in outdoor pig production'. *Vet. J.*, 171, 367–9.

Webb, N. G. and Nilsson, C. (1983), 'Flooring and injury—an overview'. In *Farm Animal Housing and Welfare*, (eds), Baxter, S. H., Baxter, M. R. and MacCormack, J. A. D., Martinus Nijhoff, The Hague, The Netherland, pp. 226–59.

Weber, R., Keil, N.M., Fehr, M. and Horat, R. (2007), 'Piglet mortality on farms using farrowing systems with or without crates'. *Animal Welfare*, 16, 277-279.

Weeks, C. A., Brown, S. N., Richards, G. J., Wilkins, L. J. and Knowles, T. G. (2012), 'Levels of mortality in hens by end of lay on farm and in transit to slaughter in Great Britain', *Veterinary Record, In Practice*, 170, 647–51.

Weng, R. C., Edwards, S. A. and English, P. R. (1998), 'Behaviour, social interactions and lesion score of group-housed sows in relation to floor space allowance'. *Appl. Anim. Behav. Sci.*, 59, 307–16.

Würbel, H., Chapman, R. and Rutland, C. (1998), 'Effect of feed and environmental enrichment on development of stereotypic wire-gnawing in laboratory mice'. *Appl. Anim. Behav. Sci.*, 60, 69–81.

Chapter 7

Pasture systems for pigs

Silvana Pietrosemoli and James T. Green, North Carolina State University, USA

1 Introduction

Markets, through consumers, lead food demand and influence the definition and design of the agriculture production systems (McGlone 2013). Improved access to transparent and up-to-date information and increased purchasing power contribute to consumers making more selective purchasing decisions, demanding safe and high-quality farm products while giving considerations to animal welfare and environmental stewardship in the production process.

Pastured pork represents a buying alternative to consumers seeking products generated in systems perceived as more environmentally and animal welfare friendly than those from confinement production systems. Surveys designed to determine customer criteria leading to buying selection indicated that pig living conditions, especially those that offered access to pasture, were a determining criterion in buying at premium prices often twice the commodity price (Norwood and Lusk 2011).

Pasture-based pig production systems have the potentiality to differentiate their products, opening the door to the creation of niche markets (Barlocco 2005; Grosso et al. 2011). Breeds or genotypes, meat taste, quality and freshness, family-farm raised, locally

http://dx.doi.org/10.19103/AS.2017.0013.21

raised, natural or humane are some of the attributes stated by the participants of the pastured pork supply chain (Honeyman 2004).

This chapter summarizes information related to pasture pig production systems and describes the pasture pig system developed at the Center for Environmental Farming Systems (CEFS), Goldsboro, North Carolina.

2 Characteristics of pasture pig systems

In pasture production systems, pigs are kept on paddocks where they can forage, have access to water, receive supplemental feed and are provided with shelter (Miao, Glatz and Ru 2004). In these systems maintaining soil vegetative cover is a primary objective. Pigs on pasture have the opportunity to express their natural behaviour such as exploring their surroundings, socializing, grazing, rooting (Bonde and Sørensen 2004; Laister and Konrad 2005) and allowing for better animal welfare (Barlocco 2005; Laister and Konrad 2005; Lindgren et al. 2014). Pasture pigs have access to diverse feed resources: grasses, legumes, herbs, roots, tubers, fruits, mast crops and small animals including invertebrates, reptiles, mammals and birds that contribute to their diet (Edwards 2003; Rodriguez-Estevez et al. 2009c). The use of electric fences, portable shelters and equipment facilitate the implementation of modern practices and technologies, and have contributed to the renewed interest that farmers are showing to these production systems. Outdoor systems have provided performance indicators similar to those obtained indoor (Gentry et al. 2004; Miao, Glatz and Ru 2004), and up to 50% reductions in production costs (Grosso et al. 2011).

Pigs on pasture are exposed to environmental conditions that could cause thermal stress, either by excessive cold or by heat, and have a broader range of activity than indoor pigs, and those could be reflected in increased (+8 to +15%) maintenance energy requirements (Edwards 2003; Stern and Andresen 2003) and poorer feed conversion (Edwards 2005; Kongsted et al. 2015). In that sense, Edwards (2003) indicated that for certain northern European countries, feed allowance for outdoor pigs could be increased by up to 15% of those for indoor-raised pigs to counteract the effects of higher nutritional needs.

Health and welfare of pigs on pasture are under pressure by the potential exposition to weather conditions, predators, wild animals, parasites and pathogens (Hovi, Sundrum and Thamsborg 2003; Heyer, Andersson and Lundstrom 2006; Roepstorff et al. 2011; Lindgren et al. 2014). Special attention should be devoted to rats and mice control, as these frequent visitors to pasture pig systems could be important reservoirs of pathogens such as *Salmonella, Leptospira, Yersinia, Brucella suis, Erysipelothrix rhusiopathiae* and *Brachyspira hyodysenteriae* (Bonde and Sørensen 2004). Nevertheless, higher health status (fewer respiratory problems, skin lesions including abscesses and hernias, tail wounds, lesser incidence of stomach ulceration and body injuries, lower veterinary and medicine costs) has been associated with outdoor pigs (Guy et al. 2002b; Edwards 2005; Lindgren et al. 2014). Furthermore, gestating sows kept indoor had, respectively, higher skin temperature (32.1 vs 30.8°C) and respiratory frequency (31.04 vs 26.07 mov/min) than those kept on pasture; pasture pig experienced better thermal conditions and consequently lower heat stress (Nazareno et al. 2012). Outdoor sows showed lower risk of lameness than indoor sows (Knage-Rasmussen et al. 2014).

Success in sustainable farming depends on the appropriate selection of the animal to be managed on pasture (Hilimire 2011). Choosing the best animal for pasture pig systems could be a daunting task: Modern versus traditional breeds and purebred versus crossbred are few of the dilemmas. Modern breeds have been improved to obtain maximum performance, but traditional breeds show better acclimatization. Higher feed intake and fibre digestibility have been reported for traditional breeds and their crosses in comparison with modern breeds (Ogle 2006). In addition to their adaptation to local conditions, including climate and available feed resources, traditional breeds are characterized by their robustness and resistance to parasites and pathogens (FAO 2007; Kongsted et al. 2015). Furthermore, due to the eating qualities of their meat, traditional breeds have gained the appreciation of chefs, cured meat processors and end sale customers (Barone et al. 2015; Lebret et al. 2015; Pugliese and Sirtori 2012).

A distinctive trait of traditional breeds is their intramuscular fat content, higher than that found in modern pig genotypes (Lebret et al. 2014; Pugliese and Sirtori 2012). Differences in fatty acid deposition have been observed between modern and traditional breeds, which tend to deposit more saturated fatty acids (Edwards 2005). Crosses between modern and traditional breeds resulted in improved performance (average daily gain (ADG), feed conversion rate, prolificacy) (Barlocco 2005; Kelly et al. 2007; Sirtori et al. 2011). Duroc crosses are of particular interest as they show increased productivity without impacting meat quality while maintaining the level of intramuscular fat, water-holding capacity and improved tenderness (Edwards 2005; Sirtori et al. 2011; Pugliese and Sirtori 2012). Additionally, traditional genotypes have been found to be more adapted to grazing than modern genotypes, but no effect of animal genetics has been evidenced on rooting behaviour (Kongsted and Jakobsen 2015). In addition to market-valued performance traits, breeding programmes for pasture pigs should include welfare-related traits such as maternal aptitude, leg conformation, disease resistance, foraging abilities and reproductive attributes (Gourdine, De Greef and Rydhmer 2010).

Around the world there is a wide range of pasture-based pork systems. All production phases can be conducted on pasture, but in some northern European countries, Great Britain and Brazil, most pasture-based units are oriented to outdoor farrowing, and growing and finishing stages are conducted mainly indoor, sometimes with access to outdoor areas. In some Mediterranean countries, forest and woodland – mainly oaks (*Quercus* spp.) and chestnut (*Castanea sativa*) species – are important resources for traditional silvopastoral systems. That is the case for Iberian pigs grown in 'La Dehesa' of Spain; Alentejano pigs from 'El Montado' in Portugal; Corsica, Basque and Gascon breeds in France; Cinta Senese and Nero Siciliano in Italy; and Greek pigs in Greece.

3 Pasture pig systems and the environment

Managing pigs on pastures may contribute to soil fertility improvement (Andresen, Ciszuk and Ohlander 2001; Rodriguez-Estevez et al. 2009c; Ocampo-Duran 2011), soil conservation (Ocampo-Duran 2011), carbon sequestration (Dourmad and Casabianca 2013), and weed and pest control (Benjamins 2002; Nunn et al. 2007; Buehrer and Grieshop 2014), and increases biodiversity while producing high-quality protein for human consumption. Organic matter and nutrients from hog manure can be utilized by existing vegetation in pastures. Nonetheless, if the rate of deposition of the nutrients is higher than the rate of plant utilization, some of these nutrients could accumulate, with the risk of being transported out of the field either by runoff or by leaching (Webb et al. 2014).

As with other livestock species, grazing pigs can involve environmental impact risks. If the natural behaviour of pigs is not properly managed, pigs can damage the vegetation, disturb the soil surface, create trails in the paddocks and deposit a large amount of nutrients in selected areas. These behavioural manifestations could generate bare soil areas (Eriksen 2001; Eriksen et al. 2006a), nutrients hot spots (Williams et al. 2000; Eriksen and Kristensen 2001), heavy metal input (Menzi et al. 1998; Monteverde et al. 2011; Dourmad and Casabianca 2013) and soil compaction (Williams et al. 2000), and as a consequence nitrate leaching (Eriksen 2001; Hermanse, Eriksen and Strudsholm 2005), denitrification (Petersen, Kristensen and Eriksen 2001), ammonia volatilization (Sommer et al. 2001; Blumetto et al. 2012) and groundwater pollution (Dourmad and Casabianca 2013), thus impacting biodiversity and ecosystem stability. Additionally, pig manure could be a source of pathogens and veterinary medicine residues.

3.1 Impacts of pigs on soil and soil nutrients

The extent of soil damage and recovery is related to soil type (Vittoz and Hainard 2002). Similarly, it is necessary to consider soil type and mineralization caused by rooting (Menzi et al. 1998) and nutrient hot spots to estimate the potential of N leaching. The difference between the nutrients that enter a farming system and the nutrients leaving the system constitutes the nutrient balance, which is related to feed efficiency and animal mineral body composition. Increases in N (+130%) and P (+170%) excretion per kilogram of body weight gain have been registered for outdoor pigs when compared with pigs housed indoor in a conventional system (attributed to a supply of nutrients that surpass nutritional requirements and outdoor pigs' higher excretion rates) (Dourmad and Casabianca 2013). The N losses from farrowing to weaning in sow paddocks have been estimated as 8%, 13% and 16–35% for denitrification, ammonia volatilization and nitrate leaching, respectively (Eriksen, Petersen and Sommer 2002). The same authors reported that only 44% of N input was removed from the system as pig body component. According to Hermansen et al. (2005), the N use efficiency decreased as the pigs' length of stay in the pasture increased. Piglets kept on pasture for the whole cycle showed a 30% efficiency of N use. Similar figures were reported by Garcia-Valverde et al. (2007), who indicated for Iberian pigs fed with an herbage mixture that 26% of the N apparently digested was incorporated into the body.

Grazing with pigs did not influence N, or carbon bonded to humic acids or fulvic acids, but the P fraction increased in soils grazed by pigs compared with areas left ungrazed (Rivero et al. 2013). An increase in soil Zn in pastures managed during 12 years with pigs (6 sows/ha/yr) on pasture (mixture of alfalfa [*Medicago sativa*], red clover [*Trifolium pratense*], white clover [*Trifolium repens*] and chicory [*Cichorium intybus*]) has been reported, but the levels observed (0.88 mg Zn/kg) were only slightly higher than some crop requirements. The soil Zn content was directly related to the concentration of N-NO_3, P, K and cation-exchange capacity values (R^2 = 0.74, 0.90, 0.89, 0.85 respectively; $P < 0.001$), and was associated with feed wastage in feeding areas and animal excretory behaviour (Monteverde et al. 2011).

In continuous stocked systems, preferred areas for nutrients deposition amounted to 10% of the pasture area (Watson et al. 2003). Similarly, in stationary systems that included a barn, outdoor pen with concrete pad and a grass/clover grazing area, most of the nutrients (43%) were deposited in the wallow area (4% of the pen), with 14% of the nutrients excreted in the transfer and grazing area that represented 94% of the pen. In mobile systems with portable shelter, feeding and drinking equipment, 95% of the nutrients were deposited on 24% of the pen area corresponding with the location of the shelter, feeding and drinking equipment, while 5% of the nutrients were deposited on the grazing area that represented 76% of the pasture. In the stationary system, changes in pig excretory behaviour could have been encouraged by the use of a long and narrow outdoor area, generating a high concentration of nutrients (Salomon et al. 2007). Additionally, higher concentrations of soil nutrients have been correlated with distance from the service area (Eriksen and Kristensen 2001; Monteverde and Del Pino 2014), as well as with selected areas for excretory behaviour identified along fences (Watson et al. 2003; Bordeaux et al. 2014). Compacted soil around feeders and drinkers shows higher risk of denitrification, whereas the areas around the shelters have a higher leaching potential (Quintern 2005).

Important considerations should be given to topography and slope when deciding on the location of an operation to minimize soil erosion, runoff and soil properties deterioration (Horta et al. 2012, Bondi et al. 2015).

3.2 Impacts of pigs on ground cover

Maintenance of ground cover in pasture pig systems is crucial to trapping and recycling soil nutrients by the forages (Rachuonyo and McGlone 2005), thus preventing them from leaving the system through runoff and leaching (Eriksen et al. 2006a). Higher ground cover damage as a consequence of rooting and trampling behaviour can be observed in the service area and along fences (Salomon et al. 2005a; Quintern and Sundrum 2006; Acciaioli et al. 2012). Frequently, 'trails' connecting different functional areas of the paddocks are formed and have the potential to be converted into water channels contributing to runoff and erosion. The first effect of grazing pigs (stocking rate of 200 m^2/pig) in a Mediterranean open forest was compaction along the enclosure fence, with increases in rooting and excavation activities on the entire surface, impacting soil porosity and structure (Pistoia et al. 2012).

3.3 Estimating the environmental impact of pasture pig systems

The significant role of feeding strategies, diets and feed efficiency on environmental impact of pasture pig systems has been stressed on life cycle assessment studies, which estimate and evaluate environmental impact through the resources employed in the production process and the discharge to the environment. The most pronounced environmental effect was observed during the finishing phase (Lammers 2011; Espagnol and Demartini 2014; Rudolph 2015). Additionally, the negative environmental impact of the excessive supply of certain nutrients in diets for outdoor pigs has been reported (Dourmad and Casabianca 2013). These concerns emphasize the importance of preventing feed losses as a way to reduce soil nutrient loads (Quintern 2005) and to improve feed efficiency (Dourmad and Casabianca 2013). The potential environmental impact of outdoor organic pig production systems has been established in terms of greenhouse gas emissions (2.2 kg CO_2-eq/kg live weight), acidification (55.4 g SO_2-eq/kg live weight) and eutrophication potential (28.7 PO_4-eq/kg live weight) (Rudolph 2015).

The use of feed resources available on the farm and their incorporation into the life cycle analysis could reduce the estimation of environmental impact of extensive pig production systems (Espagnol and Demartini 2014). A wide variation on environmental impact indicators has been observed among farms, underlining the importance of conducting individual on-farm assessment (Dolman, Vrolijk and De Boer 2012). There is also the need to improve the methodologies employed to evaluate the environmental impact of pasture pig systems by incorporating the potential carbon sequestration of the vegetative ground cover in the analysis, and its impact on climate change (Dourmad and Casabianca 2013).

4 Forages for pasture pig systems

Forages are considered a vegetative ground cover, a physical barrier that protects the soil against rain and water impact, minimizing erosion, improving water infiltration, helping to trap nutrients from animal manure, reducing runoff and nutrient leaching, or as a source of nutrients in pig diets. A healthy vegetative ground cover contributes to the improvement of the animal welfare and well-being by reducing heat impact and muddy conditions (Larsen and Kongsted 1999).

4.1 Nutritional value

Forages have the potential to meet some of the nutritional needs of pigs, and can be incorporated fresh, dehydrated or ensiled (Ogle 2006; Bikker and Binnendijk 2012; Roinsard, Maupertuis and Dubois 2014). Their cost of production is often relatively low (Kanga et al. 2012), providing amino acids – mainly lysine (Garcia-Valverde et al. 2007) –, minerals (Garcia-Valverde et al. 2007) and vitamins (Edwards 2003; Martens et al. 2012). In addition, forages increase gut fill contributing to a satiety sensation in sows fed restricted diets (Larsen and Kongsted 1999; Lagreca and Marotta 2005), and enhance intestinal health (Ogle 2006; Kambassi et al. 2014).

According to Bauza and Petrocelli (2005), the main contribution of pastures to pigs is their content of protein of good nutritional value that could complement the lower lysine, tryptophan and methionine content of grains. These authors suggest the possibility of reducing the amount of protein supplement in diets for growing and finishing pigs by 25 to 50%. The addition of a mineral premix to diets of restricted-fed pigs proved unnecessary when pigs had access to pastures (Kongsted et al. 2015). Conversely, a high level of inclusion of forages in the diet could affect pig performance negatively (Régnier et al. 2013) due to high fibre levels, reducing digestibility and nutritional energy value, and the potential presence of anti-nutritional compounds (Ogle 2006; Kambashi et al. 2014).

The inclusion of fibrous materials in diets for pigs without impacting animal performance is of interest to reduce feeding costs. Difficulties that need to be overcome include satiety/gut fill, gut motility, digestion and absorption of nutrients, changes in gut microflora, anti-nutritional effects of fibre and how to minimize the effects of anti-nutritional substances (Ogle 2006; Martens et al. 2012; Patridge 2014). Adult swine, with a developed gastrointestinal tract, show a better capability to utilize fibrous diets compared to younger pigs (Ru and Bao 2004; Sehested et al. 2004; Jørgensen et al. 2007). Morphological modifications of pig gastrointestinal tract and changes in enzyme function and gut microbial diversity have been reported as consequences of diet variations (Ogle 2006; Urriola, Cervantes-Pahm and Stein 2013). Pigs raised indoor need time to get used to the pasture environment and could suffer an adaptation stress when moved outdoors (Pistoia et al. 2008). Providing a period of adaptation to forages in the diet will improve their utilization by pigs (Bauza and Petrocelli 2005; Passos et al. 2015). According to Castillo et al. (2007), gut microbiota could require an adaptation period up to six weeks to adapt to diet changes.

In pigs, foraging ability is a developed trait. While on pasture, they select which plants and plant parts they will graze (Gustafson and Stern 2003; Sehested et al. 2004; Rachuonyo, Allen and McGlone 2005). This behaviour allows pigs to balance consumption to fulfil their nutritional requirements. When pigs were introduced to a faba bean (*Vicia faba minor*) field, Giannone (2001a) observed how they selectively avoided green leaves and immature pods and focused on grazing dry plant parts. At the end of the grazing period, most dry leaves and mature pods had been harvested and no seeds were left on the ground. No rooting damage was observed. In a mixed field with grain sorghum (*Sorghum bicolor*) and sugar beet (*Beta vulgaris* L.), Giannone (2001b) noticed that pigs alternatively and almost systematically grazed the sorghum panicles and the beet leaves. Sorghum green leaves were avoided, and the beetroots were seldom consumed. Rachuonyo, Allen and McGlone (2005) observed a predilection for legumes such as alfalfa *and* white clover over grasses such as fescue (*Festuca arundinacea*) and buffalo grass (*Buchloe dactyloides*) and, among the legumes, white clover was preferred over alfalfa.

Similarly, European wild boars showed a preference for plantain (*Plantago lanceolata*) over perennial ryegrass (*Lolium perenne*) (Hodgkinson, Carcamo and Lopez 2011). Pigs go as far as selectively chewing the fibrous component of plants, absorbing the juice and discarding the fibrous parts (Andresen, Ciszuk and Ohlander 2001; Stern and Andresen 2003; Faner 2007), or selecting the heavier acorns from the ground, and removing and rejecting the nutshells as reported for Iberico pigs (Rodriguez-Estevez et al. 2009a,b). When freshly cut forage was offered to Iberian pigs, Garcia-Valverde et al. (2007) observed a low voluntary intake (12.0–15.3g DM/kg$^{0.75}$/d). The authors concluded that the selective behaviour displayed by grazing is important to attain higher levels of intake.

A wide range of forage species could be included in pasture-based operations. The criteria to select them should include nutritional quality, adaptation to soil and climate, and sufficient resiliency to resist animal grazing, trampling and rooting behaviours. Plant species, stage of maturity and season affect the content of sugars, protein and fibre, and define forages' nutritional value. The options of species to be used include natural or introduced plant species, such as grasses, legumes, forbs, herbs and root crops. Tables 1, 2 and 3 present forages that have been used or that present potential for inclusion in pasture pig systems. Pastures, orchards and woodlots are the environments that most resemble pigs' natural habitats.

The development of low-input multispecies pastures for pasture pig systems highlights the benefits of functional biodiversity. Complementarities of plant species of different phenology included in the mixture generate more efficient utilization of available nutrients, longer growing season (Bordeaux et al. 2014), reduction of seasonal variability, higher biomass production, better forage quality, reduction of the presence of weeds (Robinson 2013), diminution of nutrient losses to the environment (Robinson 2013), and improved soil quality and carbon sequestration potential. Many benefits have been reported when diversifying pastures. For example, a pasture mix of grasses, legumes and herbs will bring to the system the higher nutritional value of the legume and herb species, and the yield and grazing resistance of the grasses. Additionally, some herbs such as chicory, plantain and sulla (*Hedysarum coronarium*) have been shown to have high mineral contents (P, K, Cl, Na, Cu, Co; Goh and Bruce 2005), and anthelmintic properties for mammals (Ramirez-Restrepo and Barry 2005).

Perennial pastures have the advantage of reducing environmental and economic costs of establishment practices and the possibility to maintain higher stocking rates.

Table 1 Leguminous species with potential use in pasture pig systems

Crop	Life cycle	Morphology	Propagation method	Growing point primary location	Target stubble	Authors who have included this species in pasture pig systems
Arachis pintoi	P	P, E	S, St	S	3	Both (2003)
Glycine max	A	E	S	E	5–10	
Lespedeza cuneata	A	S	S	E	5–10	
Lotus corniculatus	P	E	S	S	8–10	Kongsted et al. (2015)
Medicago sativa	P	E	S	E	8–10	Riart (2002), Gentry et al. (2002b), (2004); Rachuonyo et al. (2005), Garcia-Valverde et al. (2007), Fortina et al. (2011), Helmerich (2014), Jakobsen et al. (2014), Kongsted et al. (2015)
Melilotus officinalis	P	E	S	E	5–10	Kongsted et al. (2015)
Pisum sativum	A	E	S	S, E	15–20	Nilzen et al. (2001), Fortina et al. (2011)
Pueraria montana	P	P	S, R	S	4–8	
Securigera varia	P	P	S, R	S	4–8	
Trifolium alexandrinum	A	E	S	S	3–6	Bochicchio et al. (2012)
Trifolium campestre	A	S	S	E	5–10	
Trifolium hybridum	A	E	S	S	8	
Trifolium incarnatum	A	E	S	E	5–10	
Trifolium pratense	B	E	S	S	5–10	Andresen and Rebbo (1999), Stern and Andresen (2003), Blumetto et al. (2012 and 2013)

Trifolium repens	P	P	S, St	S	5–10	Stern and Andresen (2003), Sehested et al. (2004), Rachuonyo et al. (2005), Leite et al. (2006a,b), Blumetto et al. (2012, 2013), Rivero et al. (2013b), Helmerich (2014), Kongsted et al. (2015)
Trifolium subterraneum	A	P	S, St	S	2.5–7.6	Garcia-Valverde et al. (2007)
Trifolium vesiculosum	A	E	S	S	5–10	
Vicia faba	A	E	S	S	5	Bochicchio et al. (2012)
Vicia sativa	A	S	S	S	5–10	Bochicchio et al. (2012)
Vicia villosa	A	S	S	S	5–10	
Vigna unguiculata	A	E, P, S	S	E	15	Gilmore (1999), Sarria et al. (2010)

Life cycle: A: Annual, B: Biannual, P: Perennial
Morphology: E: Erect, P: Prostrate, S: Semi erect
Propagation method: R: Rhizome, S: Seed, St: Stolon
Growing point primary location: S: Stem base, E: Elevates
Target stubble: Height to stop grazing, cm

Table 2 Grass species with potential use in pasture pig systems

Crop	Life cycle	Morphology	Propagation method	Growing point primary location	Target stubble	Authors who have included this species in pasture pig systems
Agrostis gigantea	P	E	S, R	S	3–8	
Andropogon gerardii	P	E	S	S	8–13	
Astrebla lappacea	P	E	S	S	5–10	
Avena sativa	A	E	S	S	5–10	Nilzen et al. (2001), Hodgkinson et al. (2008)
Axonopus scoparius	P	S, P	S	S	3–8	
Bothriochloa ischaemum	P	E	S	S	5–8	Rachuonyo et al. (2001)
Bothriochloa pertusa	P	E	S, St	S	5 8	
Brachiaria humidicola	A	E	S	S	15	
Brachiaria mutica	P	P	S, St	S	>20	
Bromus catharticus	P	E	S	S	5–10	Riart (2002)
Bromus inermis	P	E	S	S	10	
Bromus valdivianus	P	P	S	S	5–10	Hodgkinson et al. (2008)
Buchloe dactyloides	P	S	S, St	S	15	Rachuonyo et al. (2005)
Chloris gayana	P	P	S, St	S	10	
Cynodon dactylon	P	E, S, P	S, R, St	S	3–8	Both (2003), Pietrosemoli and Green (2009); Pietrosemoli et al. (2012), Passos et al. (2015)
Cynodon nlemfuensis	P	P	St	S	15–25	Santos Ricalde and Lean (2002, 2006), Nazareno et al. (2012)
Dactylis glomerata	P	E	S	S	5–10	Garcia-Valverde et al. (2007)
Digitaria eriantha	P	P	S, St	S	15–25	
Digitaria sanguinalis	A	S	S, St	S	5–10	
Festuca pratensis	P	E	S	S	3–5	Andresen and Rebbo (1999), Andresen et al. (2001), Stern and Andresen (2003)
Festuca rubra	P	E	S	S	5	Kongsted et al. (2015)
Hordeum vulgare	A	E	S	S	5–10	Nilzen et al. (2001), Quintern (2005)
Lolium multiflorum	A	E	S	S	5–10	Quintern (2005), Pietrosemoli et al. (2012), Bordeaux et al. (2014)

Lolium perenne	P	E	S	S	5–10	Sehested et al. (2004), Horsted et al. (2012), Blumetto et al. (2012,2013), Rivero et al. (2013b), Helmerich (2014), Jakobsen et al. (2014), Kongsted et al. (2015)
Panicum virgatum	P	E	S	S	8–13	Pietrosemoli et al. (2012)
Paspalum dilatatum	P	E	S	S	5–10	
Paspalum notatum	P	E	S, St	S	3–8	Both (2003)
Pennisetum clandestinum	P	P	S, St, R	S	>5	Kanga et al. (2012)
Pennisetum glaucum	A	E	S	S	15–20	
Pennisetum orientale	A	E	S	S	8–15	
Phalaris aquatica	P	E	S	S	8–10	
Phalaris arundinacea	P	E	S, R	S	5–10	
Phleum pratense	P	E	S, Corm	S	5–10	Andresen and Rebbo (1999), Stern and Andresen (2003)
Poa pratensis	P	E	S, R	S	5–8	
Schedonorus arundinaceus	P	E, S	S	S	5–10	Riart (2002), Rachuonyo et al. (2005), Garcia-Valverde et al. (2007), Pietrosemoli et al. (2012)
Secale cereale	A	E	S	S	5–10	Pietrosemoli et al. (2012), Bordeaux et al. (2014)
Setaria sphacelata	P	E	S	S	>5	
Sorghastrum nutans	P	E	S	S	8–13	
Sorghum × drummondii	A	E	S	S	8–15	Pietrosemoli et al. (2012), Bordeaux et al. (2014)
Sorghum bicolor	A	E	S	S	15–50	Passos et al. (2015)
Sorghum bicolor × S. bicolor var. sudanese	A	E	S	S	8–15	
Sorghum halepense	P	E	S, R	S	8–13	
Tripsacum dactyloides	P	E	S	S	8–13	
Triticum × Secale	A	E	S	E	5–10	

Life cycle: A: Annual, B: Biannual, P: Perennial
Morphology: E: Erect, P: Prostrate, S: Semi erect
Propagation method: R: Rhizome, S: Seed, St: Stolon
Growing point primary location: S: Stem base, E: Elevates
Target stubble: Height to stop grazing, cm

Table 3 Herb and forb species with potential use in pasture pig systems

Crop	Life cycle	Morphology	Propagation method	Growing point primary location	Target stubble	Authors who have included this species in pasture pig systems
Beta vulgaris	A	E	S	C	5–8	Giannone (2002) and Fortina et al. (2011)
Borago officinalis	A	E	S	S	7	Kongsted et al. (2015)
Brassica campestris	A	E	S	C	5–10	Quintern and Sundrum (2006)
Brassica napus	A	E	S	S	25	Livingstone et al. (1977), Gilmore (1999), Quintern (2005), Edge et al. (2005)
Brassica oleracea	A	E	S	C	10–15	Livingstone et al. (1980)
Carum carvi	P		S	S	3–7	Kongsted et al. (2015)
Cichorium intybus	P	E	S	S	8–10	Bauza (2007), Kongsted et al. (2015)
Woodland *Quercus* spp, *Castanea sativa*	A, P	E	S	E	–	Pugliese et al. (2005), Rodrigues Esteves et al. (2009, 2010), Lebret et al. (2014)
Helianthus tuberosus	P	E	S	S	15	Quintern (2005), Quintern and Sundrum (2006), Kays and Nottingham (2008), Pousset (2010), Wünsch et al. (2011), Kongsted et al. (2013)
Hypochaeris radicata	P	P	S, R	S	10–13	Quijada et al. (2012), Hodgkinson et al. (2008)
Plantago lanceolata	P	E	S	S	8–10	Hodgkinson et al. (2008), Kongsted et al. (2015)
Salix spp, *Miscanthus giganteus*	P	E	S	S	5–10	Horsted et al. (2012)
Sanguisorba minor	P	E	S	S	7	Kongsted et al. (2015)

Life cycle: A: Annual, B: Biannual, P: Perennial
Morphology: E: Erect, P: Prostrate, S: Semi erect
Propagation method: R: Rhizome, S: Seed, St: Stolon
Growing point primary location: S: Stem base, E: Elevates
Target stubble: Height to stop grazing, cm

4.2 Pasture intake

Determining supplemental feed and forage intake in pasture-based systems is complicated. The intake is affected by feeding level, fibre content of the diet, pasture botanical composition, vegetative stage of forages, forage availability, quality and palatability, animal class, season, length of occupation in the paddock, and the quantity and quality of supplemental feed, the latter an aspect more important for growing animals than for sows (Stern and Andresen 2003; Rachuonyo, Allen and McGlone 2005; Fortina et al. 2011).

Estimating forage intake by pigs varies widely, as shown in Tables 4 and 5 that present daily intake registered for pigs and sows on pasture pig systems, respectively. Accordingly, Rodriguez-Estevez et al. (2009c) observed that the grass intake per bite was equivalent to 1.4 ± 0.1 g (fresh) or 0.26 ± 0.02 g (DM). A seasonal intake pattern was observed, with higher intake during spring and fall versus summer (Rivera-Ferre et al. 2001; Edwards 2003; Hodgkinson, Lopez and Navarrete 2009). An average soil intake of 0.3 kg of soil/sow/d (0.28 to 0.41 kg soil/sow/d for summer and fall, respectively) has been estimated for grazing lactating sows (Jurjanz and Roinsard 2014).

Low forage DM intake (less than 2% of DM intake) was observed in growing pigs that had *ad libitum* access to a supplemental-cereal-based feed. Pasture pigs had 11.3% higher DM intake than similar pigs indoors, and this was reflected in lower feed efficiency (Kelly et al. 2007). Conversely, no differences in feed conversion ratio were observed between pastured and indoor pigs (Presto et al. 2007).

Inclusion of pastures in growing to finishing pig diets has allowed reduction of the supplemental feed total daily intake in the range of 10–30% (Both 2003; Bauza 2007; Faner 2007). Conversely, lengthier growing periods have been observed when supplemental feed is substituted by pastures (Bauza 2007; Fortina et al. 2011).

4.3 Pig utilization of nutrients

Growing pigs can make a limited use of nutrients from forages and, according to Carlson et al. (1999) pasture contribution could represent 10% of their DM intake. The degree of utilization would be related to factors associated with the forage itself (species, stage of maturity, management), with the animal (age, breed, physiological stage), and even with the season.

Pastures have the potential to provide 20 and 16% of the energy and 25 and 28% of the crude protein requirements of growers and finishers, respectively (Bauza and Petrocelli 2005). Fortina et al. (2011) indicated that a mixed pasture composed of forage pea (*Pisum sativum*), white clover, sugar beet and alfalfa could fulfil 50% of the nutritional requirement of crossbred finishing pigs (90–140 kg BW), whereas Kanga et al. (2012) were able to replace only 20% of the supplemental feed when kikuyu (*Pennisetum clandestinum*) grass was fed to growing pigs (27 ± 3.8 kg BW). According to Hodgkinson et al. (2009), pastures (*Lolium perenne* and *Plantago lanceolata*) could contribute up to 52% of the digestible energy requirements for European wild boars.

As reported by Rivera-Ferre et al. (2001) and Edwards (2003), good-quality pasture could supply 30–45% of energy and 45–50% of amino acid requirements for gestating sows. Gestating sows grazing star grass and supplemented with diets formulated to provide 19, 26 and 33 MJ of DE/day showed increments in weight and back fat thickness with increasing feeding level, but reduced grazing activity, time spent grazing and distance walked (Santos-Ricalde and Lean 2006). However, no effect was observed in litter

Table 4 Daily intake (kg/head) of growing and finishing pigs on pasture

Stage of production	Pasture	Intake kg/head		Reference
		Forage	Supplemental feed	
27–100 kg	Mixture: *Lolium* spp. *Trifolium repens*	0.15 kg DM	*Ad libitum*	Danielsen et al. (1999)
50–60 kg	Mixture: *Lolium* spp. *Trifolium repens*	0.5 kg fresh (0.02–0.15 kg/kg OM)	*Ad libitum*	Mowat et al. (2001)
30–70 kg	Mixture: *Medicago sativa* *Festuca arundinacea*	0.15 kg DM (spring) 0.04 kg DM (summer)	*Ad libitum*, limited in CP	Riart (2002)
70–100 kg	*Bromus catharticus*	0.51 kg DM (spring) 0.32 kg DM (summer)		
30–80 kg	Mixture: *Lolium multiflorum*, *Cichorium intybus*	0.37–0.385 kg (0.17–0.21 kg/kg DM)	Restricted 20% once a day	Bauza (2007)
80–120 kg	*Trifolium pratense*	0.70–0.80 kg DM (0.25–0.30 kg/kg DM)		
112–160 kg	Mixture: *Quercus rotundifolia, Cistus monspeliensis, C. salvifolius, Retama sphaerocarpa, Lavandula stoechas, Thymus zygis, Agrostis* sp., *Poa bulbosa, Trifolium subterraneum, T. arvense, T. cherleri, T. stellatum, Ornithopus* sp., *Diplotaxis* sp., *Echium plantagineum, Erodium* sp., *Chamaemelum mixtum, Senecio vulgaris* and *Diplotaxis* sp.	7.1–8.4 kg fresh acorn 2.0–2.7 kg fresh grass (0.0958 kg/kg ME)	No supplemental feed	Rodriguez Estevez et al. (2009)

90–141 kg	*Pisum sativum* *Trifolium* spp. *Beta vulgaris* *Medicago sativa*	1.7 kg DM 1.2 kg DM 1.4 kg DM 0.5 kg DM	To supply 50% of daily energy requirements	Fortina et al. (2011)
14–23 kg European wild boar	Mixture: *Lolium perenne* *Trifolium repens*	0.106 kg DM (0.07 kg/kg ME) 0.140 kg DM (0.10 kg/kg ME) 0.229 kg DM (0.16 kg/kg ME)	0.829 kg DM 0.832 Kg DM 0.886 kg DM	Rivero et al. (2013b)
58–90 kg	Mixture: *Medicago sativa* *Lolium perenne*	**High CP (20.5%) fed pigs:** 0.330 kg DM *alfalfa*/pig/d **Low CP (10.7%) fed pigs:** 0.470 kg DM alfalfa/pig/d	2.2 kg DM	Jakobsen et al. (2015)

Table 5 Daily intake (kg/head) of sows on pasture

Stage of production	Pasture mixture	Intake kg/head Forage	Supplemental feed	Reference
Pregnant sows	*Lolium perenne* *Trifolium repens*	3.4 to 8.2 kg fresh (spring) 6.5 to 8.1 kg fresh (summer)	1.5 kg or 3.0 kg	Rivera-Ferre et al. (2001)
Pregnant sows	*Lolium* spp. *Trifolium repens*	1.1 to 10.5 kg fresh (spring) 4.3 to 11.8 kg fresh (summer)	Commercial levels, amount not reported	Edwards (2003)
Lactating sows	*Lolium* spp. *Trifolium repens*	0.2 to 1.6 kg DM	*Ad libitum*	Jurjanz and Roinsard (2014)
Pregnant sows	*Cynodon nlemfuensis*	0.231 kg DM 0.180 kg DM 0.136 kg DM	19 MJ DE 26 MJ DE 33 MJ DE	Santos Ricalde and Lean (2006)

performance (pigs/litter, birth and 21 d weight, and mortality from farrowing to weaning) or in milk composition. Although star grass apparently did not contribute significantly to nutrient energy supply, an intermediate feeding level (26 MJ DE/d) was enough to generate weight gain. The lower feeding regime caused mobilization of body fat reserves and decreases in back fat thickness (Santos-Ricalde and Lean 2002).

Pigs could obtain energy from the nutrients contained in forages such as bermudagrass (*Cynodon dactylon*), and forage sorghum and sweet sorghum (*Sorghum bicolor*), but the utilization of the N contained in the grasses was poor (Passos et al. 2015). Similarly, Jerusalem artichoke (*Helianthus tuberosus*) represented an appropriate energy source for restricted-fed pigs on pasture, ensuring about 60% of the pigs' energy requirements. Restricted-fed pigs showed lower weight gain (–54.25%) and better feed conversion ratio (+59%) and spent more time grazing (Kongsted, Horsted and Hermansen 2013).

Older animals, especially gestating sows, can better utilize forages. With an average intake of 1.5 kg DM/sow/d, pastures could substitute 50–70% of the supplemental feed during gestation and about 25% during lactation (Bauza 2007).

4.4 Digestibility

Kanga et al. (2012) reported lower digestibility coefficients for crude protein (0.77 vs 0.72) and crude fibre (0.75 vs 0.54) for restricted-fed growing pigs (20% of concentrate voluntary intake) managed on Kikuyo grass pastures than for pigs managed indoor and that had *ad libitum* access to fresh cut Kikuyo. Similarly, the inclusion of 250 g of dried star grass in diet for sows impaired the digestibility of the diet, affecting the coefficients of total tract apparent DM digestibility (0.91 vs 0.76), organic matter (0.92 vs 0.76), crude protein (0.85 vs 0.77), neutral detergent fibre (0.78 vs 0.38), acid detergent fibre (0.70 vs 0.27), gross energy (0.91 vs 0.77) and the amount of N excreted (9.5 g/d vs 6.5 g/d) (Santos-Ricalde and Lean 2006).

Garcia-Valverde et al. (2007) reported total tract digestibility of DM of 0.65 and OM of 0.64 for a mixed pasture (*Medicago sativa*, *Medicago scutellata*, *Medicago polymorpha*, *Trifolium subterraneum*, *Trifolium resupinatum*, *Biserrula pelecinus*,

Dactylis glomerata, Festuca arundinacea, Fumaria officinalis and Quenopodium albus), and Rivera-Ferre et al. (2001) observed a range of herbage OM digestibility of 0.47 to 0.79 for a ryegrass/white clover pasture. The former authors estimated that Dehesa pasture can contribute up to 16% of the ingested metabolizable energy (ME) of Iberian pigs during the 'Montanera'.

Differences in fibre digestibility (Hodgkinson, Schmidt and Ulloa 2008) between the domestic pig (*S. scrofa domesticus*, Landrace × Large White) and the European wild boar (*Sus scrofa* L.) were found with the latter having a lower efficiency of fibre digestion. Energy digestibility values for feedstuff with low fibre content such as corn (*Zea mays*) and oat (*Avena sativa*) determined by domestic pigs could be used in diet formulation for the European wild boar, but with feed with higher fibre content such as in alfalfa, the use of these values would be of limited contribution. No differences were observed in protein digestibility between domestic pigs and European wild boar (Hodgkinson, Schmidt and Ulloa 2008). According to Quijada, Bitsch and Hodgkinson (2012), a mixed pasture composed of perennial ryegrass, roadside brome (*Bromus valdivianus* Phil), plantain and hairy cat's ear (*Hypochaeris radicata*) did satisfy between 142% (spring) and 52% (summer) of the daily digestible energy requirements for maintenance of growing European wild boars (18 to 25 kg LW) raised in a semi-extensive system.

4.5 Intake management

Some strategies could be implemented to encourage pasture consumption, such as a reduction in supplemental feed allowance (Andresen, Ciszuk and Ohlander 2001; Both 2003; Kanga et al. 2012), or the frequent provision of new grazing areas (Andresen and Redbo 1999; Stern and Andresen 2003; Gustafson and Stern 2003). The implementation of restricted feeding strategies proved to be effective and have resulted in increased forage intake and improvement of feed efficiency up to 9–15% (Both 2003; Kongsted et al. 2015). Danielsen et al. (1999) reported an increase in grass intake of 32% when a 30% restriction of the concentrate feed was implemented. These authors observed, however, a decrease in growth performance (−11% ADG). The effects of supplemental feed restriction on pasture pigs are presented in Table 6.

Bauza and Petrocelli (2005) did not observe an effect on growth rate when feed restriction up to 20% of the voluntary intake was imposed on pigs on pastures of alfalfa, perennial ryegrass, canarygrass (*Phalaris aquatica*) and white clover. But with higher levels of restriction, the decline in growth rate was almost proportional to the intensity of the restriction. The tendency can be observed in the equation $Y = -0.7364X + 105.36$ ($R^2 = 0.6123$ where Y is the growth rate expressed as a per cent of control animals growth rate and X is the per cent of supplemental feed restriction). Additionally, these authors mentioned that the restriction had more impact on finishers than growers as a consequence of the former higher energy requirements. Furthermore, to avoid a diminution in growth rate, they recommend the use of supplemental feeds with higher energy concentrations than recommended in nutritional tables, when implementing feed restriction over 25%.

The physical capacity of the gastrointestinal tract of pigs can represent an intake-limiting factor when forage is used to replace supplemental feed.

To minimize effects on performance, Bauza (2007) advised implementing supplemental feed restriction no higher than 20% for growing and finishing pigs on pasture. The same author had previously suggested the unsuitability of implementing 30% feed restriction to replacement gilts (Bauza 2005).

Table 6 Effects of supplemental feed restriction on animal performance

Stage of production	Forage/mixture *Ad libitum*	Supplemental feed restriction % intake	Animal performance relative to no feed restriction	Reference
27–100 kg	*Lolium* spp.	30%	+32% forage intake	Danielsen et al. (1999)
	Trifolium repens		−11% ADG	
50–85 kg	*Phleum pratense*	20%	+ 5% forage intake	
	Festuca pratensis		−15% ADG	
	Trifolium pratense		No differences in feed conversion rate	Stern and Andersen (2003)
	Trifolium repens		Higher grazing frequency	
			Higher rooting activity	
73–100 kg	*Arachis pintoi*	0%	45% restriction had effect on:	
	Cynodon dactylon	25%	−26% weight gain	
	Paspalum notatum	35%	−30% feed intake	Both (2003)
		45%	7% improvement feed efficiency	
			Longer time (7d) to reach market weight	
30–50 kg 50–90 kg	*Trifolium pratense*	Up to 20%	No effect on growth rate	
	Trifolium repens			Bauza and Petrocelli (2005)
	Lolium perenne			
27 kg	*Pennisetum clandestinum*	20%	Lower digestibility of CP, CF, NDF and ADF compared with pigs receiving cut fresh Kikuyo	Kanga et al. (2012)

34–105 kg	*Trifolium repens* *Lolium perenne* *Festuca rubra* *Plantago lanceolata* *Medicago sativa* *Lotus corniculatus* *Carum carvi* *Cichorium intybus* *Borago officinalis* *Melilotus officinalis* *Sanguisorba minor*	100% Danish feeding recommendation + 15% Restricted 20% wk 1–4 and 50% wk 5–10 + mineral and vitamin mix Restricted 20% wk 1-4 and 50% wk 5–10 without mineral and vitamin mix 19% CP, 8.9 MJ NE/kg once a day	−23 to −25% ADG Improved supplemental feed conversion ratio (−15 and −9%)	Kongsted et al. (2015)
58–90 kg	*Medicago sativa* *Lolium perenne*	High CP feed CP 20.5 % Low CP feed 48% reduction in CP, CP 10.7 %	**For pigs in alfalfa:** −18% ADG −14% Feed conversion rate +42 % Intake for low CP fed pig **For pigs in ryegrass:** −33% ADG −31% Feed conversion rate	Jakobsen et al. (2015)

5 Animal performance in pasture pig systems

Pastured pigs can exhibit performance parameters comparable to those obtained by pigs reared indoor (Andresen, Ciszuk and Ohlander 2001; Gentry et al. 2004; Fortina et al. 2011), but careful comparisons must be conducted because differences in climatic conditions, forage species, animal genetics, feeding regime, diet composition and farming practices can influence performance results. Climate variations, higher activity and feed wastage have been indicated as potential causes of lower feed conversion rates reported for outdoor pigs (Stern and Andresen 2003; Kelly et al. 2007). A season influence on pig performance was also observed. During summer, pasture pigs showed a 10.9% higher ADG, whereas no differences in growth rate of pigs on pasture versus indoor pigs were established in winter (Gentry et al. 2002a). In a meta-analysis study, the performance of pigs reared outdoor was compared with those reared indoor (Demori et al. 2012). Higher feed intake, and lower ADG and feed efficiency were reported for outdoor pigs. The performance of pigs managed in pasture compared with indoor pigs is presented in Table 7.

Iberian free-range pigs showed ADG of 0.76 +/−0.01 kg/d during a 2-month 'Montanera' finishing phase. The age and sex of the animals had a significant influence on ADG (Rodriguez-Estevez et al. 2011). For non-supplemented Iberian free-range pigs in the finishing phase and with a diet based on acorn and grass (4.0 ± 0.29 kg acorn and 0.7 ± 0.08 kg of grass, DM basis), Rodriguez-Estevez et al. (2010) reported a gain to feed conversion rate of 253 ± 13.1 g/kg DM, while the value for the ADG to ME ratio was equivalent to 15.3 ± 0.79 g/MJ ME.

Table 7 Performance of pigs on pasture- versus indoor-kept pigs

Stage of production	Pasture	Effect on performance compared with indoor pigs	Reference
9–114 kg	*Medicago sativa*	+ 6.4% final live weight + 15.3% feed intake − 9.8 % gain to feed ratio	Gentry et al. (2002b)
50–128 kg	Open woodland pastures *Quercus* spp., *Castanea sativa*	Lower growth rate + 200 d slaughter age	Pugliese et al. (2005)
19–97 kg	Grass	+ 2.56% age at slaughter + 85.7% forage intake No differences in daily gain − 12.5% feed efficiency	Hermansen et al. (2005)
Growing pigs	Meta-analysis, Several studies including many forage species	+ 9% feed intake − 2% ADG − 3 % feed efficiency	Demori et al. (2012)
14–21 kg European wild boar	Mixture: *Trifolium repens* *Lolium perenne*	+ 26.3% ADG + 30.3 % feed efficiency	Rivero et al. (2013b)
35–145 kg	Grass, acorn and chestnut	Lower ADG	Lebret et al. (2014)

Outdoor sow farms managed either conventionally or organically were compared by Larsen and Kongsted (1999). Records showed average stocking rates equivalent to 15.7 sows/ha, 19.1 piglets/sow/yr and piglet mortality rate of 24.3%. Evaluating the performance of sows (Landrace x Yorkshire) managed indoor or on pasture during lactation, Fajardo (2009) did not report any statistical difference for litter size, piglets born alive, litter weight at birth, litter weight at weaning, number of stillborn piglets, piglets weight gain, sows body condition score or weaning–oestrus interval.

6 Carcass and meat quality of pig on pasture

Pigs on pasture are exposed to climatic elements, can wander around the paddock for exercise (Isabel and López-Bote 2000) and have the opportunity to consume forages as part of their diets. These factors could elicit changes in the chemical, physical and sensorial characteristics of the meat (Pugliese and Sirtori 2012); adipose tissue (Pugliese et al. 2005; Lebret et al. 2014; Lebret et al. 2015); fat deposition (Lebret et al. 2014; Lebret et al. 2015); fatty acids composition (Bochicchio et al. 2012; Blumetto Velazco et al. 2013; Lebret et al. 2015); and muscle metabolism and glycolytic capacity (Fortina et al. 2011), thus influencing carcass composition and pork quality (Simantke and Sundrum 2001). Carcass and meat quality have been observed to be influenced by season. During the summer, outdoor pigs showed greater back fat thickness, larger loin eye area and darker meat, whereas during the winter no differences were observed in growth rate, carcass traits, pork quality sensory scores or shear force values among outdoor or indoor-kept pigs (Gentry et al. 2002a). Carcass and meat quality traits of pigs managed on pastures are presented in Table 8.

The potentiality of improving carcass quality through the inclusion of forages is of particular interest for pigs from heritage breeds which tend to have fatter carcasses compared with pigs from modern genotypes. An interaction between the environment and the genetic of the animals has been related to improved eating quality (Bonneau and Lebret 2010; Lebret et al. 2014; Lebret et al. 2015). Improvements in the dietetic properties of the fat without impacting the technological properties of the meat were observed in Italian Cinta Senese pigs managed in open woodland pasture (Pugliese, Bozzi and Campodoni 2005). The main change in back fat fatty acid composition attributable to the incorporation of different pastures is related to the content of linoleic, linolenic and other n-3 polyunsaturated fatty acids, in addition to vitamin E (Lebret 2008; Bochicchio et al. 2012).

The inclusion of more fibrous dietary components could negatively impact carcass yield, a consequence of a higher gastrointestinal tract weight (Zijlstra and Beltranena 2013; Pietrosemoli et al. 2016) or higher gut fill (Heyer, Andersson and Lundstrom 2006). Higher carcass yield was reported for pasture pigs compared to indoor pigs, but no difference was observed in lean meat carcass content (Presto et al. 2007). Conversely, Heyer, Andersson and Lundstrom (2006) found a lower carcass yield in outdoor pigs. Carcass and pork quality traits of outdoor-managed pigs were compared by means of meta-analysis with those of indoor pigs. The production system did not affect any of the carcass traits (hot carcass weight and yield, back fat thickness and lean meat percentage) or pork (drip and cooking loss, pH, the initial temperature, shear force, marbling, hardness, juiciness and tenderness) quality indicators evaluated (Demori et al. 2012).

Table 8 Carcass and pork traits of pigs on pasture- versus indoor-kept pigs

Pasture	Effect on carcass and pork traits compared with indoor pigs	Reference
Mixture: *Pisum sativum* *Avena sativa* *Hordeum vulgare*	Lower CP meat content Lower water-holding capacity Higher levels of PUFA	Nilzen et al. (2001)
Medicago sativa	Higher back fat thickness + 12.5% Higher loin eye area + 9% Higher score for pork flavour + 6.2% Higher intramuscular fat Lower cooking loss More reddish-pink colour score Lower shear force Lower L* values	Gentry et al. (2002b)
Medicago sativa	Higher carcass weight + 10.9% Similar or better meat quality	Gentry et al. (2004)
Open woodland pastures *Quercus ilex*, *Castanea sativa*	Higher percentage of total fat cuts Higher intramuscular fat	Pugliese et al. (2005)
Grass	Increase age at slaughter + 2.56% Higher meat content + 2.3% Increase forage intake + 85.7% No differences in daily gain Worsened feed efficiency − 12.5%	Hermansen et al. (2005)
Mixture: *Trifolium pratense* *Trifolium repens* *Lolium perenne*	Lower back fat thickness Darker meat Higher polyunsaturated and lower monounsaturated fatty acids	Blumetto Velazco et al. (2013)
Grass, acorn and chestnut	Lower back fat thickness Higher ham proportion in relation to BW	Lebret et al. (2014)

7 Pasture management

The management challenge for pasture pig systems is to design grazing systems that would provide quality forages year-round and at the same time guarantee the persistence of the forages. Therefore the management should focus on the identified effects triggered by the pigs, and should include reducing deterioration of ground cover, minimizing nutrients build up, improving nutrients distribution and lowering soil compaction. The implementation of best management practices must be tailored to edaphic and climatic realities of each farm. Table 9 presents a summary of strategies that have been implemented while managing pasture pig systems.

How well-established is a pasture at the time of pig introduction plays an important role in the persistence of the sward. Older, well-established pastures show higher survival probability (Menzi et al. 1998). While on pastures, pigs affect the forage by grazing, rooting and trampling; not all the forage produced will be consumed by the grazing animals.

The efficiency of utilization is the proportion of forage consumed or destroyed by the animals, and tends to increase as the grazing system is intensified. Strip and rotational stocking systems would generally present higher efficiency than continuous stocking systems. Bauza (2005) estimated the efficiency of utilization of pastures by pigs to be approximately 70%; similar values are hard to attain even with dairy cattle. Increasing the forage availability encouraged grazing instead of rooting behaviour (Andresen and Redbo 1999, Horsted et al. 2012).

Pigs tend to organize in small grazing groups possibly as a way to reduce competition for the dietary resources (Roepstorff and Mejer 2001). Studies conducted in the Dehesa (Spain) under semi-natural conditions showed that the size of the group could be an indicator of the concentration of supplies. Larger groups could reveal more concentrated resources (Rodriguez-Esteves et al. 2010). This behaviour may generate uneven utilization of the pasture with under- and over-grazed areas, suggesting the suitability of dividing large pastures to obtain a better utilization.

According to Quintern (2005), the criteria leading the management of pasture pig systems should include limiting the amount of nutrients imported into the system and optimizing the distribution of these nutrients onto the paddocks. The author suggests strategies directly related to the supplemental feed to be offered to the pigs: implementing a phase feeding strategy adjusting the N concentration of the ration to the nutritional requirements of the pigs, adopting restricted feeding programmes, avoiding feed waste and removing the excess nutrients deposited on the ground by crop harvesting. To improve the distribution of the nutrients, the approaches proposed by this author centre on the needs to modify the excretory behaviour of the pigs, and include periodic equipment (feeding and drinking stations, shelters and shades) movements and the provision of new grazing areas, thus proposing the suitability of establishing a rotational stocking management system. Changes in feeding and management programmes could reduce N losses via leaching up to 50% (Webb et al. 2014).

To reduce damage to the soil structure, it is necessary to avoid grazing when the soil is wet. The ideal response to wet and poaching conditions would be to remove the animals from the ground, but perhaps a more practical alternative would be the use of a 'sacrifice area' where the animals could be maintained when conditions become unfavourable.

7.1 Grazing systems for pasture pig

In grazing systems, quality and persistence of forage ground cover are related to the rest period between two successive grazing. A production system that secures the survival of the vegetative ground cover needs to be based on controlling the number of animals and the length of their stay in certain areas. Consequently, a rotational stocking management approach seems most suitable for pasture pig systems (Pistoia et al. 2012; Helmerichs 2014). Frequent defoliations deplete stored reserves, causing a decline in foliar area and a reduction in photosynthetic activity and yield. Two alternative stocking methods may be implemented for pasture pigs: continuous and rotational stocking. Continuous stocking allows animals to roam and graze in the pasture during the entire productive cycle. In rotationally managed pasture, the area is divided into smaller paddocks, and the animals have access to them sequentially based on growth and recovery rate of the forage. There are many modalities of rotational stocking such as strip stocking, leader/follower stocking, mixed-species stocking and mob stocking (Allen et al. 2011). Strip stocking generally allows for better forage utilization and less time exposure to a specific site for trampling

Table 9 Environmental impact of grazing management and stocking rates with various forage species

Forage	Management: Grazing system	Management: Stocking rate m^2/hd	Supplemental feed	Environmental impact	Reference
Mixture: *Trifolium pratense* *Phleum pratense* *Festuca pratensis*	Weekly rotational stocking for 33 d	20 m^2/hd/wk or 10 m^2/hd/wk Pigs 37 kg BW	Low 13.7% CP or High 19.1% CP, twice a day	Higher rooting behaviour with higher stocking rate	Andresen and Rebbo (1999)
Mixture: *Poa trivialis*, *Dactylis* sp. and *Bromus* sp.	Continuous stocking for 133 d	576 m^2/hd, 240 m^2/hd or 100 m^2/hd Pigs 26–70 kg BW	Restricted 20%, adjusted weekly to BW	Faster reduction in ground cover with the higher stocking rates Remaining ground cover: 76, 57 and 30%, respectively Highest parasite infestation at the high stocking rate	Thomsen et al. (2001)
Mixture: *Trifolium pratense* *Phleum pratense* *Festuca pratensis*	Weekly rotational stocking for 56 d	20 m^2/hd/wk or 10 m^2/hd/wk Pigs 36–76 kg BW	Low CP 13.9% High CP 19.7%, Twice a day	Improved soil nutrient distribution Deeper tillage work and higher wheat grain yield with higher stocking rate	Andresen et al. (2001)
Bothriochloa ischaemum	Continuous stocking for 252 d	572 m^2/hd 286 m^2/hd Gilts 207 kg BW	Low CP 12.6% High CP 14.7%	Higher SR decreased ground cover (92 vs 84%) No differences in soil NO_3	Rachuonyo et al. (2002)

Mixture: *Phleum pratense* *Festuca pratensis* *Trifolium pratense* *Trifolium repens* *Triticum aestivum* *Hordeum vulgare* *Pisum sativum* *Avena sativa* *Solanum tuberosum* *Beta vulgaris*	Strip stocking for 180 d New stripe every day 3m^2/pig (30–40 kg), 6m^2/pig (> 80 kg) or larger area every 3 weeks Huts, feeding through and wallowing areas moved every 2–3 wks	2000 m^2/hd 1000 m^2/hd Pigs 25–105 kg 10000 m^2/sow	0.54 MJ per kg BW at 25 kg, to 0.30 MJ kg BW at 110 kg 16 to 18% CP	Creation of areas of ammonia emissions was avoided	Gustafson and Svenson (2003)
Mixture: *Phleum pratense* *Festuca pratensis* *Trifolium pratense* *Trifolium repens*	Daily strip stocking for 42 d	35 m^2/hd/wk Pigs 50–82 kg BW	16.7 % CP 11.9 MJ ME/kg Twice a day Feed allowance changed every week according to BW	Feed restriction (20%) increased grazing (12%) and rooting (46%) behaviour Improved distribution of soil nutrient	Stern and Andresen (2003)
Mixture: *Lolium perenne* *Trifolium repens*	Continuous stocking for 155 d, alternate Sows/heifers or mixed sows+heifers	Mixed species grazing 1471 m^2/heifer, 603 m^2/sow or 1818 m^2/heifer + 1429 m^2/sow	16.2% CP 13.04 MJ/kg ME	Alternate and mixed system: Improved ground cover and herbage quality Changes in botanical composition Positive on heifer BWG	Sehested et al. (2004)

(*Continued*)

Table 9 (*Continued*)

Forage	Management: Grazing system	Management: Stocking rate m^2/hd	Supplemental feed	Environmental impact	Reference
Mixture: *Brassica rapa* *Hordeum vulgare* *Lolium* sp. *Helianthus tuberosus* *Solanum tuberosum*	Continuous stocking + static equipment vs Weekly strip stocking + movements of equipment every 3 wks, for 63 d	19.2 m^2/hd/wk Pigs 30–120 kg BW	19.7–21.5% CP Restricted 20% (weekly adjustment) Fed twice/d	Weekly allowance of new land and equipment movements improved distribution of soil nutrients	Quintern and Sundrum (2006)
Mixture: Grass-clover	Continuous stocking for 124 d, monthly movements of huts and troughs	20.5 m^2/hd 77.6 m^2/hd 110.8 m^2/hd Pigs 18–26 kg BW	Restricted 20% or *Ad libitum*	Season affected ground cover deterioration Risk of nutrient loss No effect on extractable P Improved soil nutrients distribution	Eriksen et al. (2006a)
Mixture: Grass-clover	Continuous or split paddock stocked for 70 d (pregnant sows) 63 d (lactating sows) Ringed vs unringed	**Pregnant sows** 364 m^2/hd 182 m^2/hd **Lactating sows** 328 m^2/hd 164 m^2/hd 18.2 m^2/sow/wk	**Pregnant sows** diet restricted 30% to 1.75 kg feed/d **Lactating sows** *Ad libitum* 9 kg feed/d	Less pronounced rooting in ringed sows paddocks Higher ground cover on ringed sows paddocks 38 vs 14% and 81 vs 64% for pregnant and lactating sows, respectively Ring had no effect with 364 m^2/sow stocking rate	Eriksen et al. (2006b)

Trifolium repens	Continuous, alternated (occupation 14d + rest 14d) Rotational (occup 7d + rest 21d) stocking for 54 d	37.5 m^2/hd/wk Pigs 27–90 kg BW	*Ad libitum*	Improved soil nutrient distribution Improved pasture persistence	Leite et al. (2006a,b)
Apple orchard floor vegetation	Continuous stocking for 2 d	46.45 m^2/hd 24.4 m^2/hd Pigs 45 kg BW	*Ad libitum*	−58 % undesirable vegetation	Nunn et al. (2007)
Cynodon dactylon	Weekly rotational stocking for 56 d	126.5 m^2/hd/wk 84.3 m^2/hd/wk 50.6 m^2/hd/wk Pregnant sows 294 kg	Restricted, 14% CP 3.2 kg/d	Ground cover at animal removal: 82, 81 and 71%, respectively Severe soil compaction with the highest SR	Pietrosemoli and Green (2009)
Cynodon dactylon	Continuous stocking for 91 d	270 m^2/hd 135 m^2/hd 90 m^2/hd 68 m^2/hd Pigs 18–119 kg BW	*Ad libitum*, 15% CP	Stocking rate affected the ground cover Remaining ground cover 93, 85, 80 and 68%, respectively	Pietrosemoli, Green and Vibart (2009)
Mixture: *Festuca arundinacea* *Bromus catharticus*	Continuous stocking for 168 d	Pigs 25–40 kg 83 and 41 m^2/hd, Pigs 80–105 kg 233 and 116 m^2/hd	*Ad libitum*, Commercial feed	Stocking rate and animal category had impact on ground cover, soil N and soil compaction	Campagna et al. (2011)

(*Continued*)

Table 9 (*Continued*)

	Management				
Forage	Grazing system	Stocking rate m^2/hd	Supplemental feed	Environmental impact	Reference
Mixture: *Trifolium pratense*, *Trifolium repens* *Lolium perenne*	Weekly rotational stocking for 84 d	30.8 m^2/hd/wk Pigs 42–107 kg BW	*Ad libitum*, 3110–3196 Kcal ED/kg 3.15 kg feed/hd/d + 0.258 kg forage/hd/d	Persistence of ground cover Increase in soil content of NH_4 and NO_3 Higher nutrients loading on service area Low runoff and NH_3 emissions Centralizing the service area could reduce the risk of nutrients loss	Blumetto et al. (2012)
Mixture: *Salix viminalis*, *S. viminalis* x *S. schwerinii*, *Lolium perenne*, *Miscanthus giganteus* *Populus* sp., *Betula*	Continuous stocking for 54 d	367 m^2/hd or 117 m^2/hd Pigs 52–100 kg BW	As indoor + 10% and as Indoor + 20% (winter) 16.7% CP, 12.3 MJ/kg Once/d	Concentrated excretory behaviour in willow area Lower nitrate leaching risk in willow area More rooting damage in spring	Horsted et al. (2012)
Cynodon plestostachyum	Continuous stocking for 8 d	75 m^2/hd Sows	Restricted 3250 Kcal EM/d 2.2 kg/sow/d	Sows on pasture showed more comfort indicators	Nazareno et al. (2012)

Schedonorus arundinaceus	Continuous, weekly rotational or strip stocking for 84 d	Continuous 200 m²/hd Rotational Wk 1–8 44.5 m²/hd/wk Wk 9–12 66.7 m²/hd/wk Strip stocking Wk 1–8 25 m²/hd/wk Wk 9–12 50 m²/hd/wk Pigs 23–85 kg BW	*Ad libitum,* 16% CP, Supplemental feed intake 1.96 kg/hd/d	Lesser ground cover in continuous system Rotational system showed lower soil NO_3, P, K, Mn and Cu No differences among grazing systems in paddocks botanical composition	Pietrosemoli and Green (2012a,b) Pietrosemoli, Luginbuhl and Green (2012)
Sorghum bicolor Mixture: *Secale cereale* + *Lolium multiflorum*	Continuous stocking for 84 d, weekly movement of shelters and drinker stations	135 m²/hd Pigs 29–93 kg BW	*Ad libitum* 16% CP	No effect of movements of equipment on ground cover or soil nutrients distribution Reduction in soil compaction	Bordeaux et al. (2014)
Malus domestica, *Prunus avium,* *Pyrus communis*	Post-harvest rotational grazing for 2 d	675 m²/hd Pigs 73 kg BW	Restricted fed 20–30%	Decreased grasses, forbs and fallen fruits Increased bare soil by 57% Lower pest damaged fruits in subsequent harvest	Buehrer and Grieshop (2014)
Mixture: *Medicago sativa* *Lolium perenne*	Weekly strip stocking for 40 d	83.3 m²/hd/week Pigs 58–90 kg BW	High CP 20.5% Low CP 10.7%	Low CP increased rooting, more intense effect on ryegrass	Jacobsen et al. (2014)

(*Continued*)

Table 9 (*Continued*)

Forage	Management: Grazing system	Management: Stocking rate m²/hd	Supplemental feed	Environmental impact	Reference
Mixture:	Weekly strip stocking for 77 d	15.2 m²/hd/week	High CP 20.5%	Increased rooting and grazing in new allotted area	Helmerich (2014)
Medicago sativa		Pigs 58–90 kg BW	Low CP 10.7%	Lower urinating frequency in new allotted area	
Trifolium repens			Once a day, 2.23 kg/d		
Lolium perenne					
Mixture:	Weekly strip stocking for 70 d	**Herbage mixture**	Allowance	Final ground cover wk 10: 27%, 15% and 8%	Kongsted and Jakobsen (2015)
Trifolium repens		Wk 1–4: 125 m²/hd	100% Danish recommendation + 15%	Faster reduction in ground cover in paddocks with restricted feed pigs	
Lolium perenne		Wk 5: 156m²/hd	Restricted	Rooted area represented 81%, 92% and 96% of the bare soil area	
Festuca rubra		Wk 6: 187.5m²/hd	20% wk 1–4 and 50% wk 5–10	3–4 m² uprooted/hd/d	
Plantago lanceolata		Wk 7: 218.8m²/hd	+ Mineral and vitamin mix		
Medicago sativa		Wk 8: 250 m²/hd	Restricted		
Lotus corniculatus		Wk 9: 338 m²/hd	20% wk 1–4 and 50% wk 5–10		
Carum carvi		Wk 10: 338 m²/hd	Without mineral and vitamin mix		
Cichorium intybus		**Chicory**	19% CP		
Borago officinalis		Wk 9: 20 m²/hd	8.9 MJ NE/kg once a day		
Melilotus officinalis		Wk 10: 34 m²/hd			
Sanguisorba minor		Pigs 34–95 kg BW			

and lounging, and it provides more control over animal grazing of new plant regrowth (Bauza 2005).

In addition to the stocking method, some other tactics can be employed to improve the functionality of these systems: 1) conducting strategic movements of service structures (shelters, shades, feeding and drinking stations) would allow for a better distribution of 'heavy use areas', manure and hence of soil nutrients (Benfalk et al. 2005; Hermansen, Eriksen and Strudsholm 2005; Horta et al. 2012), and lessening soil compaction (Bordeaux et al. 2014); 2) the use of nose rings; 3) the provision of diets with a high fibre content; and 4) the establishment of special rooting areas in the pasture with buried feedstuff to encourage rooting (Bornett, Edge and Edwards 2003; Edge et al. 2004; Edge, Bulman and Edwards 2005).

The excretory behaviour of pigs differed between stationary and mobile systems. In both systems, pigs created sections of soil with high accumulation of nutrients and elevated mineral N content (representing 24 and 4% of the area for the mobile and stationary system, respectively). Higher concentration of nutrients was recorded for the mobile system (Salomon et al. 2007). Similarly, an elevated concentration of nutrients was observed in the wallowing area. The hot spots showed a concentration of Zn up to levels above regulations (Salomon et al. 2005b).

The benefits of rotational stocking systems will be reflected in ground cover maintenance (Vittoz and Hainard 2002; Stern and Andresen 2003; Quintern 2005), a better distribution of manure and consequently of soil nutrients (Salomon et al. 2005b; Quintern 2005; Quintern and Sundrum, 2006), and in a reduction of the risk of parasite contamination (Lindgren et al. 2014). More ground cover was observed when sows were provided with nose rings 38 vs 14% and 81 vs 64% for gestating and lactating sows, respectively (Eriksen et al. 2006b). The use of nose rings is controversial because it is related to welfare issues. An alternate stocking system, in which the pasture was divided into two equal subunits grazed alternately by sows and heifers at weekly intervals, showed an improvement in ground cover in pastures managed with unringed gestating sows (Sehested et al. 2004).

The natural inclination of pigs is to spend more than 80% of their time along fences or drinking/feeding stations (Andresen and Redbo 1999), and their motivation to explore new surroundings advocated for strip stocking systems as a way to improve soil nutrient distribution (Quintern 2005), and to avoid localized emissions of ammonia (Gustafson and Svensson 2003). Three stocking methods for pastures for pigs were compared in Brazil. The stocking method (continuous, alternate and rotational) did not have an effect on animal performance (Leite et al. 2006b), but pigs under the alternate system spent more time eating supplemental feed and less time grazing compared to the pigs in the other two management systems (Leite et al. 2006a).

During growing cycles of 12 weeks, implementing periods of occupation of one week and stocking rates equivalent to 187m^2/pigs (permanent access: 16.6 m^2/pig as service area + weekly access to grazing plots measuring 14.2 m^2/pig with an initial pig body weight of 46 ±5.03 kg), forage persistence was achieved with slight changes in soil nutrients, and minimal losses due to runoff and NH_3 emissions. Therefore, this management system represents a viable alternative to raising pigs on pasture without compromising environmental quality (Blumetto et al. 2012).

Comparing the intake and the grazing behaviour of European wild boars under continuous (during five days) or rotational (daily access to a new strip) stocking systems for five days, Rivero et al. (2013a) did not find differences among systems. The time dedicated to playing was greater in the continuous system. The authors observed more animal activity during the first three hours of grazing and suggested devising a stocking system whereby wild boars would have access to the pasture for a limited time of the day to reduce vegetative ground cover damage by rooting and trampling (Rivero, Lopez and Hodgkinson 2013a).

7.2 Stocking rate

Whether pigs are managed on pasture or in forestland, the intensity of grazing, rooting and trampling, and the amount of nutrients loaded into the system are affected by animal density (number of animals per area), animal size and the duration of time that they stay in a given area (Menzi et al. 1998; Quintern 2005; Bondi et al. 2015).

The degree of the impact of pigs on the vegetative ground cover and on the soil has been associated with high stocking rates (Rachuonyo, Pond and McGlone 2002; Acciaioli et al. 2012; Bondi et al. 2015). The establishment of high stocking rates implies the use of large amounts of supplemental feed. The damage exerted to the vegetative ground cover by the pigs limits the capacity of the pasture to utilize the nutrients deposited on the soil, thus increasing the risk of nutrients and sediment loss (Eriksen et al. 2006a; Campagna et al. 2011; Sikala 2012). The implementation of high stocking rates could also have health implications. Animals managed using a high stocking rate would be more prone to gastrointestinal parasitic infestation (Roepstorff et al. 1999; Thomsen et al. 2001).

Campagna et al. (2011) compared the effects of low (4000 kg LW/ha) and high (8000 kg LW/ha) stocking rates and animal category using growers (25–40 kg; 83 vs 41 m^2/pig, respectively) and finishers (80–105 kg; 233 vs 116 m^2/pig, respectively) managed on a tall fescue and bromegrass (*Bromus catharticus* Vahl) pasture. Stocking rate had an impact on ground cover (62 vs 25% for the low and high stocking rate, respectively). The greatest impact on ground cover was observed on paddocks with growers managed under the high stocking rate. Animal category also impacted ground cover and nutrients deposition. Younger, more active animals generated more damage (65 and 45% ground cover, respectively), while older/bigger animals increased available N deposition on the soil. Soil compaction was similar for both animal categories. Within low stocking rates, finishers caused more compaction than growers. Similarly, total deterioration of the ground cover was registered with stocking rates equivalent to 100 m^2 per growing-finishing with animals weighing 20 to 100 kg (Hermansen 2005).

The evolution of soil properties using two wild boar stocking rates and two vegetative cover (woodland and olive plantation) was followed by Macci et al. (2012). Low stocking densities favoured soil organic matter accumulation and biological activity, while high densities were associated with soil degradation. Woodland exerted a beneficial influence protecting the soil compared to the olive grove, probably due to higher organic matter deposition (through leaf litter). Similarly, a high stocking rate of Cinta Senese and Yorkshire pigs (40–50 vs 1–2 pigs/ha) on soils of a Mediterranean woodland (*Castanea sativa, Fraxinus ornus, Robinia pseudoacacia, Quercus ilex, Q. cerris, Q. petraea, Laurus nobilis, Hedera helix, Ruscus aculeatus* and *Rubus ulmifolius)* deteriorated all soil chemical and structural properties occasioning loss of soil fertility and quality, an effect that was not observed with the low stocking rate. Soil deterioration, such as decrease in carbon content, lower chemical parameters linked to OM and increased mineralization, was greater in steep slope areas whereas soil compaction was not affected by this factor (Bondi et al. 2015).

The stocking rates (285.7 and 571.4 m^2/sow, respectively) at which sows were managed during the gestation period affected animal performance, with a higher number of pigs weaned and a lower mortality rate for the higher stocking rate (8.4 vs 7.1 pigs/sow and 18.1 vs 25.7%, respectively, Rachuonyo, Pond and McGlone 2002). Conversely, Sueskun-Ospina and Ocampo-Duran (2015) found no effect of gestating sows stocking rate (450 vs 150 m^2/sow) on weaning–oestrus interval (5.19 ± 1.35 days), fertility rate (79.15 ± 1.31 %), piglets mortality rate (24.7 ± 0.45 %), number of piglets born (11.5) and weaned (8.33), and weight of piglets at birth (1.32 kg) or at weaning (8.8 kg).

Higher N content in soils from paddocks managed with higher stocking rates (25, 18 and 12 sows/ha) was reported by Williams et al. (2000), with N surpluses of 576, 398 and 265 kg N/ha, respectively. According to Quintern (2005), a stocking rate equivalent to 10 pigs/ha/y allows for an acceptable degree of environmental impact.

7.3 Mixed stocking systems

Utilizing the distinctive grazing and browsing behaviour of various animal species, mixed herds could have a positive impact on the control of undesirable plant species, the diversification of farm income and the overall improvement of pasture productivity. Additionally, mixed stocking (system where different animal species graze the same pasture not necessarily simultaneously) could improve the efficiency of utilization of the pastures (Sehested et al. 2004) and reduce gastrointestinal parasite problems as well (Roepstorff et al. 1999). The digestive physiology of pigs, which are monogastrics, and their selective foraging behaviour make them perfect grazing companions for herbivores and ruminant animal species, mainly because the nutritive requirements of these animal species bear little similarities. Mixed stocking (cattle, horses and pigs) is a strategy that could also be employed for nature conservation purposes (Beinlich and Poschlod 2002).

Mixed stocking strategies provide the opportunity to manage the animals together or sequentially (Doran 2015). Grazing gestating sows and heifers in a mixed stocking system resulted in increased weight gain for both species when grazed together as compared to grazing them alone (Søegaard et al. 2000; Sehested et al. 2004). In addition, a reduction in heifer gastrointestinal parasite load was reported for heifers managed with sows (Sehested et al. 2004). The botanical composition of the pasture and the quality of the forage, however, were affected when evaluating mixed alternatives. Better quality indicators were reported in pastures where heifers grazed alone due to the selective grazing behaviour of the sows which concentrated their foraging on clovers, leaving more fibrous material such as stems behind (Sehested et al. 2004). In the same study, a higher amount of refused forage was observed in paddocks managed sequentially with sows and heifers, and in which sows were the first grazing group compared to both species grazed together simultaneously. In Greece, Mpampouli et al. (2009) observed that ground cover declined (62.5 vs 39.5%) and botanical composition shifted with a decrease in species richness (10.5 vs 8.0%) among open *Quercus frainetto* forest plots grazed only by small ruminants versus plots grazed by mixed groups of goats, sheep and wild boars.

7.4 Hogging down pastures

Pigs can be managed to effectively graze grain crops such as corn or combinations with other crops. Gilmore (1999) indicated that pastures planted only with corn present nutritional limitations for pig systems, and recommended pumpkin (*Cucurbita pepo*), soya bean or cowpea (*Vigna unguiculata*) seeded simultaneously or in alternate rows with the corn, cereal rye (*Secale cereale*), velvet bean (*Mucuna pruriens*), rape (*Brassica napus*) or soya bean. This same author proposed a stocking rate equivalent to 35–40 pigs (68 kg BW)/ha during 1 month, and recommended to only provide the hogs access to an area that they can clean up in 10 days.

Hogging down faba bean (*Vicia faba minor*) resulted in a viable strategy, saving harvesting costs and seed losses in the field estimated at 10% of the harvest (Giannone 2001a). Hogs receiving no supplement feed and grazing a mixture of grain sorghum (86%, physiological maturity stage) and sugar beet (14%) alternatively consumed the sorghum panicles and beet leaves, but seldom ate the beet roots or the sorghum green leaves (Giannone 2001b).

Finishing Cinta Senese pigs, under a 30% DM feed restriction, were managed in four pastures containing two different combinations of forages: 1. grazing mixtures: Vetch (*Vicia sativa*) + berseem clover (*Trifolium alexandrinum*) + oat, and 2. gleaning/hogging down pastures including sorghum, fava bean and barley (*Hordeum vulgare*). Pigs from both kinds of pastures showed similar growth performance. At harvest (110–150 kg BW), the back fat fatty acid composition of the pigs varied according to pasture type, with a higher content of fatty acid n-3 and n-6 found in pigs managed on the grazed pastures (Bochicchio et al. 2012).

7.5 Undesirable species control

The rooting behaviour of pigs could replace mechanical soil tillage thus contributing to the control of undesirable species while simultaneously enriching the soil with the nutrients contained in the manure. Pigs could be especially successful controlling plants with vegetative reproductive habits, and have the advantage of being able to reach areas not accessible to machinery.

Pig rooting was comparable to mechanical tillage, especially when the ground was moist. Wheat grain yield was recorded in plots previously rotationally managed with pigs (36–76 kg BW) at two stocking rates (10m^2/pig/wk vs 20 m^2/pig/wk, pigs). Higher yield was obtained with the higher stocking rate. The nutrient budget showed a balance between N input and its output in pig ADG and subsequent wheat grain at the lower stocking rate and the lower level of protein on supplemental feed (13.9 vs 19.7%). Better growth rate and feed conversion were obtained at the highest level of protein (Andresen et al., 2001). Conversely, no effect of previous pig stocking rates (37, 74, 111 and 148 head/ha bermudagrass, 18–119 kg BW) on subsequent forage (cereal rye + annual ryegrass and sudangrass) yield was recorded by Renner et al. (2011).

To control weeds and till the soil, Schivera (2009) recommended stocking rates equivalent to 13 474 kg BW/ha for 7–10 days. Grazing pigs were an effective way to control brambles (*Rubus* spp.) and ferns (Henney 2012). Similarly, using pigs during land preparation in rice fields resulted in a reduction (62%) of *Cyperus rotundus* (Kathiresan, 2012). Effective control of couch grass (*Elymus repens*), creeping thistle (*Cirsium arvense*), and lambsquarter (*Chenopodium album*) in vegetable gardens and dock (*Rumex crispus*, *R. obtusifolius*) in pastures has been reported by Robinson (2013) who advised that on occasion, there would be a need to re-educate pigs to 'appreciate' the edible value of plants species to which currently we do not assign nutritional importance, thus allowing for a change of paradigm to the weed concept.

The implementation of restricted feeding practice could encourage rooting activity (almost doubling the frequency showed by *ad libitum* fed pigs) and impacting the vegetative ground cover (Menzi et al. 1998; Stern and Andresen 2003; Kongsted and Jakobsen 2015). When managed appropriately, this increment in rooting activity could be of benefit when soil tillage is the ultimate goal (Kongsted and Jakobsen 2015).

Potential issues related to the use of pigs to control undesirable vegetation would be potential toxic compounds found in some plants (Henney 2012), and the eventual dispersion of unwanted species by seeds that could have maintained their viability after passing through pig digestive tracts (O'Connor and Kelly 2012).

8 Case study: pasture pig system developed at the Center for Environmental Farming Systems (CEFS)

Researchers at the CEFS (Pietrosemoli et al. 2012) primarily focused on maintaining vegetative ground cover and reducing the accumulation of nutrients on pastures of tall fescue, switchgrass (*Panicum virgatum*), bermudagrass, sorghum, annual ryegrass (*Lolium multiflorum*) and cereal rye under different stocking rates and grazing methods with either gestating sows or growing to finishing pigs.

Results showed that stocking rates over 50 growing to finishing pigs/ha have a negative effect on the maintenance of ground cover of tall fescue and bermudagrass grazed continuously for 12 weeks. Under rotational management, three sows' (231 kg BW) stocking rates (10, 15 and 25 sows/ha) were compared during three seasons. Severe soil compaction was documented in paddocks managed with 25 sows/ha after one grazing cycle (8 weeks). To maintain the vegetative ground cover, the stocking rate for bermudagrass pastures should not be higher than 15 gestating sows/ha managed with a rotational stocking with weekly movements.

Sorghum-sudangrass or cereal rye + annual ryegrass managed continuously for 12 weeks at a stocking rate of 74 pigs/ha (29–93 kg BW) showed a reduction in soil compaction when shelters and feeding and drinking stations were moved weekly. Ground cover was similar regardless of equipment management. However, ground cover deterioration was quicker on sorghum-sudangrass (summer months) as compared with the cereal rye + annual ryegrass (spring months). The movement of the equipment had no effect on soil nutrients distribution (Bordeaux et al. 2014).

Renner (2011) found that nutrients deposited on bermudagrass which had been used for finishing two groups of pigs could be removed in subsequent crops of cereal rye with annual ryegrass in winter followed by sorghum-sudangrass the subsequent summer.

The ground cover after 12 weeks was best (+14%) and soil nutrient concentrations were lower (1.3 and 21% less P and K, respectively) in tall fescue paddocks under rotational stocking equivalent to 50 pigs/ha (23–85 kg BW) with fixed or portable equipment (shelter, shades and drinking stations), compared to pastures grazed continuously (Pietrosemoli and Green 2012a,b). Animal performance (ADG: 0.732 kg/pig/d, supplemental feed intake: 1.96 kg DM/pig/d, gain to feed ratio: 0.37 kg gain/kg feed) (Pietrosemoli and Green, 2012a) and pasture botanical composition (tall fescue 65%, other grasses 30.3% and broadleaf species 4.7%, Pietrosemoli, Luginbuhl and Green 2012) did not differ among stocking methods after two groups of growing to finishing pigs. Rotational with portable equipment management resulted in a higher workload.

As a result of these experiences, a pasture pig system (3 sows + 2 litters/yr + 1 boar stocked on 1.2 ha) has been proposed, with the primary intent of maintaining at least 75% ground cover within the pastures at all times. All production phases (breeding, gestation, farrowing-lactation and growing-finishing) are conducted on pasture (see Fig. 1). Lactating sows and the boars are managed in individual pastures, whereas all other production phases are managed as a group of animals. Shelters, shade, feeding

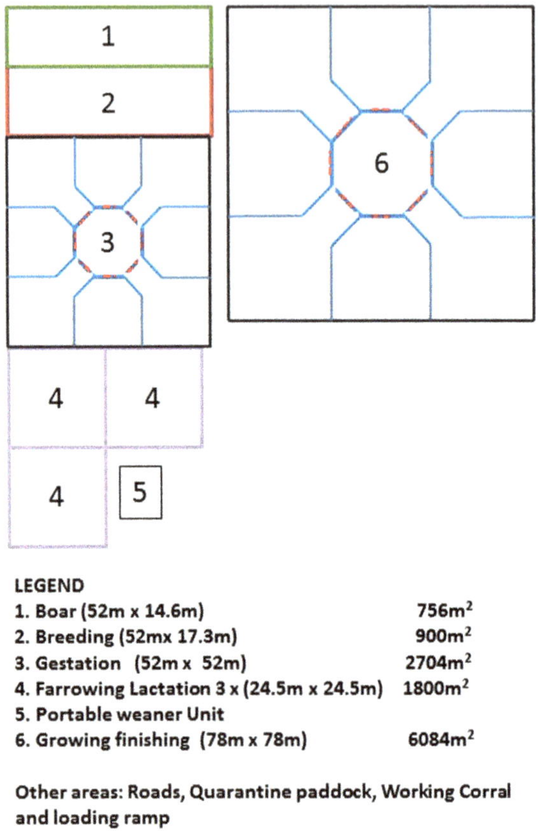

Figure 1 Farm layout for a farrow-to-finishing operation with 3 sows, 1 boar and 2 litters/sow/year stocked on 1.2 ha.

and drinking equipment are portable and located over hard perforated platforms. Electric fences are used to delimit the pastures using polywire for internal paddock subdivisions. Animals should be hardy and adapted to management on pasture. Up-to-date herd genetic, reproductive, sanitary, and nutritional principles and management practices can be implemented. Feeding programmes oriented to fulfil nutritional requirements need to be adjusted to physiological status and animal category. It is very important to maintain a daily routine in management to avoid unnecessary stress to the animals.

Farrowing-lactation paddocks and portable units for weaners are managed under continuous stocking system. Boar and breeding paddocks could be managed either in continuous or in an alternate (dividing the area into two subplots using them sequentially) stocking system, whereas gestation and growing to finishing pastures are rotationally stocked. To facilitate the management of the pastures, the occupation period of the subplots in the rotational stocking system is set at seven days. In the continuous stocking system, shelters, shade and drinking stations are moved every three weeks. Each pasture is occupied for the length of the physiological stage or productive cycle, with two farrowing per sow/year and 6 to 8 weeks of weaning age.

The three sows are managed as one farrowing group. After weaning, sows are moved to the breeding pasture while the piglets are stationed in a portable weaner unit where they will stay for one month. The sows will remain in the breeding pasture for four weeks, and will then be moved to the gestating area where they will stay for twelve weeks. One week before the expected farrowing date, the sows will be relocated to the individual farrowing-lactation pastures. Weaners will be moved to the growing-finishing paddocks where they will be kept until reaching market weight at roughly 100–130 kg.

The pastures under rotational stocking management are divided into nine sections as suggested by Pietrosemoli and Green (2015). Pigs have permanent access to the central service area, and to one or more of the other eight sections where the feeder is located during the grazing period (see Fig. 2). The shelter and the drinking station are found in the central service area. Electric fences provide flexibility to the system, allowing for successive changes in the number of paddocks available per pasture (see Fig. 3). The rotational system presented can be adapted to different sizes according to the availability of land, but our experience showed easier management with pastures of 2030 to 4060 m^2.

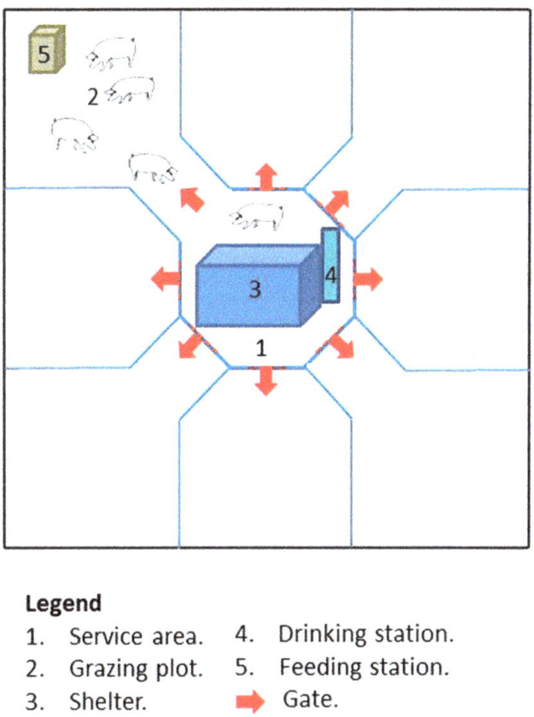

Figure 2 General diagram of a rotationally managed pasture. Arrows indicate animals' weekly movements.

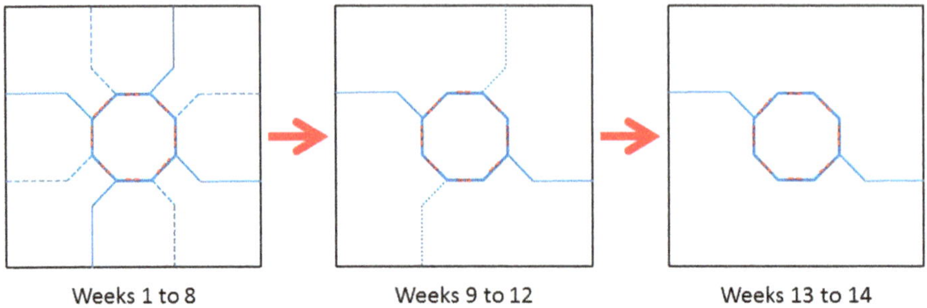

Figure 3 Example of how a paddock can evolve by removing internal fences (dotted lines). This example illustrates how the paddock shape and size change over 14 weeks of occupation.

The use of perennial grasses which have stolons and/or rhizomes or are tolerant of close grazing (bermudagrass and tall fescue) has proven to be effective in maintaining the ground cover in North Carolina temperate environment. Considering the potential need to renovate 20% of the grazing area per year, annual forages could be used as companion crop (wheat, barley, oats, rye, crabgrass [*Digitaria Haller*], sorghum or sorghum-sudangrass, millet, annual ryegrass, rape and other brassicas) for temporary cover.

The estimates of nutrients output per year by the respective phases of production on a farrow-to-finish operation stocked at three sows (producing 20 pigs/yr each) and one boar on 1.2 ha can be seen in Fig. 4. The plant-available nutrients (PA N, PA P, PA K) were estimated using coefficients for mineralization and other losses according to the North Carolina Nutrient Planning Guidelines (NRCS 2007). With this system, few nutrients

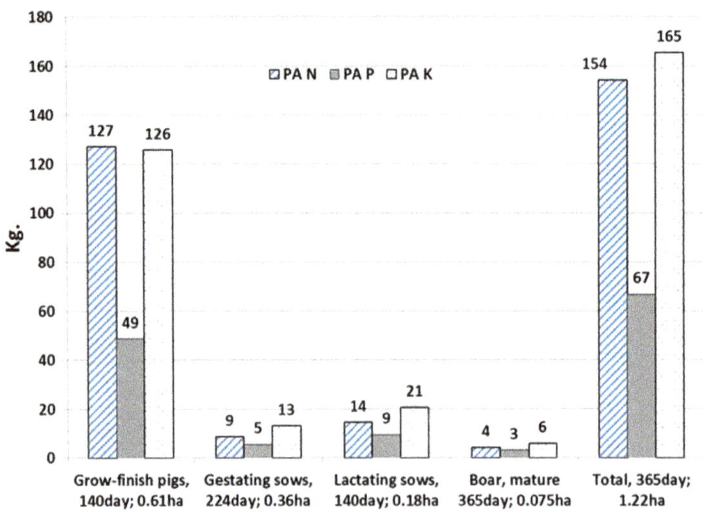

Figure 4 Nutrient output expressed as plant available (PA) for a farrow to finish operation with one boar and two farrows of three sows producing 10 pigs each. Coefficients for mineralization of excreta were 0.4, 0.9 and 0.9 for N, P and K, respectively.

would be removed from the site because there is no crop removal. Therefore, soil nutrient monitoring would be advisable to avoid excessive nutrients accumulation and environmental issues. When necessary, crops can be utilized to take up the excess of nutrients deposited by the pigs, and the area dedicated to other farming activities.

9 Conclusion

Properly managed, pasture pig systems can sustain natural resources, improve the quality of life of producers, support rural community development and enhance farmers' profit (Silva Filha and Barbosa 2011). In pasture pig systems, reasonable management can be accomplished with minimal resources and capital, and the market price per pound of pork is typically 40–60% higher than for conventionally raised hogs (Andresen and Redbo 1999; Andresen 2000). These systems are an option for small and limited-resource farmers because of their low start-up costs, representing one-third to one-fifth of the investment required for indoor facilities, and reduced operating costs (Grosso et al. 2011). Nevertheless, the system can be intensified adopting some of the same practices that are implemented in confinement operations.

A plan to minimize the potential environmental impact of pasture-based pig operations should include strategies to maintain at least 75% ground cover with vegetation, to avoid the accumulation of nutrients in the soil, to improve the distribution of deposited nutrients, to minimize soil compaction and to reduce runoff.

Ground cover maintenance and the establishment of vegetated filter or buffer strips reduce the loss of soil and nutrients offsite. Ground cover conservation strategies should involve the establishment of appropriate stocking rates and length of stay of pigs per paddock; periodic movements of shelters, shades and feeding structures; and the use of hard perforated platforms under feeders and waterers.

The incorporation of forages of good quality to the feeding programme of pasture pig represents an alternative to reduce the production costs.

Improved management and housing conditions could help in reducing welfare issues of pasture pigs, such as parasite infections and thermal stress.

The intention of this chapter is to present evidence that pigs can be reasonably managed on pastures while protecting natural resources and producing satisfactory animal performance and welfare. The case study was designed to illustrate a system that could be employed to adapt and shape the model to individual circumstances.

10 Future trends

In spite of the existence of a general understanding of pasture pig systems, there is limited information related to the animal management practices (stocking methods, use of environmental enrichment, feed management) that allow for maintenance of vegetation during the various phases of production (especially during gestation and growing to finishing) at a wide range of stocking rates.

Another area requiring the focus of research would be nutrient loading and distribution patterns of the manure and urine within the pasture and how that affects nutrient losses and subsequent crop removal, and supplemental nutrients application rates and patterns.

Although there is much information available related to outdoor swine production, most of the existing research has been conducted in small paddocks, and the results would be hard to predict when extrapolating the information to commercial-scale paddock size. Therefore, it is necessary to conduct pasture pig research on a size that easily permits the reproduction of the system by producers.

The incorporation of local resources to pasture pig systems will be reflected in the reduction of production costs and better sustainability of the system. Future research should be oriented towards the incorporation of alternative feedstuff as supplemental feed for pasture pig systems.

There is limited knowledge on multispecies plant mixtures for pasture pig systems. More information on the management of mixed-species pastures and its effect on the stand quality and persistence, the animal and the environment is essential to better understand the productive potential of these pastures.

Overall, research should focus on best management practices for pasture pig systems that will allow for pasture quality and persistence, animal performance, carcass and pork quality, and environmental sustainability.

11 Where to look for further information

Detailed information can be found in:

Andresen, N.(2000). The foraging pig. *Resource Utilisation, Interaction, Performance and Behaviour of Pigs in Cropping Systems, Agraria, 227*.

Edwards, S. A. (2003). Intake of nutrients from pasture by pigs. *Proc. of the Nutrition Society*. pp. 62, 257–65.

Edwards, S. A. (2005). Product quality attributes associated with outdoor pig production. *Livestock Production Science*, 94, 5–14.

Miao, Z. H., Glatz, P. C. and Ru, Y. J. (2004). Review of Production, Husbandry and Sustainability of Free-range Pig Production Systems. *Asian-Aust. J. Anim. Sci.* 17(11), 1615–34.

Thorton, K. (1988). Outdoor pig production. Farming press limited. 224pp. ISBN-13: 9780852361788

Some research centres that are generating information for pasture pig systems:

1. Aarhus University, Department of Agroecology, Denmark
2. Swedish University of Agricultural Science, Center for Sustainable Agriculture, Uppsala, Sweden
3. INRA, Livestock Production Systems, France
4. EMBRAPA, Brazil
5. INTA, Argentina
6. Kassel University, Faculty of Organic and Agricultural Sciences, Germany
7. Universidad de Córdoba, Departamento de Producción Animal, Spain
8. University of Newcastle, Department of Agriculture, UK
9. Universita degli Studi di Firenze, Department of Agricultural, Food and Forestry Systems, Italy
10. Universidad de la Republica, Facultad de Agronomía, Uruguay

12 Acknowledgements

Research conducted at the Center for Environmental farming Systems (**CEFS**) was supported by the United States Department of Agriculture – Natural Resources Conservation Service (**USDA-NRCS**), Sustainable Agriculture Research and Education (**USDA-SARE**) program and the **Kellogg Foundation**. We thank **CEFS** swine unit staff for support on field activities. Finally, thanks to Dr Jean-Marie Luginbuhl for his time spent proof reading, listening and advising.

13 References

Acciaioli, A., Grifoni, F., Fontana, G., Esposito, S. and Franci, O. (2012). Evaluation of forest damage derived from the rearing of Apulo-Calabrese pig. In: De Pedro, E. J. (ed.), Cabezas, A. B. (ed.). *7th International Symposium on the Mediterranean Pig*. [online] Zaragoza: CIHEAM. Options Méditerranéennes: Série A. Séminaires Méditerranéens; n. 101, 133–6. Available at: http://om.ciheam.org/om/pdf/a101/00006667.pdf [Accessed 4 May 2015].

Allen, V. G., Batello, C., Berretta, E. J., Hodgson, J., Kothmann, M., Li, X., McIvor, J., Milne, J., Morris, C., Peeters, A. and Sanderson, M. (2011). An international terminology for grazing lands and grazing animals. *Grass and Forage Science*, 66, 2–28.

Andresen, N. (2000). The foraging pig. *Resource Utilisation, Interaction, Performance and Behaviour of Pigs in Cropping Systems, Agraria*, 227.

Andresen N., Ciszuk P. and Ohlander L. (2001). Pigs on grassland – animal growth rate, tillage work and effects in the following winter wheat crop. *Biological Agriculture and Horticulture*, 18 (4), 327–43. DOI:10.1080/01448765.2001.9754896.

Andresen, N. and Redbo, I. (1999). Foraging behaviour of growing pigs on grassland in relation to stocking rate and feed crude protein level. *Applied Animal Behaviour Science*, 62, 183–97.

Barlocco, N. (2005). Alimentación de cerdos en Crecimiento y engorde en pastoreo permanente. In: Bauza, R. (ed.). *Utilización de pasturas en la alimentación de cerdos*. [online] Montevideo: Universidad de la Republica. Uruguay. 15–22. Available at: http://upc.edu.uy/images/documents/extension/Jornada-Taller_Pasturas_dic05.pdf [Accessed 8 June 2016].

Barone, C. M. A., Di Matteo, R., Rillo, L., Rossetti, C. E., Pagano F. and Matassino, D. (2015). Pork quality of autochthonous genotype Casertana, crossbred Casertana x Duroc and hybrid Pen ar Lan in relation to farming systems. *Agronomy Research*, 13(4), 900–6.

Bauza, R. (2007). Alimentos alternativos para animales monogástricos. In: *IX Encuentro de Nutrición y Producción en Animales Monogástricos*. [pdf] Montevideo: Universidad de la Republica, Facultad de Agronomia. 47–55. Available at: http://upc.edu.uy/ixe/category/23-memorias-del-encuentro [Accessed 8 June 2016] ISBN: 978-9974-0-0399-6.

Bauza, R. (2005). Utilización de pasturas en la alimentación de reproductores. In: Bauza, R. (ed.). *Utilización de pasturas en la alimentación de cerdos*. [pdf] Montevideo: Universidad de la Republica. Facultad de Agronomia. 5–14. Available at: http://upc.edu.uy/images/documents/extension/Jornada-Taller_Pasturas_dic05.pdf [Accessed 8 June 2016].

Bauza, R. and Petrocelli, H. (2005). Uso de pasturas en el crecimiento-terminación de cerdos: pastoreo con acceso restringido. In: Bauza, R. (ed.). *Utilización de pasturas en la alimentación de cerdos*. [pdf] Montevideo: Universidad de la Republica. Facultad de Agronomia. 23–32. Available at: http://upc.edu.uy/images/documents/extension/Jornada-Taller_Pasturas_dic05.pdf [Accessed 8 June 2016].

Beinlich, B. and Poschlod, P. (2002). Low intensity pig pastures as an alternative approach to habitat management. In: Redecker, B., Härdtle, W., Finck, P., Riecken, U. and Schröder, P. (eds). *Pasture Landscapes and Nature Conservation*, pp. 219–26. Publisher Springer Berlin Heidelberg New York. DOI 10.1007/978-3-642-55953-2_16.

Benfalk, C., Lindgren, K., Lindahl, C. and Rungren M. (2005). Mobile and stationary systems for organic pigs – Animal behavior in outdoor pens. In: Köpke, U., Niggli, U., Neuhoff, D., Cornish, P., Lockeretz, W. and Willer, H. (eds). *Researching Sustainable Systems.* [pdf] ISOFAR, pp. 242–5. Available at: http://orgprints.org/4313/4/4313-Benfalk_etal_4p_revised-ed.pdf [Accessed 8 June 2016.]

Benjamins, D. (2002). *Oxford Sandy and Black pigs as a method of weed control. When do they stop being an asset and start becoming a problem?* BSc Thesis. De Montfort University, Riseholme School of Agriculture, UK. 43p.

Bikker, P. and Binnendijk, G. P. (2012). *Grass silage in diets for organic growing-finishing pigs*; Report 603; Wageningen UR Livestock Research: Lelystad, Holland, 22p.

Bochicchio, D., Comellini, M., Marchetto, G., Goracci, J. and Della Casa, G. (2012). Il pascolo nell' allevamento del suini biológico mediterraneo. In: Zanoli, R. and Vairo, D. (ed.). *Agricoltura biológica: modelo sostenibile per un Mediterraneo in transizione.* [pdf] Ancona. Grab-it, 16–21. ISBN 978-88-9077-400-3. Available at: http://www.enea.it/it/pubblicazioni/pdf-volumi/V2011CongressoRIRAB.pdf [Accessed 14 June 2016].

Bonde, M. and Sørensen, J. T. (2004). Herd health management in organic pig production using a quality assurance system based on Hazard Analysis and Critical Control Points. *NJAS – Wageningen Journal of Life Sciences*, 52(2), 133–43. doi:10.1016/S1573-5214(04)80009-5.

Bondi, G., Peruzzi, E., Macci, C., Masciandaro, G. and Pistoia, A. (2015). Changes in soil organic matter associated with pig rearing: Influence of stocking densities and land gradient on forest soils in central Italy. *Agriculture, Ecosystems and Environment*. 211, 32–42. http://dx.doi.org/10.1016/jagee.2015.05.003.

Bonneau, M. and Lebret, B. (2010). Production systems and influence on eating quality of pork. *Meat Science*. 84(2), 293–300. doi: 10.1016/j.meatsci.2009.03.013.

Bordeaux, C., Grossman, J., White, J., Osmond, D., Poore, M. and Pietrosemoli, S. (2014). Effects of rotational infrastructure within pasture-raised pig operations on ground cover, soil nutrient distribution, and bulk density. *Journal of Soil and Water Conservation*. 69(2), 120–30 doi:10.2489/jswc.69.2.120.

Bornett, H. L. I., Edge, H. L. and Edwards, S. A. (2003). Alternatives to nose-ringing in outdoor sows: 1. The provision of a sacrificial rooting area. *Applied Animal Behaviour Science*, 83 (4), 267–76.

Both, MdC. (2003). *Comportamento e produção de suínos mantidos em pastagem e submetidos a diferentes níveis de restrição alimentar.* PHD Thesis. Universidade Federal do Rio Grande do Sul, 127p. Available at: http://www.lume.ufrgs.br/bitstream/handle/10183/5267/000468021.pdf?sequence=1 [14 June 2016].

Blumetto Velazco, O. R., Calvet Sanz, S., Estellés Barber, F. and Villagrá García, A. (2013). Comparison of extensive and intensive pig production systems in Uruguay in terms of ethologic, physiologic and meat quality parameters. *Revista Brasileira de Zootecnia*, 42(7), 521–9. Available at: https://dx.doi.org/10.1590/S1516-35982013000700009.

Blumetto, O., Calvet, S., Estellés, F., Villagrá, A. and Torres, A. G. (2012). Caracterización productiva y ambiental de un Sistema semi-extensivo de engorde de cerdos en condiciones de sequía en Uruguay. *Información Técnica Económica Agraria*, 108 (3), 256–74.

Buehrer, K. A. and Grieshop, M. J. (2014). Postharvest grazing of hogs in organic fruit orchards for weed, fruit, and insect pest management. *Organic Agriculture*, 4,223–32.

Campagna, D., Dichio, L., Ausilio, A., Bessón, P. A., Silva, P. and Spinollo, L. (2011). Efectos de la carga animal sobre el tapiz vegetál y el recurso suelo en un sistema de producción porcina al aire libre. In: Barlocco, N. and Vadell, A. (eds). *Producción de Cerdos a Campo. Aportes para el desarrollo de tecnologías apropiadas para la producción familiar.* [pdf] Montevideo, Uruguay. Universidad de la Republica. Facultad de Agronomia. 92–101. Available at: http://www.upc.edu.uy/images/documents/extension/Publicacion%2015%20anos%20UPC.pdf.

Carlson, D., Laerke, H. N., Poulsen, H. D. and Jorgensen, H. (1999). Roughages for growing pigs, with emphasis on chemical composition, ingestion and faecal digestibility. *Acta Agriculture Scandinavica*, 49A,129–36.

Castillo, M, Martín-Orúe, S. M., Anguita, M, Pérez, J. F. and Gasa J., (2007). Adaptation of gut microbiota to corn physical structure and different types of dietary fibre, *Livestock Science*, 109, 149—152.

Danielsen, V.; Hansen, L. L.; Møller, F.; Bejerholm, C. and Nielsen, S. (1999). Production results and sensory meat quality of pigs fed different amounts of concentrate and *ad libitum* clover grass or clover grass silage. In Hermansen, J. E., Lund, V. and Thuen, E. (eds), *Ecological Animal Husbandry in the Nordic Countries*, Danish Research Centre for Organic Farming: Tjele, Denmark, pp. 79–86.

Demori, A. B., Lovatto, P. A., Andretta, I., Kipper, M., Lehnen, C. R. and Remus, A. (2012). Criação intensiva de suínos em confinamento ou ao ar livre: estudo meta-analítico do desempenho zootécnico nas fases de crescimento e terminação e avaliação de carcaça e carne no Longissimus dorsi. *Ciência Rural*, 42(7), 1294–9. Available at: https://dx.doi.org/10.1590/S0103-84782012000700025.

Dolman, M. A., Vrolijk, H. C. J. and De Boer, I. J. M. (2012). Exploring variation I economic, environmental and societal performance among Dutch fattening pig farms. *Livestock Science*, 149, 143–54.

Doran, M. (2015). Multi-Species Grazing Systems. In: Ellsworth, S. (ed.). *Outdoor Hog Production.* Best Practices for Resource Conservation in the San Francisco Bay Area. [pdf] Available at: http://www.acrcd.org/Portals/0/Multi%20Species%20Grazing.pdf [18 November 2015.]

Dourmad, J-Y. and Casabianca, F. (2013). Effect of husbandry systems on the environmental impact of pig production. *Acta agriculturae Slovenica*, Supplement 4, 197–204. ISSN: 1581-9175 Available at: http://prodinra.inra.fr/record/253477 [10 June 2016].

Edge, H. L., Bornett, H. L. I., Newton, E. and Edwards, S. A. (2004). Alternatives to nose-ringing in outdoor sows: 2.The provision of edible or inedible overground enrichment. *Animal Welfare*, 13(2), 233–7.

Edge, H. L., Bulman, C. A. and Edwards, S. A. (2005). Alternatives to nose-ringing in outdoor sows: the provision of root crops. *Applied Animal Behavioural Science*, 92 (1–2), 15–26.

Edwards, S. A. (2003). Intake of nutrients from pasture by pigs. *Proceedings of the Nutrition Society.* 62, 257–65.

Edwards, S. A. (2005). Product quality attributes associated with outdoor pig production. *Livestock Production Science*, 94, 5–14.

Eriksen, J. (2001). Implications of grazing by sows for nitrate leaching from grassland and the succeeding cereal crop. *Grass & Forage Science*, 56(4), 317–22.

Eriksen, J., Hermansen, J. E., Strudsholm, K. and Kristensen, K. (2006a). Potential loss of nutrients from different rearing strategies for fattening pigs on pasture. *Soil Use Management* 22, 256–66.

Eriksen, J. and Kristensen, K. (2001). Nutrient excretion by outdoor pigs: a case study of distribution, utilisation and potential for environmental impact. *Soil Use and Management*, 17, 21–9.

Eriksen, J., Petersen, S. O. and Sommer, S. G. (2002). The fate of nitrogen in outdoor pig production. *Agronomie*, 22(7–8), 863–7. doi:10.1051/agro: 2002045.

Eriksen, J., Studnitz, M., Strudsholm, K., Kongsted, A. G. and Hermansen, J. E. (2006b). Effect of nose ringing and stocking rate of pregnant and lactating outdoor sows on exploratory behavior, grass cover and nutrient loss potential. *Livestock Science*, 104, 91–102.

Espagnol, S. and Demartini, J. (2014). In Rita Schenck, R. and Huizenga, D. (eds). Proceedings of the *9th International Conference on Life Cycle Assessment in the Agri-Food Sector (LCA Food 2014)*, 8–10 October 2014, San Francisco, USA. ACLCA, Vashon, WA, USA. 364–71. Available at: http://lcacenter.org/lcafood2014/proceedings/LCA_Food_2014_Proceedings.pdf ISBN: 978-0-9882145-7-6.

Fajardo, D. S. (2009). *Evaluación de dos sistemas de instalaciones y manejo para la etapa de lactancia, comparando la producción porcina tradicional vs la producción al aire libre.* BSc Thesis. Universidad de la Salle, Colombia. 125p.

Faner, C. L. (2007). La pastura de alfalfa como fuente de alimentación para cerdos en crecimiento y terminación. *Boletín AACP*. Available at: www.produccion-animal.com.ar.

FAO. (2007). Animal genetic resources and resistance to disease. In: Rischkowsky, B. and Pillin, D. (eds), *The State of the World's Animal Genetic Resources for Food and Agriculture*, FAO, Rome, Italy, pp. 101–12.

Fortina, R., Brugiapaglia, A., Tassone, S., Malfatto, V. and Cavallero, A. (2011). Pasture-Based Swine Management: Behaviour and Performances of Growing-Finishing Pigs. *Animal Science and Biotechnologies*, 44 (1), 37–40.

García-Valverde, R., Nieto, R., Lachica, M. and Aguilera, J. F. (2007). Effects of herbage ingestion on the digestion site and nitrogen balance in heavy Iberian pigs fed on an acorn-based diet. *Livestock Science*, 112, 63–77.

Gentry J. G., McGlone J. J., Blanton J. R. and Miller M. F. (2002a). Alternative housing systems for pigs: influences on growth, composition, and pork quality. *Journal of Animal Science*, 80(7), 1781–90.

Gentry, J. G., McGlone, J. J, Miller, M. F. and Blanton, J. R. (2002b). Diverse birth and rearing environment effects on pig growth and meat quality. *Journal of Animal Science*, 80(7), 1707–15.

Gentry, J. G., McGlone, J. J., Miller, M. F. and Blanton, J. R. (2004). Environmental effects on pig performance, meat quality, and muscle characteristics. *Journal of Animal Science*, 82(1), 209–17.

Giannone, M. (2001a). Pascoli di favino per i suini biologici. *Rivista di suinicoltura*. 42, 125–30.

Giannone, M. (2001b). Sorgo e barbabietola per i suini biologici. *Rivista di suinicoltura*. 42, 38–42.

Gilmore, J. (1999). Making hogs work for you. *Small Farm Today*. 16(5), 67–9. Available at: http://aims.fao.org/serials/c_b72837fe [Accessed 12 June 2016].

Goh, K. M. and Bruce, G. E. (2005). Comparison of biomass production and biological nitrogen fixation of multi-species pastures (mixed herb leys) with perennial ryegrass-white clover pasture with and without irrigation in Canterbury, New Zealand. Agriculture, *Ecosystems and Environment*, 110, 230–40.

Gourdine, J. L., De Greef, K. H. and Rydhmer, L. (2010). Breeding for welfare in outdoor pig production: A simulation study. *Livestock Science*. Volume 132(1–3), 26–34. doi:10.1016/j.livsci.2010.04.017

Green, J. T., Benson, G. and Pietrosemoli, S. (2014). Nutrient and vegetation Management in outdoor hog production systems. [pdf] Available at: https://cefs.ncsu.edu/resources/nutrient-and-vegetation-management-in-outdoor-hog-production-systems-2014/. [Accessed 23 June 2016].

Grosso, L., Barbieri, S., Ferrante, V. and Ferrari, P. (2011). La producción porcina a campo en Italia. In: Barlocco, N. and Vadell, A. (eds). *Producción de Cerdos a Campo. Aportes para el desarrollo de tecnologías apropiadas para la producción familiar*. Montevideo. Universidad de la república. Facultad de Agronomía. 110–15. Available at: http://www.upc.edu.uy/images/documents/extension/Publicacion%2015%20anos%20UPC.pdf

Gustafson, G. M. and Stern, S. (2003). Two strategies for meeting energy demands of growing pigs at pasture. *Livestock Production Science* 80, 167–74.

Gustafson, G. M. and Svensson, L. (2003). Ammonia emissions from pigs on deep litter beds and in a strip grazing system. *Journal of Sustainable Agriculture*, 23(2), 41–51. DOI:10.1300/j064v23n02_05.

Guy, J. H., Rowlinson, P., Chadwick, J. P. and Ellis, M. (2002a). Growth performance and carcass characteristics of two genotypes of growing-finishing pig in three different housing systems. *Animal Science*, 74, 493–502.

Guy, J. H., Rowlinson, P., Chadwick J. P. and Ellis, M. (2002b).Health conditions of two genotypes of growing-finishing pig in three different housing systems: implications for welfare. *Livestock Production Science*, 75, 233–43. doi:10.1016/S0301-6226(01)00327-X.

Helmerichs, J. (2014). *Paddock management systems for organic growing pigs Effect of land allocation strategies on foraging activity and excretory behavior*. MSc Thesis. Aarhus University. Department Of Agroecology. 45p.

Henney, J. (2012). *An evaluation of the use of pigs as a method of bracken control*. [pdf] MSc. Grazing Animals Project Heritage Grazing Traineeship. 40p. Available at: http://www.grazinganimalsproject.org.uk/habitat_land_management.html?publication=1;154 [Accessed 23 June 2016].

Hermansen, J. E. (2005). Integration of organic pig production into land use. In: Sundrum, A. and Friedrich Weißmann, F. (eds). *Organic Pig Production in Free Range Systems*. [pdf] Braunschweig. FAL *Agricultural Research*. 281, 3–12. ISSN 0376-0723 ISBN 3-86576-005-8. Available at: http://orgprints.org/6094/1/6094.pdf [Accessed 4 June 2016].

Hermansen, J. E., Eriksen, J. and Strudsholm, K. (2005). Full or partial outdoor rearing of slaughter pigs. Effects on performance, carcass quality and nutrient load. In: *Researching Sustainable Systems*. [pdf] Adelaide. ISOFAR. 250–3. Available at:http://http://http://orgprints.org/4368/4/4368-Hermansen_etal_4p_revised-ed.pdf [Accessed 4 June 2016].

Heyer, A. H., Andersson, K. and Lundstrom, K. (2006). Performance, carcass and technological meat quality of pigs in indoor and outdoor production systems. *Acta Agriculturae Scand* Section A. 56, 55–64. DOI: 10.1080/09064700600670904.

Hilimire, K. (2011). Integrated Crop/Livestock Agriculture in the United States: A Review. *Journal of Sustainable Agriculture*, 35(4), 376–93, DOI: 10.1080/10440046.2011.562042.

Honeyman, M. S. (2004). Extensive bedded indoor and outdoor pig production systems in USA: current trends and effects on animal care and product quality. *Livestock Production Science*, 94(1–2), 15–24.

Hodgkinson, S. M., Cárcamo, A. and López, I. (2011). Selective grazing of *Lolium perenne* and *Plantago laceolata* by growing European wild boar (*Sus scrofa* L.) in a semi-extensive system. *Livestock Science*, 140, 268–74.

Hodgkinson, S. M., López, I. F. and Navarrete, S. (2009). Ingestion of energy, protein and amino acids from pasture by grazing European wild boar (*Sus scrofa* L.) in a semi-extensive production system. *Livestock Science* 122, 222–6.

Hodgkinson, S., Schmidt, M. and Ulloa, N. (2008). Comparison of the digestible energy content of maize, oats and alfalfa between the European wild boar (*Sus scrofa* L.) and Landrace x Large White (*Sus scrofa domesticus*). *Animal Feed Science and Technology* 144, 167–73.

Horsted, K., Kongsted, A. G., Jorgesen, U. and Sorensen, J. (2012). Combined production of free-range pigs and energy crops-animal behavior and crop damages. *Livestock Science*. 150, 200–8. doi:10.1016/j.livsci.2012.09.006.

Horta, C., Batista, M., Roque, N. and Almeida, J. (2012). Environmental impact of pig production: soil P forms evolution, spatial distribution and P losses in drainage waters. *Revista de Ciencias Agrarias*. 35(1), 36–48. ISNN: 0871-018 X.

Hovi, M., Sundrum, A. and Thamsborg, S. M. (2003). Animal health and welfare in organic livestock production in Europe: current state and future challenges. *Livestock Production Science*, 80, 41–53.

Isabel, B. and López Bote, C. (2000). Ácidos grasos, antioxidantes naturales y ejercicio: factores que determinan la calidad de los productos del cerdo ibérico. In: *II Jornadas El cerdo ibérico y sus productos*. Salamanca. Consejeria de Agricultura y ganadería, 2–10.

Jakobsen, M., Kongsted, A. G. and Hermansen, J. E. (2015). Foraging behaviour, nutrient intake from pasture and performance of free-range growing pigs in relation to feed CP level in two organic cropping systems. *Animal*, 9(12), 2006–16. doi:10.1017/S1751731115001585.

Jørgensen, H., Serena, A., Hedemann, M. S. and Bach-Knudsen, K. E. (2007). The fermentative capacity of growing pigs and adult sows fed diets with contrasting type and level of dietary fibre. *Livestock Science*, 109 (1–3), 111–14.

Jurjanz, S. and Roinsard, A. (2014). Ingestion d'herbe et de sol par des truies au pâturage. Alter Agri, 125, 27–8.

Kambashi, B., Picron, P., Boudry, C., Théwis, A., Kiatoko, H. and Bindelle, J. (2014). Nutritive value of tropical forage plants fed to pigs in the Western provinces of the Democratic Republic of the Congo. *Animal Feed Science & Technology*, 191, 47–56. DOI:10.1016/j.anifeedsci.2014.01.012.

Kanga, J. S., Kanengoni, A. T., Makgothi, O. G. and Baloyi, J. J. (2012). Estimating pasture intake and nutrient digestibility of growing pigs fed a concentrate-forage diet by n-alkane and acid-insoluble ash markers. *Tropical Animal Health and Production*, 44(7), 1797–802. doi:10.1007/s11250-012-0141-1.

Kathiresan, R. (2012). Utility tag, farming elements and ITK for sustainable management of weeds in changing climate. *Pakistan Journal of Weed Science Research*, 18, 271–82, Special Issue.

Kays, S. J. and Nottingham, S. F. (2008). Value in human and animal diets. In: Kay, S. J. and Nottingham, S. F. (eds), *Biology and Chemistry of Jerusalem Artichoke Helianthus tuberosus L*. CRC Press, Boston, 2007, pp. 97–125. DOI: 10.1201/9781420044966.ch6.

Kelly, H. R. C., Browning, H. M., EL Day, J., Martins, A., Pearce, G. P., Stopes, C. and Edwards, S. A. (2007). Effect of breed type, housing and feeding system on performance of growing pigs managed under organic conditions. *Journal of the Science of Food and Agriculture*, 87(15), 2794–2800.

Knage-Rasmussen, K. M., Houe, H., Rousing, T. and Sørensen, J. T. (2014). Herd- and sow-related risk factors for lameness in organic and conventional sow herds. *Animal*, 8(1), 121–7.

Kongsted, A. G., Horsted, K. and Hermansen, J. E. (2013). Free-range pigs foraging on Jerusalem artichokes (*Helianthus tuberosus* L.) – Effect of feeding strategy on growth, feed conversion and animal behaviour. *Acta Agriculturae Scandinavica*, 63, 76–83.

Kongsted, A. G. and Jakobsen, M. (2015). Effect of genotype and level of supplementary concentrate on foraging activity and vegetation cover in an organic free-range pig system. Acta Agriculturae Scandinavica: Section A, Animal Science. Sep-Dec 2015, Vol. 65 Issue 3/4, p 139–47. 9p.

Kongsted, A. G., Nørgaard, J. V., Jensen, S. K., Lauridsen, C., Juul-Madsen, H. R., Norup; L. R., Engberg, R. M., Horsted, K. and Hermansen, J. E. (2015). Influence of genotype and feeding strategy on pig performance, plasma concentrations of micro nutrients, immune responses and faecal microbiota composition of growing-finishing pigs in a forage-based system. *Livestock Science*, 178, 263–71.

Lagreca, L. and Marotta, E. (2005). Comportamiento y características ingestivas de forraje en la cerda a campo. In: *III Encuentro Latinoamericano de Especialistas en Sistemas de Producción Porcina a Campo*. [pdf] Marco Juarez. INTA. Available at: http://www.produccion-animal.com.ar/produccion_porcina/00-produccion_porcina_general/24-comportamiento_ingestivo.pdf [10 June 2016]

Laister, S. and Konrad, S. (2005). Behaviour, performance and carcass quality of three genotypes of growing-finishing pigs in outdoor pig production in Austria: A pilot study. In: *Organic Pig Production in Free Range Systems*. Eds. Sundrum, A., Weißmann, F. Landbauforschung Völkerode, Special Issue 281. P. 13.

Lammers, P. J. (2011). Life-cycle assessment of farrow-to-finish pig production systems: A review. *CAB Reviews: Perspectives in Agriculture, Veterinary Science, Nutrition and Natural Resources*, 6, 1–8.

Largerkvist, C. J., Carlsson, F. and Viske, D. (2006). Swedish consumer preferences for animal welfare and biotech: A choice experiment, *AgBio Forum*, 9, 51–8.

Larsen, V. A. and Kongsted, A. G. (1999). Sows on pasture. In: Hermansen, J. E. and Erling Thuen (eds). *Ecological Animal Husbandry in the Nordic Countries*. [pdf]Horsens. NJF 303, 99–105. Available at: http://www.darcof.dk/publication/rapport/dar_2.pdf. [Accessed 24 October 2008].

Lebret, B. (2008). Effects of feeding and rearing systems on growth, carcass composition and meat quality in pigs. *Animal*. 2. 10. 1548–58.

Lebret, B., Dourmad, J. Y., Mourot, J., Pollet, P. Y. and Gondret, F. (2014). Production performance, carcass composition and adipose tissue traits of heavy pigs: influence of breed and production system. *Journal of Animal Science*, 92(8), 3543–56.

Lebret, B., Ecolan, P., Bonhomme, N., Méteau, K. and Prunier, A. (2015). Influence of production system in local and conventional pig breeds on stress indicators at slaughter, muscle and meat traits and pork eating quality. *Animal*, 9(08), 1404–13. doi:10.1017/S1751731115000609.

Leite, D., Silva, M., Medeiros, R., Saibro, J., Pavan, M. A., Zanella, J. A. and Barrey, M. A. (2006a). Comportamento de suínos submetidos a diferentes sistemas de pastejo em pastagem de trevo-branco. *R. Bras. Zootec.*, 35(4), 1774–9. Available at: http://www.scielo.br/scielo.php?script=sci_arttext&pid=S1516-35982006000600027&lng=en&nrm=iso [Accessed 14 June 2016]. http://dx.doi.org/10.1590/S1516-35982006000600027.

Leite, D., Silva, M., Medeiros, R., Saibro, J., Pavan, M. and Barrey, M. (2006b). Efeito de diferentes sistemas de pastejo sobre o desempenho de suínos mantidos em pastagem de trevo-branco (*Trifolium repens* L.). *R. Bras. Zootec.*, 35(3), 792–6. Available at: http://www.scielo.br/scielo.php?script=sci_arttext&pid=S1516-35982006000300022&lng=en&nrm=iso>. [Accessed 14 June 2016]. http://dx.doi.org/10.1590/S1516-35982006000300022.

Lindgren, K., Bochicchio, D., Hegelund, L., Leeb, C., Mejer, H., Roepstorff, A. and Sundrum, A., (2014). Animal health and welfare in production systems for organic fattening pigs. *Organic Agriculture*, 4(2), 135–47.

Livingstone, R. M., Baird, B. A. and Atkinson T. (1980). Cabbage (*Brassica oleracea*) in the diet of growing – finishing pigs. *Animal Feed Science and Technology*, 5(1), 69–75. doi:10.1016/0377-8401(80)90012-7.

Livingstone, R. M., Jones, A. S. and Mennie, I. 1977. Swedes (*Brassica napus*) for growing pigs: Chemical composition and use as a replacement for barley in the diet. *Animal Feed Science and Technology*, 2(1), 31–40.

Macci, C., Doni, S., Bondi, G., Davini, D., Masciandaro, G. and Pistoia, A. (2012). Effects of wild boar (*Sus scrofa*) grazing on soil properties in Mediterranean environment. *CATENA*, 98, 79–86.

Martens, S. D., Tiemann, T. T., Bindelle, J., Peters, M. and Lascano, C. E., (2012). Alternative plant protein sources for pigs and chickens in the tropics – nutritional value and constraints: a review. *Journal of Agriculture and Rural Development in the Tropics*, 113, 101–23.

McGlone, J. J. (2013). The future of pork production in the world: Towards sustainable, welfare-positive systems. *Animal*, 3(2), 401–15; doi:10.3390/ani3020401.

Menzi, H., Stauffer, W., Zihlmann, U. and Weisskopft, P. (1998). Environmental impacts of outdoor pig production. In: Martinez, J. and Maudet, M. N. (eds), 8th International Conference on the *Network on Recycling of Agricultural, Municipal and Industrial Residues in Agriculture* (RAMIRAN 98). Rennes. *FAO ESCORENA* Editions Quae, 2, 31–9. ISBN: 978-2-85362-531-9.

Miao, Z. H., Glatz, P. C. and Ru, Y. J. (2004). Review of production, husbandry and sustainability of free-range pig production systems. *Asian Australasian Journal of Animal Sciences*, 17(11), 1615–34.

Monteverde, S. and Del Pino, A. (2014). Variabilidad espacial en las propiedades químicas del suelo en un sistema de producción con cerdos en pastoreo. *Revista Ciencias Agronómicas* 23, 037–045.

Monteverde, S., Del Pino, A., Piaggio, J. and Nay, H. (2011). Metales pesados en la producción de cerdos a campo. In: *VII Jornadas técnicas de la Facultad de Veterinaria. UDELAR.* 190. Available at: http://www.fvet.edu.uy/?q=tecnolog%C3%ADa-agropecuaria/publicaciones [Accessed 27 August 2013].

Mowat, D., Watson C. A., Mayes R. W., Kelly H., Browning H. and Edwards S. A. (2001). Herbage intake of growing pigs in an outdoor organic production system. *British Society of Animal Science Annual Meeting, York*, 169.

Mpampouli, S., Abraham, E. M., Kyriazopoulos, A., Fotiadis, G., Parissi, Z. M. and siouvaras, C. N. (2009). The effects of wild boar grazing on the floristic diversity of a silvopastoral oak system. In Alternative functions of grassland. Proceedings of the *15th European Grassland Federation Symposium*, Brno, Czech Republic, 7–9 September. Editors Cagaš , B.; Macháč Machác Czech Repu*Grassland Science in Europe*, 14, 192–5. ISBN 978-80-86908-15-1 Record Number 20093293375.

Nazareno, A. C., Da Silva, I. J. O., Nunes, M. L. A., De Castro, A. C., Miranda, K. O. S., and Trabachini, A. (2012). Caracterização bioclimática de sistemas ao ar livre e confinado para a criação de matrizes suínas gestantes. *Revista Brasileira de Engenharia Agrícola e Ambiental*, 16(3), 314–19. Available at: http://www.scielo.br/scielo.php?script=sci_arttext&pid=S1415- 43662012000300013&lng=en&nrm=iso. https://dx.doi.org/10.1590/S1415-43662012000300013 [Accessed 19 June 2016].

Nilzén, V., Babol, J., Dutta, P. C., Lundeheim, N., Enfält, A-C. and Lundström K. (2001). Free range rearing of pigs with access to pasture grazing – effect on fatty acid composition and lipid oxidation products. *Meat Science*, 58(3), 267–75. [Accessed 12 April 2016] doi:10.1016/S0309-1740(00)00164-9.

Norwood, F. B. and Lusk, J. L. (2011). A calibrated auction-conjoint valuation method: valuing pork and eggs produced under differing animal welfare conditions. *Journal of environmental Economics and Management*, 62, 80–94.

NRCS. (2007). Conservation planning guidelines for outdoor swine operations. Available at: https://efotg.sc.egov.usda.gov/references/Delete/2008-7-12/OSOTECHNOTE.pdf [Accessed 25 July 2016].

Nunn, L., Embree, C. G., Hebb, D., Bishop, S. D. and Nichols, D. (2007). Rotationally grazing hogs for orchard floor management in organic apple orchards ISHS. *Acta Horticulturae*, 737(9), 71–78, http://dx.doi.org/10.17660/ActaHortic.2007.737.9 *ISSN*: 0567-7572 [Accessed 19 June 2016].

Ocampo-Duran, A. (2011). Producción porcina tropical: cuando el alimento viene de las alturas. In: Barlocco, N. and Vadell, A. (ed.). *Producción de Cerdos a Campo. Aportes para el desarrollo de tecnologías apropiadas para la producción familiar.* [pdf] Montevideo. Universidad de la Republica, Facultad de Agronomia. 116–20 Available at: http://www.upc.edu.uy/images/documents/extension/Publicacion%2015%20anos%20UPC.pdf. [Accessed 12 April 2016]

O'Connor, S-J. and Kelly, D. (2012). Seed dispersal of matai (*Prumnopitys taxifolia*) by feral pigs (*Sus scrofa*). *New Zealand Journal of Ecology*, 36(2), 228–31. Available at: http://www.newzealandecology.org/nzje/[Accessed 7 August 2016].

Ogle, B. (2006). Forage for pigs: nutritional, physiological and practical implications. In: *Forages for Pigs and Rabbits.* [pdf] Phnom Penh MEKARN-CelAgrid. Available at: http://www.mekarn.org/proprf/ogle.htm [Accessed 21 August 2012].

Passos, A. A., Andrade, C., Phillips, C. E., Coffey, M. T. and Kim, S. W. (2015). Nutrient value of spray field forages fed to pigs and the use of feed enzymes to enhance nutrient digestibility. *Journal of Animal Science*, 93(4), 1721–8.

Patridge, G. (2014). Challenges in feeding high-fiber diets to pigs. [online] Pig International. Available at: www.WATTAgNet.com [Accessed 21 June 2016].

Petersen, S. O., Kristensen, K. and Eriksen, J. (2001). Denitrification losses from outdoor piglet production. *Journal of Environmental Quality*, 30(3), 1051–8.

Pietrosemoli, S. and Green, J. T. (2015). Designing Pasture Subdivisions for Practical Management of Hogs. [pdf] Available at: https://cefs.ncsu.edu/resources/designing-pasture-subdivisions-for-practical-management-of-hogs-2015/[Accessed 26 June 2016].

Pietrosemoli, S. and Green, J. T. (2009). Efecto de la carga animal de cerdas adultas en la cobertura vegetal de pasto Bermuda (*Cynodon dactylon*) durante el invierno. *ALPA*, 17(1), 447–51.

Pietrosemoli, S. and Green, J. T. (2012a). Effect of outdoor swine management systems on tall fescue (*Festuca arundinacea* L.) ground cover and animal performance. *Animal Science*, 90, E-Suppl. 3/*Journal of Dairy Science*, 95, E-Suppl. 2: 33.

Pietrosemoli, S. and Green, J. T. (2012b). Soil nutrients in tall fescue (*Festuca arundinacea* L.) paddocks managed under different outdoor hog systems. *Animal Science*, 90, E-Suppl. 3/*Journal of Dairy Science* 95, E-Suppl. 2: 32.

Pietrosemoli, S., Luginbuhl, J-M. and Green, J. T. (2012). Effect of outdoor swine management systems on the botanical composition of tall fescue (*Festuca arundinacea* L.) paddocks. *Animal Science*, 90, E-Suppl. 3/*Journal of Dairy Science*, 95, E-Suppl. 2: 33.

Pietrosemoli, S., Green, J. T. and Vibart, R. (2009). Effects of stocking rate of weaned to finishing pigs on bermudagrass ground cover. *Journal of Animal Science*, 87, E-Suppl. 2/*Journal of Dairy Science*, 92, E-Suppl. 1, 449.

Pietrosemoli, S., Green, J., Bordeaux, C., Menius, L. and Curtis, J. (2012). Conservation practices in outdoor hog production systems: Findings and Recommendations from CEFS. [pdf] Available at: http://www.cefs.ncsu.edu/publications/conservation_practices_2012.pdf [Accessed 26 June 2016].

Pietrosemoli, S., Moron-Fuenmayor, O. E., Paez, A. and Villamide, M. J. (2016). Effect of including sweet potato (*Ipomoea batatas* Lam) meal in finishing pig diets on growth performance, carcass traits and pork quality. *Animal Science Journal*. 10pp. Doi:10.1111/asj.12546.

Pistoia, A., Bondi, G., Balestri, G., Man, I. D. and Masciandaro, G. (2012). Soil degradation by grazing pig in Mediterranean environment. In: De Pedro, E. J. and Cabezas, A. B. (eds). *7th International Symposium on the Mediterranean Pig*. [pdf] Zaragoza: CIHEAM. *Options Méditerranéennes: Série A. Séminaires Méditerranéens*; 101, 187–91. Available at: http://om.ciheam.org/om/pdf/a101/00006677.pdf [Accessed 4 May 2015].

Pistoia, A., Poli, P., Casarosa, L., Balestri, G. and Ferruzzi, G. (2008). Utilizzo del pascolo boschivo nell allevamento biológico del suino. In: Paolo Pignatteli, P., Ferrante, V., Mattiello, S., Barbieri, S. and Lolli, S. (eds). *Zootecnia biologica e ricerca: Investire insieme. VI Convegno Nazionale dell Associazione Italiana di Zootecnia biologica e biodinamica.* [pdf] Arezzo. ZooBioDi, 68–73. ISBN 978-88-903475-0-4. Available at: http://www.zoobiodi.it/doc/ATTI_2008_DEFINITIVI.pdf [Accessed 16 May 2015].

Pousset, J. (2010). Culture biologique du topinambour. [pdf] Groupement régional d'agriculture biologique de Basse-Normandie. *Biodoc* n° 22. Available at: http://www.bio-normandie.org/wp-content/uploads/2009/09/Biodoc-n-22-Culture-biologique-du-topinambour-_2_1.pdf [Accessed 21 June 2016].

Presto, M. H., Andersson, H. K., Wallgren, P. and Lindberg, J. E. (2007). Influence of dietary amino acid level on performance, carcass quality and health of organic pigs reared indoors and outdoors. *cta Agriculturae Scandinavica Section A*, 57, 61–72.

Pugliese, C. and Sirtori, F. (2012). Quality of meat and meat products produced from southern European pig breeds. *Meat Science*, 90(3), 511–18.

Pugliese, C., Bozzi, R., Campodoni, G., Acciaioli, A., Franci, O. and Gandini, G. (2005). Performance of Cinta Senese pigs reared outdoors and indoors: 1. Meat and subcutaneous fat characteristics. *Meat Science*, 69(3), 459–64. http://dx.doi.org/10.1016/j.meatsci.2004.09.001.

Quijada, R. P., Bitsch, N. I. and Hodgkinson, S. M. (2012). Digestible energy content of pasture species in growing European wild boar (*Sus scrofa* L.). *Journal of Animal Physiology and Animal Nutrition*, 96(3), 421–7. DOI:10.1111/j.1439-0396.2011.01158.x.

Quintern, M. (2005). Integration of organic pig production within crop rotation: Implications on nutrient losses In: *Organic Pig Production in Free Range Systems*. Eds Sundrum, A.; Weißmann, F., Special Issue 281.

Quintern, M. and Sundrum, A. (2006). Ecological risks of outdoor pig fattening in organic farming and strategies for their reduction - results of a field experiment in the centre of Germany. *Agriculture, Ecosystems and Environment*, 117(4), 238–50.

Rachuonyo, H. A., Allen, V. G. and McGlone, J. J. (2005). Behavior, preference for, and use of alfalfa, tall fescue, white clover, and buffalograss by pregnant gilts in an outdoor production system. *Journal of Animal Science*, 83(9), 2225–34.

Rachuonyo, H. A. and McGlone, J. J. (2005). Impact of outdoor gestating gilts on soil nutrients, vegetative cover, rooting damage and pig performance. *Journal of Sustainable Agriculture*, 29(3), 69–87.

Rachuonyo, H. A., Pond, W. G. and McGlone, J. J. (2002). Effects of stocking rate and crude protein intake during gestation on ground cover, soil-nitrate concentration, and sow and litter performance in an outdoor swine production system. *Journal of Animal Science*, 80(6), 1451–61.

Ramírez-Restrepo, C. A. and Barry, T. N. (2005). Alternative temperate forages containing secondary compounds for improving sustainable productivity in grazing ruminants. *Animal Feed Science and Technology*, 120(3–4), 179–201.

Régnier, C., Bocage, B., Archimède, H., Noblet, J. and Renaudeau, D., (2013). Digestive utilization of tropical foliages of cassava, sweet potatoes, wild cocoyam and Erythrina in Creole growing pigs. *Animal Feed Science and Technology*, 180(1), 44–54.

Renner, B. M. (2011). *The effect of stocking rate history on soil nutrient levels and forage nutrient uptake in pasture hog production systems*. MSc. Thesis. North Carolina State University, Crop Science. 82 p.

Renner, B., Pietrosemoli, S., Luginbuhl, J-M., Raczkowski, C., Green, J. T. and Grossman, J. (2011). Effect of stocking rate on forage production, soil compaction and root numbers in a swine pasture system. *Journal of Animal Science*, 89, E-Suppl. 1/*Journal of Dairy Science*, 94, E-Suppl. 1, 315.

Rivera-Ferre, M. G., Edwards, S. A., Mayes, R. W., Riddoch, I. and Hovell, F. D. D. (2001).The effect of season and level of concentrate on the voluntary intake and digestibility of herbage by outdoor sows. *Animal Science*, 72, 501–10.

Rivero, C., Cabrales, E., Santana, G., Rivas, M., Pulido, M., Rey, J. C., Lobo, D., Lozano, Z. and Araque, H. (2013). Efecto del pastoreo de cerdos sobre las fracciones de nitrógeno, carbono y fósforo del suelo. *Revista Temas Agrarios*, 18(1), 23–33. ISSN 0122-7610.

Rivero, J., López, I. and Hodgkinson, S. (2013a). Pasture consumption and grazing behaviour of European wild boar (*Sus scrofa* L.) under continuous and rotational grazing systems, *Livestock Science*, 154(1–3), 175–83.

Rivero, M. J., López, I. F. and Hodgkinson, S. M. (2013b). Pasture dry matter consumption in European wild boars (*Sus scrofa* L.) as affected by herbage allowance. *Journal of Animal Science*, 91(4), 1758–64 doi:10.2527/jas.2012-5289

Robinson, J. (2013). *Towards a framework for agroecological pig production: maximizing functional roles for pigs within food and farm systems*. MSc. Thesis. Norwegian University of Life Sciences. 105 pp.

Rodríguez-Estévez, V., García, A. and Gómez, A. G. (2009a). Characteristics of the acorns selected by free range Iberian pigs during the montanera season. *Livestock Science*, 122(2–3), 169–76.

Rodríguez-Estévez, V., García, A. and Gómez, A. G. (2009b). Intrinsic factors of acorns that influence the efficiency of their consumption by Iberian pigs. *Livestock Science*, 122(2–3), 281–5.

Rodríguez-Estévez, V., García, A., Peña, F. and Gómez, A. G. (2009c). Foraging of Iberian pigs grazing natural pasture in the dehesa. *Livestock Science*, 120(1), 135–43.

Rodríguez-Estévez, V., Sánchez-Rodríguez, M., García, A. and Gómez-Castro, A. G. (2010). Feed conversion rate and estimated energy balance of free grazing Iberian pigs. *Livestock Science*, 132(1), 152–6.

Rodríguez-Estévez, V., Sánchez-Rodríguez, M., García, A. and Gómez-Castro, A. G. (2011). Average daily weight gain of Iberian fattening pigs when grazing natural resources. *Livestock Science*, 137(1), 292–5.

Rodríguez-Estévez, V., Sánchez-Rodríguez, M., Gómez-Castro, A. G. and Edwards, S. A. (2010). Group sizes and resting locations of free range pigs when grazing in a natural environment. *Applied Animal Behaviour Science*, 127(1), 28–36.

Roepstorff, A. and Mejer, H. (2001). Strategies for parasite control in organic pigs. In The 5th NAHWOA Workshop, Rodding, 11–13 November. 72–8.

Roepstorff, A., Mejer, H., Nejsum, P. and Thamsborg, S. M. (2011). Helminth parasites in pigs: New challenges in pig production and current research highlights. *Veterinary Parasitology*, 180(1), 72–81.

Roepstorff, A., Monrad, J., Sehested, J. and Nansen, P. (1999). Mixed grazing with sows and heifers: parasitological aspects. In Ecological animal husbandry in the Nordic countries. Eds Hermansen, J. E., Lund, V. and Thuen, E. Book Proceedings from *NJF-seminar No. 303*, Horsens, Denmark, pp. 41–4 *DARCOF Report No. 2* Record Number 20013092887.

Roinsard, A., Maupertuis, F. and Dubois, A. (2014). De l'enrubannage pour les truies gestantes. Atler Agri, 125, 25–8.

Ru, Y. J. and Bao, Y. M. (2004).Feeding dry sows ad libitum with high fibre diets. *Asian-Australasian Association of Animal Science*, 17(2), 283–300.

Rudolph, G. (2015). Animal health and welfare and environmental impact of different husbandry systems in organic pig farming in selected European countries. University of Natural Resources and Life Sciences. Vienna. PHD Thesis.130 pags. Available at: https://forschung.boku.ac.at/fis/suchen.hochschulschriften_info?sprache_in=en&ansicht_in=&menue_id_in=206&hochschulschrift_id_in=10773.

Salomon, E., Benfalk, C., Geng, Q., Lindahl, C., Lindgren, K. and Tore'n, A. (2005a). Outdoor pig systems in organic agriculture- animal environment, plant nutrient management and working environment. *NJF Report 1*, 203–6. ISSN 1653-2015. Swedish Institute of Agricultural and Environmental Engineering (JTI).

Salomon, E., Benfalk, C., Lindahl, C. and Lindgren, K. (2005b). Mobile and stationary systems for pigs – Nutrient excretion, distribution on outdoor areas and environmental impact. In *Researching Sustainable Systems. ISOFAR*, P 254–7.

Salomon, E., Hkerhielm, H., Lindahl, C. and Lindgren, K. (2007). Outdoor pig fattening at two Swedish organic farms-Spatial and temporal load of nutrients and potential environmental impact. *Agriculture, Ecosystems and Environment*, 121(4), 407–18.

Santos-Ricalde, R. H. and Lean, I. J. (2002). Effect of feed intake during pregnancy on productive performance and grazing behavior of primiparous sows kept in an outdoor system under tropical conditions. *Livestock Production Science*, 77(1), 13–21.

Santos-Ricalde, R. H. and Lean, I. J. (2006). Digestibility of dried star grass (*Cynodon nlemfuensis*) and consumption of fresh star grass in primiparous pregnant sows. *Animal Feed Science and Technology*, 129, 12–22. doi:10.1016/j.anifeedsci.2005.11.012.

Sarria, P., Montoya, C., Yusti, L. M., Orejuela, I., Guevara, M., Cruz, A. C., Arredondo, J., Londoño, A. and Peters, M. (2010). Nutritive value of leaf meal of cowpea (*Vigna unguiculata* (L) Walp.) for growing pigs. *Livestock Research for Rural Development*, 22 (6) [Online] Available at: http://www.lrrd.org/lrrd22/6/sarr22110.htm [Accessed 8 April 2016].

Schivera, D. (2009). Raising Organic Pigs. [pdf] MOFGA Fact Sheet #16. 3 p. Available at: http://www.mofga.org/Portals/2/Fact%20Sheets/FS%2016%20Organic%20Pigs.pdf [Accessed 23 June 2016].

Sehested, J., Søegaard, K., Danielsen, V., Roepstorff, A. and Monrad, J. (2004). Grazing with heifers and sows alone or mixed: herbage quality, sward structure and animal weight gain. *Livestock Production Science*, 88 (3), 223–38.

Sikala, K. (2012). Free-range organic pigs foraging Jerusalem artichokes *(Helianthus tuberosus)*- effect of feeding strategy on feed intake, feed conversion, growth, animal behavior, nutrient balances and profitability. MSc. Thesis. Aarhus University, Department of agroecology. 70p.

Silva Filha, O. L. and Barbosa, É.J. R. (2011). Como se produz suínos locais na região Nordeste do brasil. In: Barlocco, N. and Vadell, A. (eds). *Producción de Cerdos a Campo. Aportes para el desarrollo de tecnologías apropiadas para la producción familiar.*[pdf] Montevideo. Universidad de la Republica. Facultad de Agronomia, 126–8. Available at: http://www.upc.edu.uy/images/documents/extension/Publicacion%2015%20anos%20UPC.pdf.

Simantke, C. and Sundrum, A. (2001). Auch Schweine shatzen Raufutter. *Bioland*, 3, 22–3.

Sirtori, F., Crovetti, A., Meo Zilio, D., Pugliese, C., Acciaioli, A., Campodoni, G., Bozzi, R. and Franci, O. (2011). Effect of sire breed and rearing system on growth, carcass composition and meat traits of Cinta Senese crossbred pigs, *Italian Journal of Animal Science*, 10(4), p.e47. http://dx.doi.org/10.4081/ijas.2011.e47.

Søegaard, K., Sehested, J. and Danielsen, V. (2000). Mixed grazing with heifers and pregnant sows. *Grassland Science in Europe*, 5, 359–61.

Sommer, S. G., Sogaard, H. T., Moller, H. B. and Morsing, S. (2001). Ammonia volatilization from sows on grassland. *Atmospheric Environment*, 35(11), 2023–32.

Stern, S. and Andresen, N. (2003). Performance, site preferences, foraging and excretory behaviour in relation to feed allowance of growing pigs on pasture. *Livestock Production Science*, 79(2), 257–65.

Suescún-Ospina, S. T. and Ocampo-Duran, A. (2015). Comportamiento y desempeño productivo de cerdas de cría a campo abierto en condiciones de Piedemonte, Orinoquia Colombiana. *Revista Orinoquia*. 19(1), 35–40. ISSN: 0121-3709.

Thomsen, L. E., Mejer, H., Wendt, S., Roepstorff, A. and Hindsbo, O. (2001). The influence of stocking rate on transmission of helminth parasites in pigs on permanent pasture during two consecutive summers. *Veterinary Parasitology*, 99(2), 129–46. doi:10.1016/S0304-4017(01)00454-X.

Urriola, P. E., Cervantes-Pahm, S. K. and Stein, H. H. (2013). Fiber in swine nutrition, In: Chiba, L. I. (ed.), *Sustainable Swine Nutrition*, pp. 255–76. John Wiley&Sons, Inc., USA.

Verbeke, W., Pérez-Cueto, F. J. A., de Barcellos, M. D., Krystallis, A. and Grunert, K. G. (2010). European citizen and consumer attitudes and preferences regarding beef and pork, *Meat Science*, 84 (2), 284–92.

Vittoz, P. and Hainard, P. (2002). Impact of free-range pigs on mountain pastures in the Swiss Jura. *Applied Vegetation Science*, 5(2), 247–54.

Watson, C., Atkins, T., Bento, S., Edwards, A. and Edwards, S. (2003). Appropriateness of nutrient budgets for environmental risk assessment: a case study of outdoor pig production. *European Journal of Agronomy*, 20(1), 117–26.

Webb, J., Broomfield, M., Jones, S. and Donovan, B. (2014). Ammonia and odour emissions from UK pig farms and nitrogen leaching from outdoor pig production. A review. *Science of the Total Environment*, 470, 865–75.

Williams, J. R., Chambers, B. J., Hartley, A. R., Ellies, S. and Guse H. S. (2000). Nitrogen losses from outdoor pig farming systems. *Soil Use and Management*, 16(4), 237–43.

Wünsch, K., Maier, S., Gruber, S. and Claupein, W. (2011). Harvest of Jerusalem artichoke (*Helianthus tuberosus* L.) with grazing pigs compared to mechanical harvest. *Journal Für Kulturpflanzen*, 63 (6), 179–85. ISSN 0027-7479.

Zijlstra, R. T. and Beltranena, E. (2013). Alternative feedstuffs in swine diets, In: Chiba, L. I. (ed.), *Sustainable Swine Nutrition*, pp. 229–53. John Wiley&Sons, Inc., USA.

Chapter 8

Welfare of gilts and pregnant sows

Sandra Edwards, Newcastle University, UK

1 Introduction

The wild ancestors and wild relatives of the modern domestic sow live in small stable family groups. Many of the behaviour patterns that they show have been conserved in modern sows which, when placed back into a semi-natural environment, spend 0.6 of the daylight hours grazing and rooting (Stolba and Wood-Gush, 1989). In addition to feeding, drinking and excretory behaviours, maintenance behaviours include those with thermoregulatory function, such as nest building and wallowing, and others with skin care functions, such as rubbing. In this complex environment, exploratory behaviours, including locomotion, orientation to stimuli, nosing and manipulation of objects, also occupy significant periods of time (0.1–0.2 of the day). The stable social situation is reflected in the spatial association of stable subgroups, low incidence of agonistic behaviours and social facilitation of behaviour within the group. In intensive farmed conditions, circumstances for the pregnant sow are greatly changed. Food is provided in a concentrated form, and foraging behaviours are no longer functional in increasing energy and nutrient supply. Long-term stable group structure is disrupted, space allowance is greatly reduced and the environment is frequently utilitarian and barren. This situation restricts expression of the

http://dx.doi.org/10.19103/AS.2017.0013.22

behaviour patterns developed by sows in their evolutionary history and may give rise to behaviours considered abnormal and indicative of reduced welfare.

Traditional systems, in which small groups of pregnant sows were housed in outdoor paddocks or in covered straw yards, fell into disfavour as herd sizes increased, and it became more difficult to manage them. The 1960s saw the large-scale development and adoption of individual gestation stall and tether housing systems for sows, and these rapidly became the norm in many pig-producing countries. Such systems offered the advantages of low space requirement and ease of management. Sows could no longer fight and individual feeding could ensure that the nutritional needs of all animals were precisely met without competition. With the small space allowances and enclosed buildings associated with individual housing, automated air temperature control was possible, but provision of bedding and daily cleaning-out presented a difficult and laborious manual task. In consequence, this was automated in many buildings by using fully or partly slatted floors, through which all excreta passed for storage away from the animals as a slurry. The slurry could then be mechanically removed from the building at any convenient time. The resulting form of housing offered a relatively low cost, simply managed system for pig production enterprises under all farming conditions (Fig. 1). However, the large-scale commercial use of such systems became a focus for public concern because of the associated indications of welfare impairment for the animals.

2 Welfare issues of individual confinement systems

Following concerns publicised in widely read articles and books, such as Ruth Harrison's *Animal Machines* published in 1964, scientific studies of the consequences of different housing systems during gestation multiplied during the 1980s. Comparison of sows

Figure 1 Gestation stall housing which allows protected individual feeding, but restricts space and foraging opportunity. Photo courtesy of Prairie Swine Centre, Canada.

housed in stalls with those housed in small groups of 4–5 animals in either cubicles (pens in which animals can move freely between stalls used for feeding/resting and a communal dunging/exercise area) or in larger pens with separate lying, activity and feeding areas showed that confined animals spent more time lying down, had shorter activity periods and spent significant periods of time chewing on the bars of the pen (Jensen, 1980a). When straw was provided, sows housed in stalls ate it all quickly, whereas those in loose housing showed a variety of different behaviour patterns associated with nest building and exploration. These early behavioural indications of problems with confined housing were followed by more detailed experimental studies comparing the development of behaviour of groups of animals from a common background when they were put into different housing systems. Similar results were obtained in comparisons of stalls and group housing (Svendsen and Bengtsson, 1983) or tethers and loose housing (Vestergaard and Hansen, 1984). It was again shown that confined animals, in either system, were less active, showed less manipulation of straw and spent more time bar and chain biting and drinking. Behavioural differences persisted when the animals were taken to farrowing accommodation, with those previously housed in stalls or tethers being more restless at farrowing and having longer delivery intervals. Despite being unable to physically interact, tethered animals also showed more aggressive behaviours directed at neighbouring animals.

To assess in more detail the significance of the differences in behaviour in different housing systems for animal welfare, a series of experiments was carried out in Australia. A preliminary study compared non-pregnant gilts in five different housing systems: individually tethered, housed in pairs, indoor group pens, outdoor concrete yards with simple shelter and outdoor paddocks (Barnett et al., 1984). As in previous studies, outdoor groups were more active, whilst indoor groups, especially those in tethers and pairs, spent more time sham chewing, bar biting and drinking. Assessment of animal welfare by measuring the diurnal levels of blood corticosteroids showed that animals of the indoor group pens had lowest levels, whilst those housed in pairs in limited space appeared to exhibit a chronic stress response. This was reflected in higher cortisol levels, a disrupted diurnal pattern of blood cortisol and slower response to and recovery from a transport stressor. A subsequent study with pregnant pigs compared tethers, stalls, indoor groups and paddocks (Barnett et al., 1985). Again outdoor pigs were found to be more active, spending about 0.2 of the daylight hours in rooting and grazing. Although animals in stalls showed increased oral/nasal behaviours, it was those in tethers which showed lower activity and evidence of a chronic physiological stress response. The extent of the increase in corticosteroids in tethered animals was sufficient to induce metabolic changes, reflected in higher blood glucose and urea levels (Barnett et al., 1985) and reduced responsiveness of the immune system to an external challenge (Barnett et al., 1987a).

In a later study (Barnett et al., 1987a) it was noted that sows in tethers, whilst performing less exploratory behaviour than those in groups, actually had more aggressive interactions with neighbours and more retaliatory behaviour in such interactions. This has led to an investigation of the possibility that reduced welfare of tethered animals was at least partially due to social stress arising from being in enforced close proximity to a neighbouring animal and might be improved by altering tether design. When the partition between adjacent tethered sows was changed from one of vertical bars to a wire mesh, the total number of interactions was decreased, aggressive interactions were virtually eliminated and corticosteroid levels were similar to those of stall or group-housed animals (Barnett et al., 1987b). Barnett et al. (1989) therefore suggested that stall housing was less stressful than tethers because the increased opportunity for back and forward movement reduced the amount of head-to-head contact between pigs. An effect of stall design on the nature of interactions and stress responses was subsequently found (Barnett et al., 1991), but this was in contrast to the previous theory. Animals in stalls with horizontal bars spent less time concurrently in front of the stall with a neighbour but, despite lower levels of aggression and head-to-head contact, had higher cortisol levels than those in stalls with vertical bars. Thus, the extent to which the physiological indicators of stress associated with confinement housing can be explained by social factors is still incomplete.

3 Nature and significance of stereotyped behaviour in gestating sows

As described previously, the majority of studies have shown a marked increase in abnormal oral behaviours in confined sows. In many cases such behaviours develop into stereotypies – repetitive behaviours which are fixed in pattern of performance and have no obvious function. The range and frequency of abnormal oral/nasal behaviours were described by Cronin and Wiepkema (1984) who studied in detail 36 neck-tethered sows. All behaviours which they considered to be abnormal were oral in nature. These totalled 11 (of 50) recorded actions (chewing/biting, sucking, mouth stretching, palate grinding, tongue flicking, licking, nibbling, nosing, rooting, pressing with rooting disc and pause) and were directed to 5 (of 9) possible substrates (trough, floor, bars, chain and nil). Up to 0.55 of stereotyped behaviours were performed with nil substrate, that is 'sham' behaviours. The mean number of fixed routines increased with stage of pregnancy until day 80 and then declined.

In a similar detailed study, Stolba et al. (1983) showed a strong increase in stereotyped behaviour over parities. Young sows showed high levels of 'drowsy' inactivity but also spent considerable time manipulating the limited amount of straw given. In second- and third-parity animals, time spent in straw-related behaviours decreased, but locomotory and investigative behaviours directed at the pen components increased and stereotyped behaviours started to develop. Older sows showed further increases in stereotyped behaviour, but all behavioural sequences became less variable with increasing age, and a higher proportion of behaviours were self-directed, that is 'sham' behaviours.

The development of stereotyped behaviour in individual tethered sows was traced by Cronin (1985). He showed that animals tethered for the first-time pass through a number of consistent behavioural stages. First comes a period of initial escape behaviour, in which the sow pulls on the tether, threshes about and screams. After 2–14 min, during which time escape attempts become shorter and less vigorous, this gives way to a period of inactivity. The animal lies immobile for long periods, making whining vocalisations. This phase typically lasts for about one day but with a range of 2 hours to 16 days. The animal then becomes more active again, displaying investigatory and aggressive behaviour. Initially a wide variety of such behaviours are performed, but over time there is a reduction in range of behaviour and basic stereotypies begin to develop over 8–55 days. Based on these observations Cronin (1985) suggested that the development of stereotypies was due to restraint and loss of control. However, an alternative hypothesis was developing at this time as a result of other studies.

4 Hunger in the pregnant sow

A number of studies had noted that abnormal oral behaviours occurred most frequently in the period around feeding time (Cronin, 1985; Blackshaw and McVeigh, 1984). Detailed studies carried out at this time showed that the type of behaviour varied in relation to the time of feeding. Head weaving, bar biting and snout rubbing were more commonly observed immediately prior to feeding, especially in older sows, whilst rooting, drinker manipulation and polydipsia were more common after feeding. Some sows spent as much as 40–60% of the first hour drinking. Other behaviour types (vacuum chewing, chain manipulation) were not associated with period of feeding. These observations led to the suggestion that stereotypies arose from the persistence of feeding motivation because commercially provided rations did not supply enough proprioceptive feedback, for example, stomach distension (Rushen, 1984, 1985). If stereotypies were due to frustration of feeding behaviour and not lack of environmental stimulation (as suggested for example by Stolba et al., 1983), the different forms of abnormal oral behaviour could be explained as motivationally appropriate, representing appetitive and consummatory phases of feeding. This suggestion was more in accordance with the generic models of behaviour being developed (e.g. Hughes and Duncan, 1988) which proposed that stereotypies result from the persistence of sequences of appetitive behaviour in restrictive environments which block negative feedback.

Further support for the pivotal role of feeding in causation of stereotyped behaviour came from studies of the consequences of altering the level or nature of feed. Gilts and sows tethered in unbedded pens spent more time standing and more time in repetitive behaviour when given low feed levels than those given high feed levels (Appleby and Lawrence, 1987). A subsequent study (Terlouw et al., 1991) showed that a similar effect could be shown in loose-housed animals (in an unbedded cubicle system). The incidence of chain manipulation and polydipsia was much more greatly influenced by feed level than by the degree of freedom of movement allowed by the housing system.

The role of gut fill in reducing stereotypies has been indicated in a number of studies. The provision of straw decreased the level of activity of tethered sows and the overall time spent in pen-directed oral activities (Fraser, 1975). In stall-housed sows, sham chewing, bar biting and apathetic behaviour were all reduced by provision of straw (Sambraus and Schunke, 1982). Providing additional roughage in the form of oat husks had a similar effect in stall housed sows (Broom and Potter, 1984). Even when less energy was consumed, feeding a diet which gave greater gastric distension reduced the occurrence of post-feeding oral behaviours (Brouns et al., 1994; van der Peet-Schering et al., 2003; de Leeuw et al., 2004, 2005).

Such findings led Lawrence and Terlouw (1993) to suggest that oral stereotypies result from the modification of foraging behaviour in highly food-motivated sows kept in behaviourally restrictive environments. Outdoor sows given lower feed levels increase the time spent foraging (Edwards et al., 1993a). In the absence of normal rooting substrates, such as soil, modified behaviours are performed. In many housing systems, straw provides an alternative substrate for rooting. Sows that were fed restricted diet spent more time rooting in their bedding than those fed on satiating high-fibre diets (Brouns et al., 1994). In a more critical analysis of the role of straw, Fraser (1975) showed that provision of long straw as bedding, allowing sows to manipulate it, reduced both the time spent standing and the proportion of standing time which was engaged in pen-directed oral activities, whereas providing the same quantity of straw chopped up in the food only reduced activity time and not the proportion of abnormal behaviours in this time. Providing long straw in the trough rather than as full bedding gave an intermediate result. This suggested that both nutritional and behavioural aspects of a fibrous substrate were important in alleviating the welfare challenges of restricted feeding. Subsequently, an interaction between feed level and availability of straw in the performance of abnormal oral behaviours was demonstrated. Sows fed on low levels and deprived of straw performed four times as much chain manipulation than those either given straw or fed at higher levels but without straw (Spoolder et al., 1995).

However, feeding is not the only activity influencing occurrence of abnormal behaviours in confined sows, since there are big individual differences between animals fed in the same way. This suggests that differences in individual coping style of the animals, as well as other external influences, are also important. Repetitive behaviour of newly tethered gilts was positively correlated with the level of chain manipulation performed by their neighbours (Appleby et al., 1989). This was most apparent in susceptible low-fed animals and may reflect an additive role of other environmental stressors such as disturbance from noise. The presence of an observer can also markedly increase the occurrence of abnormal behaviours (Sambraus and Schunke, 1982).

5 Pressure to adopt group housing systems for pregnant sows

Results such as those discussed previously have given rise to increasing public concern worldwide about the welfare of sows in confinement systems. In addition to the increased incidence of abnormal oral behaviours, other identified problems included a higher incidence of lameness and leg weakness (Svendsen and Bengtsson, 1983; de Koning, 1985), of urinary tract infections (Muirhead, 1983) and of prolonged farrowing and stillbirths (Nielsen et al., 1974; Svendsen and Bengtsson, 1983). Sows confined in pregnancy were also shown to have an impaired immune response (Metz and Oosterlee, 1980; Barnett et al., 1987a), and this may be one of the reasons for some reports of increased incidence of reproductive tract and mammary infections in the sow at farrowing, and of infectious disease in suckling piglets from sows housed in such systems (Backstrom, 1973; Svendsen and Bengtsson, 1983). However, such indications from these early studies (SVC, 1997) have not always been supported by more recent critical scientific review of the effects of system on sow health (Sow Housing Task Force, 2005).

As a consequence of public pressure, a number of countries have now enacted legislation which will restrict the housing of dry sows in individual confinement systems. Some countries, such as the United Kingdom, Switzerland, Sweden, Norway and Finland, have for many years had a total ban on individual confinement systems during gestation. The European Union banned tether housing from 2005 and from 2013 has restricted the use of gestation stalls to a period of four weeks after insemination. The New Zealand and Australian industries voluntarily announced similar partial stall bans in 2015 and 2017, respectively. In Canada any new pregnancy housing built from 2014 must allow for sows to be grouped, with a number of major retailers committing to stall-free supply chains by 2022. Several US states have also voted for a stall ban, and pressure from North American retailers to abolish this system is growing. Whilst such bans may solve one set of welfare problems for the pregnant sow, they have given rise to others associated with social grouping of sows.

6 Social organisation in sows

Sows are, in nature, socially living animals (Buchenauer, 1990). The wild pigs, from which modern domestic sows have evolved, live together in groups of 1–6 sows and their offspring. The group size depends on available resources such as food and sheltering cover. The typical social organisation for the species is therefore to have small groups of related animals, of mixed age and size, which have a relatively large home range and spend much of their time in foraging over this area. However, economic pressures dictate that commercial group housing systems operate with larger, often unstable, groups of unrelated, more uniform animals. These are subject to restricted space and concentrated feeding times and locations. It is the contrast of these two situations, where the biology of the animal can be at odds with the demands of the system, which gives rise to potential social problems.

Foremost amongst the problems of group housing is that of aggression between animals, which arises over competition for resources or when unfamiliar animals are grouped together. In stable groups, aggression is minimised by the use of subtle behavioural signals (Jensen, 1980b). In a detailed analysis of social interaction patterns in dry sows, he identified ten behaviour patterns. The function of these behaviours was investigated by an analysis of sequences of behaviours which occurred when two sows interacted (Jensen, 1982/3). The results indicated that head-to-head knock and head-to-body knock were aggressive attack behaviours, parallel and inverse parallel pressing were fight behaviours, nose-to-nose was a mild threat usually emitted by more dominant animals, nose-to-body and nose-to-genital appeared to be neutral behaviours associated with individual recognition and head tilt and retreat were submissive behaviours. Investigation of sows in semi-natural environments, where total space was greatly in excess of normal allowances, supported these results and identified a further pattern of behaviour called 'aiming', an upward thrust in the air with the snout, which appeared to be a mild threat behaviour (Jensen and Wood-Gush, 1984). Jensen (1982/3) proposed that in stable groups, aggression was regulated by an avoidance order. Lower ranking sows used head-tilt or retreat behaviours to avert attack in potentially aggressive situations. Comparison of sows in stalls, cubicles and pens with separate lying, dunging and feeding areas indicated that in the former two cases, where space was limited, these behaviours could not be adequately performed (Jensen, 1984). In consequence, confinement led to unsettled dominance relationships and increased aggression.

Deciding on what constitutes adequate space for group-housed sows is difficult since there are, at present, inadequate scientific data on this subject. It is relatively simple to calculate mathematical space requirements for individual animals in different postures from their physical dimensions (Petherick, 1983) and space envelopes when the animals are standing and lying (Baxter and Schwaller, 1983). Such information is useful in calculating dimensions for static situations such as resting or feeding in stalls, but this makes up only a part of the total space requirement. In addition to these static space requirements, the results previously discussed indicate that the animals require social space. Even in the absence of overt aggression, enforced proximity to another sow may in itself be stressful. Barnett et al. (1984) demonstrated a chronic stress response in gilts housed together in pairs, which was not seen in larger groups with greater total space. The preferred distance between individuals will vary, depending on the relationship of the animals involved and their current motivational state. Foraging sows in extensive conditions maintained an average distance of 3.8 m between group members and a distance of 50 m between different groups (Stolba and Wood-Gush, 1989). A distinction must always be made between space per animal and total space. With increasing group size, and constant space allowance per sow, the total pen area is increased, as is the potential for time sharing of space.

Investigation of the social structure of stable sow groups has shown this to be far from simple. In free ranging conditions, Jensen and Wood-Gush (1984) showed that the dominance order was more linear at feeding than during subsequent foraging. The number of interactions was also highest just after feeding. They suggested that a stable dominance order is not a prerequisite for low aggression level, but that this depends also on area and familiarity of the animals. In larger groups of mixed-parity sows, very linear hierarchies were found when all individuals were ranked relative to other group members by the results of paired food competition test (Brouns and Edwards, 1994). As noted in previous studies (Sambraus, 1981), rank was highly correlated to age/live weight of the animals, and the ranking obtained in pair testing was identical to that obtained by observations at feeding time within the group situation. Csermely and Wood-Gush (1986), observing group-fed sows in pens of 15, also found that the social hierarchy was the same during both feeding and non-feeding context. Most aggression was directed by high-ranking to low-ranking animals, with the next highest category being to sows of equal rank. Aggression from a subordinate to a dominant sow occurred only in non-feeding contexts. Investigation of groups of contemporary gilts, more even in age and weight, showed much less linear hierarchies (Stewart et al., 1993) in which rank was not correlated to live weight. Examination of the animals again in the second parity, when the groups had been reformed after being individually housed during lactation, showed that relative rank had changed very little. The linearity of hierarchy was influenced by the nature of the feeding regime. Animals fed individually in stalls, which were not therefore competing for food, showed a much less linear hierarchy than those which were group fed.

6.1 Establishing the dominance hierarchy

In a natural situation, where sows live in stable, mixed-parity family groups, the hierarchy can establish itself over time with a minimum of overt aggression. Young animals establish relative dominance with their contemporaries as early as formation of the teat order and, as immature growing animals, are naturally subdominant to older and larger sows within the family group. In contrast, in most commercial group housing systems, abrupt mixing of mature sows occurs at least once in each reproductive cycle. This occurs most often when groups are reformed after weaning. Sows may be regrouped with animals they were housed with a previous parity or with other animals not recently encountered. The extent to which sows recognise others from which they have been separated during a 4- to 5-week lactation period is uncertain, although Arey (1999) demonstrated that pregnant sows could be removed and returned into groups of six after a 6-week period without any major disruption to social organisation.

When unfamiliar sows are mixed, the number of interactions is high during the first 12 h, gradually decreasing over time (Edwards et al., 1993b; Csermely and Wood-Gush, 1990a,b; Luescher et al., 1990). After three days, the level of overt aggression is relatively low and focused around feeding times (Jones and Petchey, 1987; Csermely and Wood-Gush, 1986). Newly weaned sows, which show higher overall levels of activity, show more aggression than sows remixed in mid-pregnancy (Edwards et al., 1993b). Attempts to alleviate the aggression which occurs when animals are mixed have generally had little success, and there is still relatively little knowledge on the factors ameliorating aggression.

Genetic differences in aggressive temperament of sows (Lovendahl et al., 2005), whilst undoubtedly important in this context, have not yet featured significantly in modern selection programmes. The effect of sedating the sows at mixing using pharmacological agents has been examined (Csermely and Wood-Gush, 1990a,b; Luescher et al., 1990), but, as with growing pigs, this seems only to postpone the onset of aggression and does not reduce the overall level of aggression. The use of odour-masking agents has also generally proved ineffective (Barnett et al., 1993b; Luescher et al., 1990). Provision of distractions such as fresh straw, whilst sometimes recommended commercially, has not been shown to be effective in a controlled experimental situation (Botermans, 1989).

The results from such studies indicate that some degree of fighting is inevitable when unfamiliar sows are penned together. However, the incidence or severity of fighting can be influenced by pen design. One option is to provide a large area for escape and avoidance of aggressors. Total space requirement in such situations is still poorly defined, and experiments with animals in unbedded pens have given conflicting results. When sows were mixed after weaning in groups of different sizes (3, 6 or 9 in a standard pen of 22.8 m^2), sows in smaller groups, with more space per sow, avoided or withdrew from agonistic encounters more frequently and showed less severe aggression (Mujuni et al., 1985). However, in controlled experiments with gilts (Barnett et al., 1993a), reducing space allowance from 3.4 to 1.4 m^2 reduced the incidence of aggression in the first 90 min after mixing, although it had no effect on overall damage score after three days. Shape of the pen was also found to be important, since aggression was lower in a small rectangular pen and not in a small square pen. Similar results have been obtained with older sows in a comparison of groups of six mixed in pens providing 3.7 and 6.1 m^2 per sow (Edwards et al., 1993b). During the first 12 h there were more interactions in the large pen than in the smaller pen. These results appear to contradict the theory that a larger amount of space per sow reduces aggression. However, sows in the smaller pen in all cases exhibited more damage, suggesting that pen size may have more effect on severity of interaction as a result of limited escape possibility. In larger groups of 10, 30 or 80 sows in unbedded pens with floor feeding, Hemsworth et al. (2013) demonstrated a general decline in both aggression and plasma cortisol concentrations with increasing space over the range of 1.4 to 3 m^2, while there was a general increase in farrowing rate with increasing space. It is often suggested that sows in larger groups might require less space, since absolute pen size and therefore shared free space is greater. However, there were few interactions between group size and space allowance which would support this contention. In comparison of 2, 4 and 6 m^2 per sow, Greenwood et al. (2016) found no effect of space allowance on the number or duration of fights immediately after mixing, average lesion scores or subsequent reproductive performance. However, low-ranking sows had reduced lesions with greater space allowance, supporting the welfare benefit of a spacious mixing pen. Reducing space to 2 m^2 on day four, once aggression had subsided, had no adverse effects on any welfare or production measure.

The total space required to allow the interactions necessary to establish dominance with minimal injury is therefore uncertain, but it appears to be substantial. In one study where space was virtually unlimited, 0.75 of encounters resulted in chase distances of less than 2.5 m, but some sows were pursued up to 20 m following aggressive interactions (Edwards et al., 1986). Provision of such large areas would be uneconomic for mixing a small number of sows. However, it is possible that the provision of barriers which allow animals to visually separate themselves from an aggressor may partially compensate for reduction in total space. Constructing a pen with 'pop-holes' in which pigs could hide their head and neck during an aggressive interaction reduced aggression in newly weaned pigs (McGlone and Curtis, 1985). Attempts to mimic this effect in newly mixed gilts by providing partial stalls in the pen have proved unsuccessful (Luescher et al., 1990; Barnett et al., 1993a). However, the use of barriers has been successfully adopted in the design of an arena for initial mixing of unfamiliar sows (van Putten and van de Burgwal, 1990b). This Dutch design has been reported as allowing regular mixing of unfamiliar sows without serious injury, although comparative information on levels of aggression relative to alternative pen designs has not been published. In a UK study, provision of a central barrier in a pen reduced the number of interactions and was recommended, together with increased space and *ad libitum* feeding, as a component of a specialised mixing pen in which newly mixed groups could spend the first few days to minimise injury while social dominance was established (Edwards et al., 1994).

Once the dominance hierarchy has been established, the amount of space required for harmonious social interaction is still uncertain. This is probably because of the wide variation in pen designs, manure management, feeding systems and group sizes in use, with each of these factors affecting group dynamics, hygiene and the need for space. In small stable groups of six sows, with protected feeding in separate individual stalls, the quantity of skin lesions observed on the animals increased as space was reduced from 4.8 to 3.6, 2.4 and 2.0 m^2 per sow (Weng et al., 1998). A study of sows housed in groups of five with floor feeding, from day 25 of pregnancy during two consecutive parities, showed that animals had greater lesion scores and lower back fat thickness at 1.4 m^2 per sow than at greater specific allowances. However, there was no further improvement in these measures when space was increased from 2.3 to 3.32 m^2, although subsequent litter size was significantly higher at the greatest allowance (Salak-Johnson et al., 2007). In larger groups, more injuries were reported in dynamic groups in ESF housing at a space of 2.25 than 3.0 m^2 per sow, although no adverse effects on performance were seen (Remience et al., 2008). Thus, whilst good data are limited, the current EU legislation requiring a minimum of 2.4 m^2 per sow and 2.25 m^2 per gilt, with some adjustment for group size, seems sensible.

6.2 Dynamic grouping

Housing systems of lower capital cost, which are based on large group size, generally necessitate much more frequent mixing of sows, since the size of a contemporary weaning batch is only 4–5 animals for every 100 sows in the herd. Levels of aggression are higher in such dynamic groups than in smaller static groups which remain unchanged after their initial formation, although the degree of aggression recorded has varied widely in different situations. There can be significant performance loss if mixing stress occurs at critical times in the reproductive cycle (Lambert et al., 1986; Bokma, 1990; te Brake and Bressers, 1990). The extent to which the stressors associated with dynamic grouping impact on performance is still open to question, with different conclusions from different studies (Simmins, 1993; Anil et al., 2006) which probably reflect the subtle differences in management which can influence the outcome (Spoolder et al., 2009).

The factors affecting levels of aggression in dynamic groups are still poorly understood. It has often been suggested that aggression is lower when mixing occurs in larger groups. Whilst one experimental study has reported significantly less overall aggression in a larger group (32 v 21 sows), this was primarily related to accessibility of resources since there was little reduction in aggression towards newly introduced sows (Bokma and Kersjes, 1988). The key to successful functioning of dynamic groups appears to lie in the utilisation of subgroup behaviour. Within a large group of sows, the animals tend to form subgroups which lie together in specific areas of the pen (Edwards et al., 1986). If plentiful total space is available, and access to feeding and watering points well planned, newly introduced subgroups can remain together on the periphery of the main group with a minimum of aggressive interaction (Edwards et al., 1986; Hunter et al., 1989). Integration, a process shown to be effective when introducing gilts into groups of mature sows with little damage, can then take place gradually over a period of time (Lenskens, 1991). If total space is more restricted, as in partially slatted systems, there are benefits in creating specially partitioned lying areas for different subgroups to use (van Putten and van der Burgwal, 1990b). By closing off such an area for some days prior to introduction, the new group can move into an unclaimed resting territory and again avoid confrontation with longer established subgroups.

The degree of stress involved in mixing depends not only on the rank of an animal, but also on the strategy which it adopts in social encounters. Observation of the outcome of fights occurring when gilts were progressively added to a large group at 4- to 6-week intervals showed that it was possible to categorise animals into three groups (Mendl et al., 1992). High-success pigs were socially active, aggressive and won the majority of their encounters. No-success pigs never won encounters, were relatively inactive and showed low involvement in social interactions. Low-success pigs, which were of middle rank, were aggressive but relatively unsuccessful in social encounters. It was this third category which exhibited indications of physiological stress, with higher basal cortisol levels and greatest peak cortisol levels in response to ACTH challenge. This again highlights the possibility of selection for sow temperament in reducing the problems experienced in large dynamic group systems. In studies of stable groups of gilts, there was no correlation between response to ACTH challenge and rank (Edwards et al., 1993b). Thus, being of low rank may not in itself be stressful if a stable social situation exists and the frequency of aggressive interactions is low.

7 Aggression in stable groups and the method of feed provision

Once groups have been formed, the major factor giving rise to aggression in group-housed sows is competition for resources. It is normal commercial practice to feed pregnant sows a small amount of concentrate once or twice daily, and it is competition for this food which is the major cause of aggression in stable groups of sows (Carter and English, 1983; Jones and Petchey, 1987; Csermely and Wood-Gush, 1986). The one factor which has, above all others, given rise to problems in the design, cost and management of group housing systems for sows is the need to ensure that all animals obtain an adequate share of food without excessive aggression. Many different practical solutions to the problem of food distribution are in commercial use, with varying implications for sow welfare.

7.1 Floor feeding systems

The traditional, and cheapest, way of feeding group-housed sows is to distribute the total allowance of food on the ground and leave each individual sow to eat as much as it can until the food is finished. This may involve feeding by hand or be automated with dispensing canisters (Csermely and Wood-Gush, 1986, 1990a,b) (Fig. 2). However, this mode of delivery can give rise to aggression between sows as they compete for limited feed, both in an outdoor (Jensen and Wood-Gush, 1984) and an indoor situation, where up to 85% of all aggression-related behaviours can be food-related (Csermely and Wood-Gush, 1986; Jones and Petchey, 1987). The speed at which a concentrate diet is eaten varies widely between sows (Edwards et al., 1988a). Old sows can consume their 2–3 kg allowance in as little as 5 min, but younger sows may take up to twice as long. The inequality in food intake is further compounded by the ability of dominant sows to defend a particular area of good food supply (Csermely and Wood-Gush, 1990a,b), leaving lower-ranking individuals only limited access to the resource. Dominant sows occupied the centre of the feed pile, defending this area where food was thickest. They spent more time defending food than feeding, initiating more agonistic interactions than lower-ranking sows. Greatest aggression was seen in the first 15 min after food provision, with threat being more common in the subsequent 15 min. In consequence, these dominant sows showed a high frequency of short feeding bouts, whereas low-ranking sows had similar bout lengths but longer interruptions between bouts. These differences in feeding behaviour can have severe consequences for some animals. Brouns and Edwards (1994) showed that when sows were floor-fed, weight gain of low-ranking animals was only 0.63 of the group average in pregnancy. Large variation in body condition can result by the end of pregnancy when such systems are adopted in commercial practice (Edwards, 1992), and the increased aggression directed at low-ranking sows in such systems results in higher levels of skin lesions (Stewart et al., 1993).

7.2 Pens with individual feeding stalls

Precise rationing of each individual animal with minimal aggression can only be guaranteed by individually confining the animals at the time of feeding. The combination of group housing

Figure 2 Group housing with floor feeding which gives aggressive competition for feed that can only be partly ameliorated by straw bedding.

with individual feeding stalls offers a high welfare option (Edwards, 1985) (Fig. 3). Overall aggression is markedly reduced when feeding stalls are used in comparison with group feeding (Lambert et al., 1986; Jones and Petchey, 1987). In the latter study, it was noted that only 0.3 of the difference was due to aggression at feeding, indicating that the stalls may be beneficial in other ways. Although desirable for the sow, provision of individual feeding stalls which are used for only a short period each day may often not be adopted commercially because of high capital cost arising from stall purchase and additional housing space.

7.3 Cubicles and partial barriers

Space allowance and cost can be greatly reduced by combining the feeding stall and lying area, as is done in systems with cubicles or free access stalls (Edwards, 1985; Hoofs, 1990) (Fig. 4). However, as discussed earlier, such systems may result in undesirable effects on the pattern of social interactions if group space is limited (Jensen, 1984). Further simplification of the pen can be made by reducing the length of feeding stalls, such that only partial barriers are used (Petherick et al., 1987). In a pilot study on such a system, where animals were fed a small amount of food twice daily, there was only a low incidence of poaching of feed and aggression at feeding was minimal. The feed bays were also used for resting and for retreat from aggression. However, there is little information about the functioning of such a system with larger groups of mixed parity, where it is likely that subdominant animals would suffer from food poaching and limited social space.

7.4 BioFix® or trickle-feed systems

One suggested way to minimise possible feed poaching by older, fast-eating sows, without the expense and inconvenience of closing all sows into full length stalls, is to

Figure 3 Group pens with separate lying and activity areas and individual feeding stalls, which allow both socialisation and protected feeding but have higher requirement for space and capital cost.

Figure 4 Free access stalls which reduce space requirement by combining the feeding and lying areas. Photo courtesy of Prairie Swine Centre, Canada.

use a 'biological fixation' (BioFix® or trickle-feed) system, in which feed is delivered by auger at the same controlled rate to each individual place. Since the sows cannot then eat at a differential speed, movement between feeding places gives no benefit and only short partitions along the trough are necessary to protect the feeding sows (Bengtsson et al., 1983; Hoofs, 1990; Hulbert and McGlone, 2006) (Fig. 5). Correct selection of the dispensing rate is essential to the success of the system, since too slow a rate will lead to restlessness in fast-eating sows, whilst too fast a rate will overwhelm the slow-eating animals. Experiments have shown that the number of aggressive interactions and changes of place during feeding increase as the dispensing rate drops below 100 g of pellets per minute. However, with rates of more than 120 g per minute, more sows have food accumulating in the trough and the number of aggressive interactions when feed dispensing stops is increased (Hoofs, 1990). Sows fed at a slow rate by the BioFix® method subsequently spent less time nosing elements of the pen than sows given their feed in one total delivery (Bengtsson et al., 1983).

7.5 Two-yard systems

An alternative method for automated flat-rate individual feeding which can be used with simpler housing designs is the 'two yard' system (Hunter, 1988). Sows enter a mechanically operated feeding station sequentially from one yard and, after feeding, exit via a side gate into a different yard. Sows which have not fed remain in the first yard to await attention. Although simple in concept, some practical problems exist. Unless the size of the two pens is changed by moving of gates during the day, both pens must be large enough to house the majority of the group and total space requirement is high. The biggest problem is the

Figure 5 Trickle-feed system which uses 'biological fixation' to reduce competition during feeding with only short stall protection.

possibility of mechanical malfunction and lack of precise information on the individual animal. Some sows pass through the feeder without stopping to eat (Hunter, 1988) and cannot then re-enter the feeder or be identified as not having fed.

7.6 Electronic sow feeding systems (ESFs)

It is well documented that the nutritional requirements of sows can vary widely depending on factors such as live weight, body condition and stage of pregnancy. No system of floor feeding or of flat rate feeding, for example, biological fixation or two-yard system, can meet the needs of every animal and some ability to differentially feed certain individuals is essential. In many practical situations, this is done by removing problem animals to an individual stall or pen for a period. However, automation of individual rationing was made possible in the early 1980s by the development of electronic sow feeders (Lambert et al., 1986; Edwards and Riley, 1986) (Fig. 6). In this system, animals are identified electronically by a device carried on a collar, ear tag or implant and must feed sequentially at one or more feeding stations controlled by a central computer. This enables large groups of animals to be kept in low cost, unspecialised housing (Brade et al., 1986). Sows in stable groups using this system soon develop a relatively stable feeding order, with dominant sows feeding at the start of the cycle and low-ranking sows waiting until a quieter time of day (Edwards et al., 1988b; Hunter et al., 1989). Sows newly introduced to the group generally begin low in the feeding order, progressively establishing themselves over time. They move up the feeding order as other longer-term group members are removed to farrow and new sows are introduced (Hunter et al., 1989). However, problems have occurred in practice with such systems. These have been mainly associated with design problems relating to the

Figure 6 Electronic feeding system where sows feed sequentially and can be individually rationed. Compartmentalisation of the lying area to encourage subgroup behaviour is also shown. Photo courtesy of Prairie Swine Centre, Canada.

feeding stall (Edwards et al., 1988b) and with the need for animals to feed sequentially. This is considered by some to be stressful for those animals frustrated in their attempts to feed and disruptive of the normal diurnal patterns of behaviour when animals feed throughout the night. It can also be a cause of aggression arising from competition for feeder entry. For this reason, the mechanical reliability of the gating system is of crucial importance (Edwards et al., 1988b), and the use of a walk-through stall design avoids the congestion at the feeder entrance which results with back-out stalls (Edwards et al., 1988b; Kroneman et al., 1993a,b). To reduce cost, simplified unprotected electronic sow feeding systems have been developed (e.g. FITMIX system) in which the complex stall and gating designs of conventional ESFs are absent. However, this results in a more competitive feeding environment to which some sows are unable to adapt (Chapinal et al., 2010).

Although simplistic conclusions about the relative welfare implications of these different housing systems are often made, it must be emphasised that the detail of system design and the management strategies adopted mean that sow welfare can vary as much between different versions of the same system as between the different systems (Edwards, 2000; Bench et al., 2013a,b).

7.7 Relationship between aggression and feed level

The behavioural problems encountered in group housing systems, like those of individual confinement housing, are frequently related to feeding practice. Increased levels of activity in low-fed animals may result in changes in social behaviour and aggression. It is a common observation that commercial units experiencing problems with high levels of aggression also have sows which are in poor body condition (Svendsen et al., 1990; Olsson et al., 1991). Providing food in a different form, or allowing the animals some other way to appropriately express their feelings of hunger, may alleviate such problems. Provision of a high-fibre diet reduced the number of skin lesions in floor-fed outdoor sows (Martin and Edwards, 1994). Spoolder et al. (1997) found no effect of feed level on aggression and skin lesions in an ESF system when the sows were bedded on straw, with sows on the lower feed level spending increased time in straw-directed foraging behaviour. Similarly, several studies of commercial group-housing systems have shown lower injury scores on farms with computerised feeding when generous occupational material (straw or hay) was provided (Weber et al., 1991; Gjein and Larssen, 1995). Vulva biting has been a major problem in many group-housing systems and particularly in ESF systems (van Putten and van de Burgwal, 1990a; Scott et al., 2009). There is good evidence that this abnormal social aggression may be related to feed provision. Dutch studies of sows in unbedded systems showed that vulva biting was most frequent after visits to the feed station when only small amounts (<1.0 kg) of concentrate were supplied, suggesting frustration at lack of satiation as the underlying cause. It has been demonstrated that a small quantity of food actually increases feeding motivation which, if the sow is left unsatiated, can be a major cause of reduced welfare (Terlouw et al., 1993). Supplying straw pellets in the feed station after a small amount of concentrate reduced the daily frequency of vulva biting and decreased the total number of wounded vulvas by 33% (Bure, 1991). Providing daily meals of chopped corn silage at the same time as the start of the feeding cycle also reduced the incidence of vulva biting from 30% to 10% (van Putten and van de Burgwal, 1990a). Where housing sows with straw bedding is not feasible, the presentation of straw or grass silage in racks may be beneficial (O'Connell et al., 2007; Stewart et al., 2008). However, such racks can become a focus for competition and aggression if poorly designed or sited (Kroneman et al., 1993a,b).

The problems associated with low feed levels and the cost of designing appropriate individual feeding systems indicate the desirability of providing a system in which the animal can feed *ad libitum*. Group-housed sows which are fed *ad libitum* do not eat at the same time, even when the possibility exists (Petherick et al., 1987) and feeding space can therefore be reduced in comparison to restricted systems. When fed *ad libitum* on conventional concentrate diets, dry sows will eat to excess and become obese, but this problem can be overcome by the use of specially formulated bulky diets of low nutrient density (Brouns et al., 1995; Ru and Bao, 2004) or low dry matter materials such as silages and by-products (Livingstone and Fowler, 1984; Edwards et al., 1992). Animals fed *ad libitum* in this way are less active, spend less time performing foraging behaviours indicative of hunger and have lower levels of skin lesions than animals competing for limited daily food (Brouns and Edwards, 1994; Stewart et al., 1993). Low-ranking animals are not disadvantaged by inability to obtain an adequate share of the food, as they can modify their feeding strategy to avoid conflict with more dominant animals by visiting the feeder at less popular times of a day. However, despite welfare advantages, this strategy is not widely adopted since the cost or handling logistics of these bulky diets can make it unattractive in many situations.

8 Extensive systems

The various problems described with intensive indoor housing systems for sows do not mean that welfare can only be good in an extensive situation, indeed the opposite can sometimes be the case (Edwards, 2005). In some countries, such as the United

Kingdom, the use of outdoor systems for pregnant sows has increased in recent years, partly in response to market demand for perceived higher welfare product. Animal welfare is enhanced by the large area of free space, the environmental complexity and the choice of physical and social environment which is possible in these circumstances (Fig. 7). However, there is still competition for resources: notably for food, which is usually scattered once daily on the ground, and for shelter from extremes of both heat and cold at different times of a year. In these circumstances, the lower-ranking individuals within the group can experience severe welfare problems.

9 Conclusion

Many aspects of the housing and feeding of pregnant sows in commercial farms are at variance with the behaviour patterns which have evolved in response to the living conditions of their ancestors. This mismatch gives rise to a range of welfare problems for the animals. In individual confinement systems of stalls and tethers, the inability to resolve conflicts arising from enforced proximity to neighbouring sows and the inability to appropriately express foraging behaviour motivated by the chronic hunger associated with concentrated diets can result in chronic stress responses and the development of abnormal, stereotyped oral behaviours. These outcomes have given rise to public concern and led to the banning or restricted use of such systems in a growing number of countries. However, the adoption of group housing systems has brought a new set of welfare challenges linked to aggression between animals. Aggression occurs when unfamiliar animals are grouped together, when space is

Figure 7 Outdoor system for pregnant sows with simple hut shelter and feeding on the ground.

inadequate for the natural submissive behaviours to be appropriately expressed or when competition for resources occurs. Foremost amongst these is the competition for food in chronically hungry animals. Many different housing systems exist as a result of the multiple possible combinations of feeding system, floor type, bedding, group size and space allowance. Although generalised conclusions about the relative welfare implications of these different housing systems can be highlighted, it is actually the detail of system design and management on a case-by-case basis which has most influence on welfare outcomes. Reducing hunger without increasing calorie intake, by the use of high-fibre diets or supplementary forages, or allowing appropriate expression of foraging behaviour by the provision of straw bedding or other suitable substrate, gives welfare benefits in all housing systems. In group housing systems, the provision of adequate floor space for avoidance of aggressive encounters, the use of specially designed mixing pens with features assisting escape and hiding from more dominant sows and the facilitation of subgrouping behaviour in larger dynamic groups by compartmentalisation of the resting area can all minimise the social stress occurring in the system. Future strategies for high welfare in group housing must also take into account the genetic predisposition of the animals and exploit the potential for selective breeding for beneficial social genetic effects.

10 Future trends

Whilst much of the research to date has focused on housing design and management, future research will need to investigate more closely the sow factors associated with the successful implementation of group housing. The genetic and experiential effects on aggressiveness, whilst studied extensively in growing pigs, have as yet received insufficient attention in the sow. Lovendahl et al. (2005) calculated heritabilities of 0.17–0.24 for performance of aggression at mixing, suggesting significant scope for genetic approaches to improve welfare. The current work on selection for social genetic effects in growing pigs (Bijma, 2011) could be usefully extrapolated to gestating sows.

Another area where more research is required relates to the health challenges experienced by the pregnant sow. The issue of lameness has been highlighted in earlier studies (Kroneman et al., 1993a,b; Gjein and Larssen, 1995), but the change to group housing systems seems to have exacerbated the problems (Maes et al., 2016). Once again it is a multifactorial issue, with both animal and environmental (floor space, group size, pen design and flooring) contributing factors which need to be better understood in order to make appropriate recommendations. A new and concerning finding relates to the prevalence of gastric ulcers in pregnant sows. A survey by the Danish Pig Research Centre in 2014 showed that more than 50% of cull sows had damage to the gastric mucosa and more than 25% showed serious ulceration. The extent to which these ulcers cause pain to affected animals in relation to their degree of severity is unknown, although only animals with more severe ulcers show inappetence and loss of condition. In humans, the condition is known to be acutely painful, and the similarity in anatomy might suggest this to also be the case in pigs. If so, the high prevalence of the condition constitutes a serious welfare problem (Edwards, 2015). It is now widely recognised that the nature and quantity of the diet given to the pregnant sow are fundamental to many of the welfare issues in these animals, and further work on diets which can both promote satiety and support gut integrity is needed.

11 Where to look for further information

A description of different housing systems for pregnant sows can be found in the following papers:

Edwards, S. A. (1998). Housing the breeding sow. *In Practice*, 20, 339–43.

A series of 8 booklets on different group housing systems with discussion of their strengths and weaknesses and example housing layouts was produced by the UK Pig Welfare Advisory Group, and can be found at – http://webarchive.nationalarchives.gov.uk/20031221024510/ and http://www.defra.gov.uk/animalh/welfare/farmed/pigs/#pwag.

A useful website illustrating different systems and housing conversions is produced by the National Sow Housing Conversion Project (NSHCP) of Canada – http://www.groupsowhousing.com/.

The results of national reviews on the subject of welfare implications of housing for pregnant sows can be found in the following papers:

Scientific Veterinary Committee (1997). *The Welfare of Intensively Kept Pigs*. EU Commission.

EFSA. (2007). Scientific Opinion of the Panel on Animal Health and Welfare on a request from the Commission on Animal health and welfare aspects of different housing and husbandry systems for adult breeding boars, pregnant, farrowing sows and unweaned piglets. *The EFSA Journal*, 572, 1–13.

Sow Housing Task Force. (2005). A comprehensive review of housing for pregnant sows. *Journal of the American Veterinary Medical Association*, 227, 1580–90.

Bench, C. J., Rioja-Lang, F. C., Hayne, S. M., and Gonyou, H. W. (2013a). Group gestation housing with individual feeding – I. How feeding regime, resource allocation, and genetic factors affect sow welfare. *Livestock Science*, 152, 208–17.

Bench, C. J., Rioja-Lang, F. C., Hayne, S. M., and Gonyou, H. W. (2013b). Group gestation housing with individual feeding – II. How space allowance, group size and composition, and flooring affect sow welfare. *Livestock Science*, 152, 218–27.

Reviews of the issues associated with hunger in the pregnant sows can be found in the following papers:

Meunier –Salaun, M. C., Edwards, S. A. and Robert, S. (2001). Effect of dietary fibre on the behaviour and health of the restricted-fed sow. *Animal Feed Science and Technology*, 90, 53–69.

D'Eath, R. B., Tolkamp, B. J., Kyriazakis, I. and Lawrence, A. B. (2009). 'Freedom from hunger' and preventing obesity: The animal welfare implications of reducing food quantity or quality. *Animal Behaviour*, 77, 275–88.

Reviews of the issues associated with mixing of the pregnant sows can be found in the following papers:

Arey, D. S. and Edwards, S. A. (1998). Factors affecting aggression in newly regrouped sows and the consequences for reproduction. *Livestock Production Science*, 56, 61–70.

Greenwood, E. C., Plush, K. J., van Wettere, W. H.E. J. and Hughes, P. E. (2014). Hierarchy formation in newly mixed, group housed sows and management strategies aimed at reducing its impact. *Applied Animal Behaviour Science*, 160, 1–11.

12 References

Anil, L., Anil, S. S., Deen, J., Baidoo, S. K. and Walker, R. D. (2006). Effect of group size and structure on the welfare and performance of pregnant sows in pens with electronic sow feeders. *Canadian Journal of Veterinary Research*, 70, 128–36.

Appleby, M. C. and Lawrence, A. B. (1987). Food restriction as a cause of stereotyped behaviour in tethered gilts. *Animal Production*, 45, 103–10.

Appleby, M. C., Lawrence, A. B. and Illius, A. W. (1989). Influence of neighbours on stereotypic behaviour of tethered sows. *Applied Animal Behaviour Science*, 24, 137–46.

Arey, D. S. (1999). Time course for the formation and disruption of social organisation in group-housed sows. *Applied Animal Behaviour Science*, 62, 199–207.

Backstrom, L. (1973). Environment and animal health in piglet production. A field study of incidences and correlations. *Acta Veterinaria Scandinavica*, Suppl. 41, 1–240.

Barnett, J. L., Cronin, G. M., Winfield, C. G. and Dewar, A. M. (1984). The welfare of adult pigs: The effects of five housing treatments on behaviour, plasma corticosteroids and injuries. *Applied Animal Behaviour Science*, 12, 209–32.

Barnett, J. L., Winfield, C. G., Cronin, G. M., Hemsworth, P. H. and Dewar, A. M. (1985). The effect of individual and group housing on behavioural and physiological responses related to welfare of pregnant pigs. *Applied Animal Behaviour Science*, 14, 149–61.

Barnett, J. L., Hemsworth, P. H., Winfield, C. G. and Fahy, V. A. (1987a). The effects of pregnancy and parity number on behavioural and physiological responses related to the welfare status of individual and group-housed pigs. *Applied Animal Behaviour Science*, 17, 229–43.

Barnett, J. L., Hemsworth, P. H. and Winfield, C. G. (1987b). The effects of design of individual stalls on the social behaviour and physiological responses related to the welfare of pregnant pigs. *Applied Animal Behaviour Science*, 18, 133–42.

Barnett, J. L., Hemsworth, P. H., Newman, E. A., McCallum, T. H. and Winfield, C. G. (1989). The effect of design of tether and stall housing on some behavioral and physiological-responses related to the welfare of pregnant pigs. *Applied Animal Behaviour Science*, 24, 1–12.

Barnett, J. L., Hemsworth, P. H., Cronin, G. M., Newman, E. A. and McCallum, T. H. (1991). Effects of design of individual cage-stalls on the behavioral and physiological-responses related to the welfare of pregnant pigs. *Applied Animal Behaviour Science*, 32, 23–33.

Barnett, J. L., Cronin, G. M., McCallum, T. H. and Newman, E. A. (1993a). Effects of pen size/shape and design on aggression and injuries when grouping unfamiliar adult pigs. *Applied Animal Behaviour Science*, 36, 111–22.

Barnett, J. L., Cronin, G. M., McCallum, T. H. and Newman, E. A. (1993b). Effects of 'chemical intervention' techniques on aggression and injuries when grouping unfamiliar adult pigs. *Applied Animal Behaviour Science*, 36, 135–48.

Baxter, M. R. and Schwaller, C. E. (1983). Space requirements for sows in confinement. In: *Farm Animal Housing and Welfare* (Baxter, S. H., Baxter, M. R. and MacCormack, J. A. D. (Eds)). Martinus Nijhoff Publishers, The Hague, The Netherlands, pp. 181–95.

Bench, C. J., Rioja-Lang, F. C., Hayne, S. M. and Gonyou, H. W. (2013a). Group gestation housing with individual feeding – I. How feeding regime, resource allocation, and genetic factors affect sow welfare. *Livestock Science*, 152, 208–17.

Bench, C. J., Rioja-Lang, F. C., Hayne, S. M. and Gonyou, H. W. (2013b). Group gestation housing with individual feeding – II. How space allowance, group size and composition, and flooring affect sow welfare. *Livestock Science*, 152, 218–27.

Bengtsson, A.-C., Svendsen, J. and Persson, G. (1983). Comparison of four types of housing for sows in gestation: Behaviour studies and hygiene studies. Report 36, Swedish University of Agricultural Sciences, Lund, Sweden.

Bijma, P. (2011). Breeding for social interaction for animal welfare. In: *Encyclopedia of Sustainability Science and Technology* (Meyers, R. A. (Ed.)). Springer, Larkspur, CA, pp. 1477–1513.

Blackshaw, J. K. and McVeigh, J. F. (1984). The behaviour of sows and gilts, housed in stalls, tethers and groups. *Proceedings of the Australian Society of Animal Production*, 15, 85–92.
Bokma, S. (1990). Housing and management of dry sows in groups in practice: Partly slatted systems. In: *Proceedings of an International Symposium on Electronic Identification in Pig Production*. RASE, Stoneleigh, UK, pp. 37–45.
Bokma, S. and Kersjes, G. J. K. (1988). The introduction of pregnant sows in an established group. In: *Proceedings of the International Congress on Applied Ethology in Farm Animals*, Skara, Sweden, pp. 166–9.
Botemans, J. A. M. (1989). The effect of straw on the aggression of sows during grouping. Internal report, Swedish University of Agricultural Sciences, Lund, Sweden.
Brade, M. A., Edwards, S. A. and Riley, J. E. (1986). The commercial application of electronic sow identification and feeding systems in the UK. In: *Proceedings of the 37th Annual Meeting EAAP*, Budapest, Hungary.
Broom, D. M. and Potter, M. J. (1984). Factors affecting the occurrence of stereotypies in stall housed dry sows. In: *Proceedings of the International Congress on Applied Ethology*, Kiel, Germany.
Brouns, F. and Edwards, S. A. (1994). Social rank and feeding behaviour of group-housed sows fed competitively or ad libitum. *Applied Animal Behaviour Science*, 39, 225–35.
Brouns, F., Edwards, S. A. and English, P. R. (1994). Effect of dietary fibre and feeding system on activity and oral behaviour of group-housed gilts. *Applied Animal Behaviour Science*, 39, 215–23.
Brouns, F., Edwards, S. A. and English, P. R. (1995). Influence of fibrous feed ingredients on voluntary intake of dry sows *Animal Feed Science and Technology*, 54, 301–13.
Buchenauer, D. (1990). Social and feeding behaviour of sows in group housing systems. In: *Proceedings of the Seminar on 'Group housing of sows', European Conference Group on the Protection of Farm Animals*, Brussels, Belgium, pp. 40–51.
Bure, R. G. (1991). The influence on vulva biting of supplying additional roughage in an electronic sow feeder. In: *Proceedings of the 42nd Annual Meeting EAAP*, Berlin, Germany.
Carter, A. J. and English, P. R. (1983). A comparison of the activity and behaviour of dry sows in different housing and penning systems. *Animal Production*, 36, 531–2.
Chapinal, N., Ruiz de la Torre, J. L., Cerisuelo, A., Gasa, J., Baucells, M. D. and Manteca, X. (2010). Aggressive Behavior in Two Different Group-Housing Systems for Pregnant Sows. *Journal of Applied Animal Welfare Science*, 13, 137–53.
Cronin, G. M. (1985). The development and significance of abnormal stereotyped behaviours in tethered sows. Doctoral thesis, University of Groningen, Groningen, The Netherlands, 146pp.
Cronin, G. M. and Wiepkema, P. R. (1984). The development and significance of abnormal stereotyped behaviours in tethered sows. *Annales de Recherche Veterinaire*, 15, 263–70.
Csermely, D. and Wood-Gush, D. G. M. (1986). Agonistic behaviour in grouped sows. I. The influence of feeding. *Biology of Behaviour*, 11, 244–52.
Csermely, D. and Wood-Gush, D. G. M. (1990a). Agonistic behaviour in grouped sows. II. How social rank affects feeding and drinking behaviour. *Bollettino di Zoologia*, 57, 55–8.
Csermely, D. and Wood-Gush, D. G. M. (1990b). Agonistic behaviour in grouped sows. III. The effect of grouping methods. *Bollettino di Zoologia*, 57, 271–5.
de Leeuw, J. A., Jongbloed, A. W. and Verstegen, M. W. A. (2004). Dietary fibre stabilizes blood glucose and insulin levels and reduces activity in sows (*Sus scrofa*). *Journal of Nutrition*, 134, 1481–6.
de Leeuw, J. A., Zonderland, J. J., Altena, H., Spoolder, H. A. M., Jongbloed, A. W. and Verstegen, M. W. A. (2005). Effects of levels and sources of dietary fermentable non-starch polysaccharides on blood glucose stability and behaviour of group-housed pregnant gilts. *Applied Animal Behaviour Science*, 94, 15–29.
De Koning, R. (1985). On the well-being of dry sows. Doctoral thesis, University of Utrecht, Utrecht, The Netherlands.
Edwards, S. A. (1985). Group housing systems for dry sows. *Farm Buildings Progress*, 80, 19–22.
Edwards, S. A. (1992). Scientific perspectives on loose housing systems for dry sows. *Pig Veterinary Journal*, 28, 40–51.

Edwards, S. A. (2000). Alternative housing for dry sows: System studies or component analysis? In: *Improving Health and Welfare in Animal Production*. EAAP Publication 102 (Blokhuis,, H. J., Ekkel,, E. D. and Wechsler,, B. (Eds)). Wageningen Pers, Wageningen, pp. 99–107.
Edwards, S. A. (2005). Product quality attributes associated with outdoor pig production. *Livestock Production Science*, 94, 5–14.
Edwards, S. A. (2015). Assessment and alleviation of pain in pig production. In: *Proceedings of the International Conference on Pig Welfare*. Danish Centre for Animal Welfare, Copenhagen, Denmark, 29–30 April.
Edwards, S. A. and Riley, J. E. (1986). The application of the electronic identification and computerised feed dispensing system in dry sow housing. *Pig News and Information*, 7, 295–8.
Edwards, S. A., Simmins, P. H., Walker, A. J. and Beckett, M. P. (1986). Behaviour of 400 sows in a single group with electronic individual feeding. In: *Proceedings of the International Symposium on Applied Ethology in Farm Animals*, Balatonfüred, Hungary.
Edwards, S. A., Armsby, A. W. and Large, J. W. (1988a). The effect of diet form on an individual feeding system with electronic identification for group housed sows. *Research and Development in Agriculture*, 5, 129–32.
Edwards, S. A., Armsby, A. W. and Large, J. W. (1988b). Effects of feed station design on the behaviour of group housed sows using and electronic individual feeding system. *Livestock Production Science*, 19, 511–22.
Edwards, S. A., Marconnet, C., Taylor, A. G. and Cadenhead, A. (1992). Voluntary intake and digestibility of distillery products for dry sows. *Animal Production*, 54, 486.
Edwards, S. A., Atkinson, K. A. and Lawrence, A. B. (1993a). The effect of food level and type on behaviour of outdoor sows. In: *Proceedings of the International Society for Applied Ethology*, Berlin, Germany.
Edwards, S. A., Mauchline, S. and Stewart, A. H. (1993b). Designing pens to minimise aggression when sows are mixed. *Farm Building Progress*, 113, 20–3.
Edwards, S. A., Mauchline, S., Marston, G. and Stewart, A. H. (1994). Agonistic behaviour amongst newly mixed sows and the effects of pen design and feeding method. *Applied Animal Behaviour Science*, 41, 272.
Gjein, H. and Larssen, R. B. (1995). Housing of pregnant sows in loose and confined systems—a field study. 2. Claw lesions: Morphology, prevalence, location and relation to age. *Acta Veterinaria Scandinavica*, 36, 433–42.
Fraser, D. (1975). The effect of straw on the behaviour of sows in tether stalls. *Animal Production*, 21, 59–68.
Greenwood, E. C., Plush, K. J., van Wettere W. H. E. J., and Hughes, P. E. (2016). Group and individual sow behaviour is altered in early gestation by space allowance in the days immediately following grouping. *Journal of Animal Science*, 94, 385–93.
Harrison, R. (1964). *Animal Machines*. Ballantine Books, New York, NY.
Hemsworth, P. H., Rice, M., Nash, J., Girri, K., Butler, L., Tilbrook, A. J. and Morrison, R. S. (2013). Effects of group size and floor space allowance on grouped sows: Aggression, stress, skin injuries and reproductive performance. *Journal of Animal Science*, 91, 4953–64.
Hoofs, A. (1990). Equipment assessment of group housing systems for sows in the Netherlands. In: *Proceedings of an International Symposium on Electronic Identification in Pig Production*. RASE, Stoneleigh, UK, pp. 77–82.
Hughes, B. O. and Duncan, I. J. H. (1988). The notion of ethological need, models of motivation and animal welfare. *Animal Behaviour*, 36, 1696–707.
Hulbert, L. E. and McGlone, J. J. (2006). Evaluation of drop versus trickle-feeding systems for crated or group-penned gestating sows. *Journal of Animal Science*, 84, 1004–14.
Hunter, E. J. (1988). Behaviour and welfare of dry sows in different housing conditions. Doctoral thesis, University of Reading, Reading, UK.
Hunter, E. J., Edwards, S. A. and Simmins, P. H. (1989). Social activity and feeder use of a dynamic group of 40 sows using a sow operated computerised feeder. *Animal Production*, 48, 643.

Jensen, P. (1980a). The impact of confinement on the behaviour of dry sows. An Ethological Study. Report 2, Department Animal Hygiene with Farrier's School, Skara, Sweden.
Jensen, P. (1980b). An ethogram of social interaction patterns in group-housed dry sows. *Applied Animal Ethology*, 6, 341–50.
Jensen, P. (1982/3). An analysis of agonistic interaction patterns in group-housed dry sows – aggression regulation through an 'avoidance order'. *Applied Animal Ethology*, 9, 47–61.
Jensen, P. (1984). Effects of confinement on social interaction patterns in dry sows. *Applied Animal Behaviour Science*, 12, 93–103.
Jensen, P. and Wood-Gush, D. G. M. (1984). Social interactions in a group of free-ranging sows. *Applied Animal Behaviour Science*, 12, 327–37.
Jones, R. H. R. and Petchey, A. M. (1987). The behaviour of dry sows in straw yards. In: *Pig Housing and the Environment* (Smith, A. T. and Lawrence, T. L. J. (Eds)). Occasional Publication no. 11, BSAP, Edinburgh, Scotland, pp. 134–6.
Kroneman, A., Vellenga, L., van der Wilt, F. J. and Vermeer, H. M. (1993a). Field research on veterinary problems in group-housed sows — A survey of lameness. *Zentralblatt für Veterinärmedizin. Reihe A* , 40, 704–12.
Kroneman, A., Vellenga, L., van der Wilt, F. J. and Vermeer, H. M. (1993b). Review of health problems in group-housed sows, with special emphasis on lameness. *The Veterinary Quarterly*, 15, 26–9.
Lambert, R. J., Ellis, M. and Rowlinson, P. (1986). An alternative sow housing/feeding system for dry sows based upon a sow-activated electronic feeder. In: *Proceedings of the 37th Annual Meeting EAAP*, Budapest, Hungary.
Lawrence, A. B. and Terlouw, E. M. C. (1993). A review of behavioral factors involved in the development and continued performance of stereotypic behaviors in pigs. *Journal of Animal Science*, 71, 2815–25.
Lenskens, P. (1991). The effect of the stepwise introduction of young gilts into an established group of sows on the aggression during grouping. Miscellaneous report 9, Swedish University of Agricultural Sciences, Lund, Sweden.
Livingstone, R. M. and Fowler, V. R. (1984). Pig feeding in the future: Back to nature? *Span*, 27, 108–10.
Lovendahl, P., Damgaard, L. H., Nielsen, B. L., Thodberg, K., Su, G. and Rydhmer, L. (2005). Aggressive behaviour of sows at mixing and maternal behaviour are heritable and genetically correlated traits. *Livestock Production Science*, 93, 73–85.
Luescher, U. A., Friendship, R. M. and McKeown, D. B. (1990). Evaluation of methods to reduce fighting among regrouped gilts. *Canadian Journal of Animal Science*, 70, 363–70.
Maes, D., Pluym, L. and Peltoniemi, O. (2016). Impact of group housing of pregnant sows on health. *Porcine Health Management*, 2, 17.
Martin, J. E. and Edwards, S. A. (1994). Feeding behaviour of outdoor sows: The effects of diet quantity and type. *Applied Animal Behaviour Science*, 41, 63–74.
McGlone, J. J. and Curtis, S. E. (1985). Behavior and performance of weanling pigs in pens equipped with hide areas. *Journal of Animal Science*, 60, 20–4.
Mendl, M., Zanella, A. J. and Broom, D. M. (1992). Physiological and reproductive correlates of behavioural strategies in female domestic pigs. *Animal Behaviour*, 44, 1107–21.
Metz, J. H. M. and Osterlee, C. C. (1980). Immunologische und ethologische Kritorien fur die artgemasse Haltung von Sauen und Ferkeln. *Kuratorium Tech Bauwesen Landwirtschaft Schrift*, 264, 39–50.
Muirhead, M. R. (1983). Pig housing and environment. *The Veterinary Record*, 113, 587–93.
Mujuni, B. M. K., McFarlane, J. M., Curtis, S. E. and Taylor, I. A. (1985). Effect of group size on social interactions in newly weaned sows. *Journal of Animal Science*, 63 suppl. 1, 164.
Nielsen, N. C., Christensen, K, Bille, N. and Larsen, J. L. (1974). Preweaning mortality in pigs. 1. Herd investigations. *Nordisk Veterinaermedicin*, 26, 137–50.
O'Connell, N. E. (2007). Influence of access to grass silage on the welfare of sows introduced to a large dynamic group. *Applied Animal Behaviour Science*, 107, 45–57.

Olsson, A. Ch., Andersson, M., Lenskens, P., Rantzer, D. and Svendsen, J. (1991). Herd studies of different electronic dry sow feeding systems. Report 75, Swedish University of Agricultural Sciences, Lund, Sweden.

Petherick, J. C. (1983). A biological basis for the design of space in livestock housing. In: *Farm Animal Housing and Welfare* (Baxter, S. H., Baxter, M. R. and McCormack, J. A. C. (Eds)). Martinus Nijhoff Publishers, Dordrecht, The Netherlands, pp. 103–20.

Petherick, J. C., Bodero, D. A. and Blackshaw, J. K. (1987). The use of partial barriers along the feed trough in a group housing system for non-lactating sows. *Farm buildings and Engineering*, 4, 32–6.

Remience, V., Wavreille, J., Canart, B., Meunier-Salaun, M., Prunier, A., Bartiaux-Thrill, N., Nicks, B. and Vandenheede, M. (2008). Effects of dry space allowance on the welfare of dry sows kept in dynamic groups and fed with an electronic sow feeder. *Applied Animal Behaviour Science*, 112, 284–96.

Ru, Y. J. and Bao, Y. M. (2004). Feeding dry sows ad libitum with high fibre diets. *Asian-Australasian Journal of Animal Sciences*, 17, 283–300.

Rushen, J. (1984). Stereotyped behaviour, adjunctive drinking and the feeding periods of tethered sows. *Animal Behaviour*, 32, 1059–67.

Rushen, J. (1985). Stereotypies, aggression and the feeding schedules of tethered sows. *Applied Animal Behaviour Science*, 14, 137–47.

Salak-Johnson, J. L., Niekamp, S. R., Rodriguez-Zas, S. L., Ellis, M. and Curtis, S. E. (2007). Space allowance for dry, pregnant sows in pens: Body condition, skin lesions, and performance. *Journal of Animal Science*, 85, 1758–69.

Sambraus, H. H. (1981). Das Sozialverhalten von Sauen bei Gruppenhaltung. *Zuchtungskunde*, 53, 147–57.

Sambraus, H. H. and Schunke, B. (1982). Verhaltensstorungen bei Zuchtsauen im Kastenstand. *Wiener Tierärztliche Monatsschrift*, 69, 200–8.

Scientific Veterinary Committee. (1997). *The Welfare of Intensively Kept Pigs*. EU Commission, Brussels, Belgium.

Scott, K., Binnendijk, G. P., Edwards, S. A., Guy, J. H., Kiezebrink, M. C. and Vermeer, H. M. (2009). Preliminary Evaluation of a prototype welfare monitoring system for sows and piglets (Welfare Quality project®). *Animal Welfare*, 18, 441–9.

Simmins, P. H. (1993). Reproductive performance of sows entering stable and dynamic groups after mating. *Animal Production*, 57, 293–8.

Sow Housing Task Force. (2005). A comprehensive review of housing for pregnant sows. *Journal of the American Veterinary Medical Association*, 227, 1580–90.

Spoolder, H. A. M., Burbidge, J. A., Edwards, S. A., Simmins, P. H. and Lawrence, A. B. (1995). Provision of straw as a foraging substrate reduces the development of excessive chain and bar manipulation in food restricted sows. *Applied Animal Behaviour Science*, 43, 249–62.

Spoolder, H. A. M., Burbidge, J. A., Edwards, S. A., Lawrence, A. B. and Simmins, P. H. (1997). Effects of food level on performance and behaviour of sows in a dynamic housing system with electronic feeding. *Animal Science*, 65, 473–82.

Spoolder, H. A. M., Geudeke, M. J., van der Peet-Schwering C. M. C., and Soede, N. M. (2009). Group housing of sows in early pregnancy: A review of success and risk factors. *Livestock Science*, 125, 1–14.

Stewart, A. H., Edwards, S. A., Brouns, F. and English, P. R. (1993). An assessment of the effect of feeding system on the production and social organisation of group housed gilts. *Animal Production*, 56, 422.

Stewart, C. L., O'Connell, N. E. and Boyle, L. (2008). Influence of access to straw provided in racks on the welfare of sows in large dynamic groups. *Applied Animal Behaviour Science*, 112, 235–47.

Stolba, A., Baker, N. and Wood-Gush, D. G. M. (1983). The characterization of stereotyped behaviour in stalled sows by information redundancy. *Behaviour*, 87, 157–82.

Stolba, A. and Wood-Gush, D. G. M. (1989). The behaviour of pigs in a semi-natural environment. *Animal Production*, 48, 419–25.

Svendsen, J. and Bengtsson, A.-Ch. (1983). Housing of sows in gestation. In: *Proceedings of the Guelph Pork Symposium*. University of Guelph, Guelph, Ontario.

Svendsen, J., Andersson, M., Olsson A. Ch., Rantzer, D. and Lundqvist, P. (1990). Group housing of sows in gestation in insulated and uninsulated buildings. Report 66, Swedish University of Agricultural Sciences, Lund, Sweden.

te Brake, J. H. A. and Bressers, H. P. M.. (1990). Applications in service management and oestrus detection. In: *Proceedings of an International Symposium on Electronic Identification in Pig Production*. RASE, Stoneleigh, UK, pp. 63–7.

Terlouw, E. M. C., Lawrence, A. B. and Illius, A. W. (1991). Influences of feeding level and physical restriction on the development of stereotypies in sows. *Animal Behaviour* 42, 981–91.

Terlouw, E. M. C., Wiersma, A., Lawrence, A. B. and Macleod, H. A. (1993). Ingestion of food facilitates the performance of stereotypies in sows. *Animal Behaviour*, 46, 939–50.

van der Peet-Schering C. M. C., Spoolder, H. A. M., Kemp, B., Binnendijk, G. P., den Hartog, L. A. and Verstegen, M. W. A. (2003). Development of stereotypic behaviour in sows fed a starch diet or non-starch polysaccharide diet during gestation and lactation over two parities. *Applied Animal Behaviour Science*, 83, 81–97.

van Putten, G. and van de Burgwal, J. A. (1990a). Vulva biting in group-housed sows: Preliminary report. *Applied Animal Behaviour Science*, 26, 181–6.

van Putten, G. and van de Burgwal, J. A. (1990b). Pig breeding in phases. In: *Electronic Identification in Pig Production. Proceedings of the International Symposium*. RASE, Stoneleigh, UK, pp. 115–20.

Vestergard, K and Hansen, L. L. (1984). Tethered versus loose sows: Ethological observations and measures of productivity. I. ethological observations during pregnancy and farrowing. *Annales de Recherches Veterinaire*, 15, 245–56.

Weber, R, Friedli, K. and Troxler, J. (1991). The influence of the computerised feeding system on the behaviour of sows and its effects with regard to injuries and alterations on their body. In: *Proceedings of the 42nd Annual Meeting EAAP*, Berlin, Germany.

Weng, R. C., Edwards, S. A. and English, P. R. (1998). Behaviour, social interactions and lesion scores of group-housed sows in relation to floor space allowance. *Applied Animal Behaviour Science*, 59, 307–16.

Chapter 9

Welfare of weaned piglets

Arlene Garcia and John J. McGlone, Texas Tech University, USA

1 Introduction

Nearly 1 billion pigs are in inventory on farms around the world. About half the pigs are found in China alone. The other regions with significant number of pigs include South East Asia, Japan, Brazil, Mexico, the United States, Canada and most of Europe. Among these geographical locations, we find three general types of production models: (1) modern, industrialized production systems [Modern]; (2) small-scale low-input, often-backyard, family operations [Backyard]; and (3) natural, organic, antibiotic-free, GMO-free systems [Natural]. Modern industrialized systems have been called factory farms by those that wish to criticize keeping pigs in buildings and pens. However, pigs were moved from mud lots to buildings to improve their welfare and their health in particular. Today, we may find very low pre-weaning mortality and high health among pigs in many industrialized systems. Likewise, some presumably high-welfare systems have high pre-weaning piglet mortality and significant numbers of sows with scratches and wounds compared with modern US industrialized systems (McGlone, 2006). We have also reported good welfare among sows in both indoor and outdoor production systems that are well managed (Johnson et al., 2001). One cannot support the idea that industrialized production systems are inherently bad for pig welfare. Likewise, one cannot assume that because sows and piglets are in more natural settings that their welfare is automatically good. Good and poor welfare are found both on industrialized and more on natural production systems.

http://dx.doi.org/10.19103/AS.2017.0013.24

Concerns and demands for improved animal welfare and animal handling systems from authorities, non-government organizations, markets and the public in general are increasing and are of utmost importance to the meat industry (Stoier et al., 2016). Today in the United States, all Modern farms must have an animal welfare audit conducted by a third party (one not associated with the farm or the market chain). Also, many Natural systems have an animal welfare assessment requirement of some kind. Backyard production systems often are not regulated or assessed.

In Europe, some farms use third-party audit programmes and some individual European countries have government oversight of pig farms. In developing countries, animal welfare audits are initiated, but they are far from comprehensive and they do not cover the majority of pigs in commercial production. We expect this to change over time. While governments vary in animal welfare requirements, multinational retail buyers of pork products now require third-party oversight of farm animal welfare in developed countries and they are slowly advancing animal welfare audit programmes in developing countries.

Requiring animal welfare third-party assessment is one step towards assuring a minimum animal welfare standard is met in the meat supply chain. However, the situation is made complex because we are unsure, in some cases, how to create the best animal welfare. Among the phases of pork production with opportunities to improve welfare, the weaning period has the greatest opportunity to positively impact most animals. By weaning period we mean from birth through a week or two after weaning. Pre-weaning mortality of piglets has dropped from over 25% 30 years ago to less than 10% today on Modern farms. Even at 10% pre-weaning mortality, over 100 million piglets die before weaning each year. The general classification of opportunities to improve piglet welfare falls in these categories (in order of the numbers of piglets impacted and the severity of the welfare issue):

- Pre-weaning mortality
- Weaning stress
- Painful practices
- Transportation

It is important to understand that poor management can lead to many welfare issues, and educating caregivers to make good judgements can greatly improve these issues. This chapter will focus on current practices that can be detrimental to piglet well-being, alternatives and/or solutions to these, and advances in technology that could improve animal well-being, profitability and sustainability.

2 Pre-weaning mortality

The causes of pre-weaning mortality are many. In an early, classic paper from 1971, Fahmy and Bernard documented the causes of mortality among 6,890 Yorkshire piglets from birth to 20 weeks of age. Mortality was 25.6% on average. Of this total mortality, 16.4% was from birth to weaning at 8 weeks of age. The causes of death among piglets were general weakness, crushing by the sow (these two comprise nearly half of all piglet deaths), scours, paralysis, rickets, anatomical abnormalities, pneumonia and anaemia. Interestingly, they reported again (in agreement with authors in the 1940s) that (1) sows with larger litters had more pre-weaning piglet deaths, (2) inbreeding of these purebred sows increased piglet mortality and (3) smaller body weight piglets have a higher risk of death. By examining

Fahmy and Bernard's (1971) study on causes of piglet death, one can realize that some problems are solved on modern farms – we have less scours and pneumonia on high health farms, no inbreeding, no rickets and we have bred against weak piglets and anatomical problems. So today, around the world, 45 years later, authors confirm that larger litters have more smaller, at-risk piglets and that crushing is a major cause of pre-weaning mortality (e.g. Thailand: Nuntapaitoon and Tummaruk, 2015; Sweden: Westin et al., 2015; Kilbride et al., 2015).

Solutions to reduce pre-weaning mortality have focused on the farrowing environment, sow genetics and management practices such as cross-fostering piglets among sows.

2.1 The farrowing environment

The farrowing environment is a housing or penning or keeping system where sows reside from a few days before farrowing through weaning. It may be a single system or a combination of systems. The most common system in Modern pork production systems is the farrowing crate in which a single sow is kept in an area that does not allow her to turn around and requires that she lie down more gently along bars (often metal). The farrowing crate prevents sows from lying down quickly and crushing piglets. While the farrowing crate saves piglet lives compared with the older-style pen, it is not the only system to reduce pre-weaning mortality.

The farrowing crate was first patented in the United States in 1963 by Ingvald Eide (patent publication date Feb 19, 1963; US 3077861), but it was in widespread use before then in the United States and Europe. The first scientific publication that showed a reduction in pre-weaning mortality was by Robertson et al. (1966) in England. The farrowing crate quickly spread among Modern pig farms such that by 1980, it was the most common system in pork production in North America and Europe. Still, people criticized the farrowing crate for restricting the sow's movement. Systems were sought that allowed sows to turn around, but that did not cause a rise in piglet mortality.

The first published viable alternative to the farrowing crate was the turn-around pen. In this system, the pen is 21% larger (due to 21% larger length) and the same width as a standard farrowing crate (2.7 m × 1.95 m) and had a V-shape to the farrowing railings. McGlone and Blecha (1987) showed this system had even lower pre-weaning mortality than the standard farrowing crate. The turn-around crate was modelled after a pork producer in Minnesota who used what was called the Compart crate (which was actually a pen). McGlone and Morrow-Tesch (1990) showed that the sloped-floor farrowing pen also had lower pre-weaning mortality as a farrowing crate. Frank Hurnick from Canada developed an ellipsoid farrowing crate that equalled the standard farrowing crate in piglet mortality (Lou and Hurnick, 1994). Greg Cronin developed what he called the Werribee Pen that had equal pre-weaning mortality as the farrowing crate (Cronin et al., 2000). Outdoor farrowing huts vary significantly in pre-weaning mortality. Johnson et al. (2001) found one style, called the English-style hut, that produced low levels of pre-weaning mortality comparable to the farrowing crate in buildings. And more indoor farrowing systems continue to be developed that do not use a restraining crate, often re-inventing elements of the early alternatives.

What the alternatives to the farrowing crate have in common is that they all occupy more floor space than a farrowing crate that might be found in Modern pig farms. Alternatives to the farrowing crates use 9 to 147% more space (average of 48% more) (McGlone, 2002). Floor space in buildings is expensive and so systems that use less floor space and less penning material will be more economical.

In conclusion, farrowing crates do reduce piglet mortality compared with old-style farrowing pens. But reasonable alternatives to the restrictive farrowing crate are available.

2.2 Sow genetics

Pig breeders have been attempting to increase the number of pigs weaned per sow for centuries. In recent decades, pig breeders have understood that the heritability of litter size is reasonably high and would respond to selection pressure (King and Gajic, 1969). So, companies and producers selected for increased numbers of pigs born and weaned. The result is that in the United States, the national average numbers of pigs weaned have risen from 7 to over 10 pigs per sow. All the while, pre-weaning mortality has remained about the same (in the range of 10–15%). Therefore, while we can increase the number of pigs weaned by selecting for litter size, any genetic effect on litter size is probably indirect. For example, sows that sit at all or for longer durations are more likely to crush a pig (moving from the sitting to the lying position). The heritability estimate for sitting is greater than zero and so indirectly one could select against sitting to lower piglet mortality (McGlone et al., 1991). Interestingly, the sloped farrowing pen that reduces piglet mortality also discourages sows from sitting. To improve pre-weaning mortality, environmental conditions and human behaviours need attention.

2.3 Cross-fostering

Piglets are opportunistic sucklers. They will nurse any sow, even if they have a preference for their biological mother. From the sow's perspective, she will accept any piglet that nurses, largely because when she lies down to nurse, she is on her side and cannot push away any piglet that nurses. Cattle and sheep, on the other hand, nurse standing up and they can easily step away if an unwanted neonate attempts to suckle. Therefore, cross-fostering is more easily accomplished in pigs than in ruminants.

Some sows have more piglets than they have functional teats. Domestic pigs of European decent show a behaviour called teat fidelity, which means that a given piglet chooses and defends a single teat. This means that if there are, for example, 12 teats and 13 piglets, then one piglet is sure to die. This extra piglet would be cross-fostered to another sow that had, for example, 11 piglets, but she has 12 functional teats. Cross-fostering can be done in the first few days after birth. The older the piglet, the lower the chance that cross-fostering will be successful.

Low-birth-weight (LBW) piglets are being born today as they have been for decades. The number of piglets per litter is increasing, but the litter weights have not increased (Bovey et al., 2014). The mortality rate for LBW piglets has been reported to be close to 30%. There is a linear increase in the number of LBW piglet mortality and sow parity. In parities 1 through 6 LBW piglets averaged 25.5% mortality versus 9.1% for high-birth-weight (HBW) piglets, but in parities 7 through 12 the mortality rate jumps up to 35.7% in LBW piglets and 11.2% in HBW piglets (Robert et al., 1995). Additionally, processing LBW piglets early in life increases mortality rate. LBW piglets tend to be weaker and many times appear lethargic, when they are processed pain may cause decreased periods of suckling (which has detrimental effect on piglets since they have low energy reserves and immunoglobulins and depend on colostrum for heat production; Herpin et al., 2002), which may lead to death. The decrease in milk consumption may cause piglets to use the udder as a heat source and leads to crushing of piglets; therefore, delaying processing in LBW piglet may improve survival rate (Bovey et al., 2014).

Sows are giving birth to more piglets than they can rear successfully and, therefore, have to be either supplemented or cross-fostered if they are LBW piglets. Many facilities have resorted to transferring piglets to other sows for the purpose of lactating foster piglets, known as nurse sows. The nurse sow is either given newborn piglets once she has finished raising her own or given 1-week-old piglets when her piglets have been weaned (Baxter et al., 2013).

Cross-fostering is stressful for both the piglets and the nurse sows. Piglets cross-fostered after the first day of life have to re-establish a teat order which can cause additional injuries due to fighting (Robert and Marineau, 2001) and disrupted nursings (Heim et al., 2012), which can lead to reduced milk intake and death. Additionally, cross-fostering later than 24 h after farrowing can cause the sows to demonstrate aggressive behaviours towards piglets that are not hers (Pedersen et al., 2008) and can also cause a temporary stoppage of milk let down (Price et al., 1994). Piglets that are cross-fostered the first day after birth are also dirtier and have an increased occurrence in carpal lesions and lameness as a result of fights (Sorensen et al., 2016). On the other hand, nurse sows have a higher tendency to have skin lesions, udder damage and swollen bursae on legs, as a result of prolonged stay in a farrowing crate (Sorensen et al., 2016). Additionally, being a nurse sow takes a toll on the sows' body because they are required to produce milk for a longer period of time, causing body condition score (BCS) to go down because body reserves are depleted throughout lactation (Spencer et al., 2003). Further, low BCS leads to shoulder lesions (Zurbrigg, 2006; Knauer et al., 2007), which can affect sow welfare.

3 Weaning stress

Wild or feral pigs will nurse piglets for three or more months. In commercial pork production pre-industrialized farms (prior to 1960), piglets were traditionally weaned at 8 weeks of age. Creep feed was provided starting at two weeks of age and by the time piglets were 8 weeks of age, they had been gradually weaned. Then, as production efficiencies were calculated, scientists and the industry learned that by weaning 'early', say at 28 days of age, the sow would produce more litters per sow per year and the time to get pigs to market was the same or better. Sow milk production peaks about three weeks after birth, so by four weeks of age, the piglet's appetite is not satisfied by mother's milk alone. The gastrointestinal tract is more developed at four weeks of age than earlier, so weaning is physiologically not problematic.

After the 28-day-of-age weaning phase, economists and pork production specialists concluded that weaning piglets in the 19–23 day range was most profitable. More litters were produced per sow per year, with lower weaning age; however, the piglet's digestive system is less able to handle weaning stress at 19 days of age than 28 days of age. This balance is struck based on economics more than other factors. If one sells pigs into the Modern market, economics may favour weaning at 21 days of age. If one sells in the Natural market, a weaning age of 28 days may be more economically viable because the market itself drives the weaning age higher. Some markets even push for weaning at 35 days of age to make it easier on the piglet to adapt to dry feed. However, one should be aware that with fast-growing, hungry piglets, and high-milking sows, delaying weaning is a metabolic challenge to the sow, with many sows losing significant body weight with advanced weaning age (beyond 21 days). When we observed the behaviour of sows in an outdoor system where sows can walk away from demanding piglets, we saw that sows spend increasing times away from the piglets over time (Johnson et al., 2001). A muscular, demanding 30-day-old piglet can demand significant milk from a lactating sow – and with larger litters, ten or more piglets can nurse with sufficient vigour to cause sows to become very thin. After weaning, sows have an improved metabolic situation with not producing milk, but they still may suffer a form of depression from the loss of piglets in her environment. Much less has been written and investigated about the stress of the sow after weaning. We know that sows experience a significant drop in feed intake and less activity after weaning than before weaning.

The stress that piglets experience at the time of weaning has been widely reported (Niekamp et al., 2007; Moeser et al., 2007; Campbell et al., 2013; Sutherland et al., 2014; Tao et al., 2016) and is known to induce detrimental effects, and sometimes even death. Early weaning adds even more stress because of the piglets' lack of maturity. Weaning may be the most stressful situation a pig experiences, and it is unavoidable in today's Modern production systems.

Prior to weaning, piglets have positive maternal experiences such as maternal milk, warmth, maternal pheromones, interactions with their siblings and other nurturing factors that upon weaning abruptly change (Garcia et. al., 2016). At weaning, pigs are removed from a nurturing environment and must learn to cope with a novel environment in which they must establish a social hierarchy by engaging in fighting, adapt to being without their mother and siblings, a different thermal environment and a different olfactory setting, among other novel/negative experiences. The stress of weaning, specifically the transition between nursing and eating solid foods, may lead to a period of underfeeding that affects growth, metabolism and metabolic changes associated with endocrine adjustments while the animal adapts to the diet change (Le Dividich and Seve, 2000). The disruption of the intestinal barrier (composed of a layer of epithelial cells that line the intestinal tract and serve as a protective line of defence against harmful microorganisms) due to the stress of weaning plays a critical role in disease susceptibility (Spreeuwenberg et al., 2001; Melin et al., 2004; Boudry et al., 2004). Antigenic agents that pass the intestinal barrier lead to inflammation, malabsorption, diarrhoea and potentially systemic disease (Livingston et al., 1995; Deitch et al., 1996; Berkes et al., 2003). Weaning also leads to an abrupt reduction in feed intake, which is extremely low during the first two days after weaning (McCraken et al., 1995) and leads to reduced weight gain, known as 'growth check' or post-weaning 'growth lag' (Tockach et al. 1992; Azian, 1992). The post-weaning growth lag affects economic performance because the time needed to reach market weight is increased (Zijlstra et al., 1996). The acute stress response of weaning can further disrupt homeostasis and compromise well-being in pigs (Niekamp et al., 2007) by increasing cortisol concentrations, altering neutrophil to lymph (N:L) ratio and causing deviations from normal behaviours (Sutherland et al., 2014; McGlone et al., 2016).

3.1 Segregated early weaning

Complete removal of the effects of weaning stress in commercial settings is multifactorial, but we may be able to mediate some of the effects. Segregated early weaning is a common practice to reduce vertical transmission of diseases. It involves removing weaned piglets from the sows and rearing them separately from other age groups, but weaning piglets too early can also have negative effects on growth and well-being. Increasing weaning age from 12 to 21.5 days increases weight sold per pig weaned by 1.80 ± 0.12 kg for each day increased in weaning age, with linear improvements in wean-to-finish growth and productivity that is likely due to a function of both weight and physical maturity at weaning (Main et al., 2004).

3.2 Intermittent suckling

The introduction of intermittent suckling (IS) prior to weaning may help reduce stress at weaning and additionally may help prepare them for the consumption of solid feed before weaning. Providing creep feed to piglets when they are with the sow may not tempt piglets to consume enough due to the satiation of hunger with the sow's milk. However, when piglets are separated from the sow for a certain period (6 to 8 hours), their hunger might be kindled to consume starter feed and thus, prepare them for the change in diet from liquid to solid feed. IS has several benefits, including an increase in eating behaviours shortly after weaning (Berkeveld et al., 2007), increase in average daily gain shortly after weaning (Kuller et al., 2004), less pacing and squealing at weaning that may suggest piglets are less stressed (Brown et al., 2016) and, additionally, IS causes increases in LH secretions (Langendijk et al., 2007), causing sows to ovulate and return to oestrus sooner (Brown et al., 2016). Nonetheless, IS does cause an increase in cortisol when first introduced to piglets and when N:L is increased it did not exhibit any different character than control pigs which are not exposed to IS and weaned at the same time (Turpin et al., 2016). Preparing piglets for weaning with IS is a good alternative to abrupt weaning and has several benefits, such as reducing piglet mortality by allowing the sow to be away from the piglets (possibly reducing avoidance postures and preventing piglet crushing), allowing the sows to recover from the high demand of milk production and increasing the number of litters born each year due to a faster onset of heat.

4 Painful practices: castration and ear notching/tagging

4.1 Physical castration

Piglet processing usually consists of castration, tail docking, corner/needle teeth clipping or grinding, ear notching, medication injections and identification transponder injections (in some facilities). Among these painful procedures, castration is the most painful in that it causes the most behavioural and physiological changes.

Physical castration is a common procedure performed on young piglets usually at 1–3 days of age and is a standard practice in many countries, including the United States. Physical castration is done to diminish boar taint which causes an unpleasant smell and taste in the meat of sexually mature male pigs and can be aversive to consumers. The undesirable characteristics of boar taint in the meat of uncastrated male pigs are caused by a pheromone known as androstenone and a substance produced in the gut of pigs known as skatole. In order to prevent boar taint when male pigs reach sexual maturity, they are castrated early in life. Castration is typically performed without any type of analgesia (Lumb 2007; Hewson et al., 2007; Rault et al., 2011; Hannson et al., 2011) and, therefore, it is a painful (McGlone and Hellman, 1988; Lumb, 2007; Fredriksen et al., 2011) and stressful procedure in young piglets. Various studies have shown that physical castration leads to changes in behaviour which may be indicative of pain (Wemelsfelder and van Putten, 1985; McGlone and Hellman, 1988; McGlone et al., 1993; Taylor and Weary, 2000; Taylor et al., 2001). Additionally, physical castration can lead to acute changes in physiology including the activation of the hypothalamic-pituitary-adrenal axis and activation of the sympathetic nervous system (White et al., 1995; Prunier et al., 2001). Physiological changes include increased levels of adrenocorticotropic hormone (ACTH), cortisol and lactate after castration, indicative of stress and tissue damage (Prunier et al., 2005).

The use of general and local anaesthetics may be limited by regulations and economics of the swine industry. Most importantly, the use of anaesthetics is not approved in food animals by the FDA in the United States and their use could potentially create problems if these substances are detected in meat. The use of some gaseous anaesthetics may require a gas evacuation system be in place (Prunier et al., 2005), and the use of general anaesthesia has been linked with malignant hyerperthermia in some breeds of pigs (Prunier et al., 2005).

4.2 Alternatives to physical castration

Alternatives to castration include the use of chemical compounds to destroy testicular tissue (Giri et al., 2002; Ljaz et al., 2000), the use of exogenous hormones to downregulate the hypothalamic-pituitary-gonadal axis (Schneider et al., 1998; Daxenberger et al., 2001), immunocastration (IC) (Norman and Litwack, 1997; Hillier, 1994), sperm sexing (von Borell et al., 2009), raising entire males (von Borell et al., 2009), marketing pigs at a younger age and genetic selection against boar taint.

Local destruction of testicular tissue by chemical compounds such as acids and salts may also be painful. Although many authors claim that it castration causes little pain (Prunier et al., 2006), swelling of the testes or scrotum have been observed (Giri et al., 2002), necrosis and slow wound healing (Fordyce et al., 1989).

Downregulation of the hypothalamic-pituitary-gonadal axis by exogenous hormones and IC are two alternatives to physical castration using steroid agonists or antagonists. The latter of the two is known as Improvest® and is used in several countries. The mode of action of IC is through the use of an incomplete version of GnRF (an analogue) in combination with diphtheria toxoid carrier protein for the GnRF analogue (a conjugate). Together, these produce an immunological response in the pig against both the exogenous and endogenous GnRF. The resulting decrease in the levels of FSH and LH further suppresses reproductive function (Esbenshade and Britt, 1985). Without LH and FSH stimuli the gonads will not develop and the animal will stay in a pre-pubertal condition, or the gonads will atrophy in the post-pubertal animal. Currently, Improvest® is approved in more than 60 countries, but there is no market for it in the United States due to low acceptance by consumers. Other problems with its use include laborious administration, high cost and the risk of self-injection in humans (Prunier et al., 2006).

There are several alternatives to castration, but so far none has really been permanently adopted, and physical castration continues to be the more prevalent method of castration in the United States and in some other countries. Therefore, pain during and after physical castration still remains a valid concern. The use of analgesia and/or local anaesthesia to physically castrate piglets has been adopted by several countries and may be one of the most viable ways to reduce pain during and after castration until better alternatives are found and adopted worldwide.

4.3 Ear notching/ear tags

The need for identification of individual animals is a key point for management, traceability, trade control and disease eradication (Caja et al., 2005), but has to be individualized, permanent, simple to apply and read, welfare appropriate and tamper-proof (Merks et al., 1990; McKean, 2001). The use of ear notching for identification has greatly decreased but was in the past the most common method to standardize purebred and commercial swine (Rea, 1986). It was typically performed by cutting deep notches into the ears approximately 6.5 mm deep (Widoski and Torry, 2002). Some small producers may still use ear notching as a means of identifying purebred pigs. Little research has been conducted to assess the degree of pain and distress associated with ear notching (Widoski and Torrey, 2002), or even comparing ear notching to ear tagging (Marchant-Forde et al., 2009). However, ear notching can cause pain and irritation, indicated by head shaking and scratching of the ears with hind limbs (Noonan et al., 1994). Marchant-Forde et al. (2009) reported that ear-notched pigs have worse wound scores and higher cortisol concentrations than ear-tagged pigs, and emit calls higher in frequency than control pigs. Most recently, the use of ear tags has supplanted the use of ear notching, but little is known about the pain caused due to ear tagging.

4.4 Alternatives to ear notching/ear tags

Identifying individual animals is a must and therefore, the need for an efficient method of identification led to the development of radio frequency identification (RFID). RFIDs can be placed either under the skin as ear tags (usually injected in the auricle base of the ear base) or intraperitoneally (IP). There are three main types of tags used for animal identification – boluses, ear tags and injectable glass tags – and can be grouped into low or high frequency (Voulodimos et al., 2010). Although RFIDs are probably less painful than ear notching and more efficient (because improperly notched pigs lead to animal misidentification) they still have pit falls. Conventional ear tags can be soiled with faeces, can fade and be hard to read, can get pulled off or chewed on and can also lead to misidentification of pigs; however, RFID ear tags have very similar problems. Firstly, the person trying to identify the animal must be close enough to scan the tag and if it is damaged, identifying the animal can be a problem. Out of the three placements for RFIDs, intraperitonally injected transponders were not problematic and are an efficient method for traceability of pigs (Caja et al., 2005). Intaperitonally placed RFIDs require injection of the transponder into the abdominal cavity between the intestines, a task that requires trained personnel (as the injection must be performed on the left, ventral side of the animal, at ~1 cm from the ventral line and 2cm caudal to the navel) because misplacement into the intestines and urinary bladder is possible (Caja et al., 2005). However, Babot et al. (2006) found that losses of IP transponders were very low. They further reported that on-farm losses averaged 1.6% for ear tags and1.8% for IP injected transponders, but electronic failures seem to occur in ear tags and not IP injected transponders. Thus, Babot et al. (2006) suggested that IP injected transponders give the best results compared to SQ injections and electronic ear tags (damage to the tags and losses during transport to slaughter). An additional advantage of RFIDs is that they offer automatic reading in dynamic conditions, but they cost 7 to 10 times more than conventional ear tags and must include the purchase of a reading device (Babot et al., 2010). Furthermore, electronic devices must also remain functional at the slaughter line, and removal of these devices from the carcass within 5 seconds after slaughter is considered acceptable – a task which may be difficult due to slaughtering conditions of pigs that may include using fire, hot water and a high line speed (Merks and Lambooy, 1990).

Although there are not many reports in the literature referring to negative effects on animal welfare due to the use of RFIDs, Caja et al. (2005) and Babot et al. (2006) both agree that these do not have a negative effect either on animal welfare (as no inflammatory reactions were observed and animals healed quickly) or on performance.

Other novel methods of identification include the use of auricular vein patterns (Wayne et al., unpublished), biometrics such as retinal images (Rusk et al., 2006) and facial recognition (Corkery et al., 2007). Harrel and colleagues used an OptiReader™ and Veinlite™ LED to image auricular veins in pigs and found that each pig has a distinct auricular vein pattern. They imaged vein patterns of 96 pigs, resulting in a match rate of 96.7% with an imaging error of 8.3% (mainly due to the colour of the pig, black ears were harder to match). They concluded that further research is needed to determine the uniqueness of each pig's auricular vein pattern and the proper distance for imaging. The use of retinal images is a technology that has been reported to be a viable method of identification for beef and sheep (Rusk et al., 2006; Allen et al., 2008; Adell et al., 2012). Facial recognition is a common biometric used in humans and is based on location and shape of facial attributes and overall analysis of the face image (Corkery et al., 2007). Corkery et al. (2007) found that the recognition rate for sheep was 95.3% to 96%, and is non-invasive and inexpensive.

5 Painful practices: tail docking and teeth clipping/resection

5.1 Tail docking

Piglets can many times become victims of tail biting, and therefore this is the main reason that tail docking is an accepted procedure in the swine industry. However, tail docking is painful and the use of anaesthetics and analgesics is not a common practice to reduce pain during or after tail docking. It is important to understand that tail docking reduces the incidence of tail biting, but does not eliminate tail biting entirely (Paoli et al., 2016). Harley et al. (2014) reported that out of 99% of pigs' tail docked in a study conducted in Ireland, 72.5% still presented mild tail-bite lesions at the abattoir and 2.5% of those were severe lesions.

Tail biting can result in infections and severe injuries (causing acute pain) to the tail, such as tails being bitten to the rump (leading to the victim having to be euthanized; Sutherland and Tucker, 2011). Additionally, severe tail biting can lead to reduced pig performance (van Staavern et al., 2016) and also lead to financial losses after meat inspection at the abattoir as a result of carcass condemnations due to lesions associated with tail biting (Harley et al., 2014).

Tail biting decreases by increasing docking length (Thodberg et al., 2010), but, in general, about 1.5 to 2 cm is left of pigtails (Hunter et al., 2001; Sutherland et al., 2011). Tail docking is usually performed with cutting pliers, knives, scalpel blades or hot docking irons. Although there is a benefit to tail docking due to the reduced incidence of tail biting, it is still considered a painful procedure. Tail docking can cause both acute and possibly chronic stress (Herskin et al., 2015) in addition to physiological and behavioural changes in piglets

(Sutherland and Tucker, 2011). Prunier et al. (2005) reported that adrenocorticotropin hormone (ACTH) and lactate concentrations do not differ between tail-docked pigs and control pigs at one day of age. The cortisol response also does not differ between docked and control pigs at 1 or 6 days of age when a cautery iron is used (Prunier et al., 2005; Sutherland et al., 2008). Yet Sutherland et al. (2008) reported higher cortisol concentrations at 60 minutes and at six days of age when pig tails were docked with cutting pliers compared to control pigs. Hot iron cautery docking is indicated to be less stressful compared with docking by cutting off the tail (Sutherland et al., 2008), but can lead to slower healing (Sutherland et al., 2009) and development of neuroanatomical changes, such as neuromas in the outermost tail tip (Herskin et al., 2015; Sandercock et al., 2016).

Pigs have a need for manipulable materials to satisfy a range of behavioural needs (intrinsically motivated exploratory and foraging behaviour) and if these are not met there is an increased incidence of tail-biting behaviours, especially in weaner and rearing pigs (EFSA, 2014). Tail biting outbreaks in swine facilities can be multifactorial (Sutherland et al., 2008), including lack of manipulable material, poor climate, poor air quality, poor health, nutritional deficiencies, competition for resources, social instability, high stocking density, poor pen layout and high genetic potential for lean tissue growth (Fig. 1; EFSA, 2014).

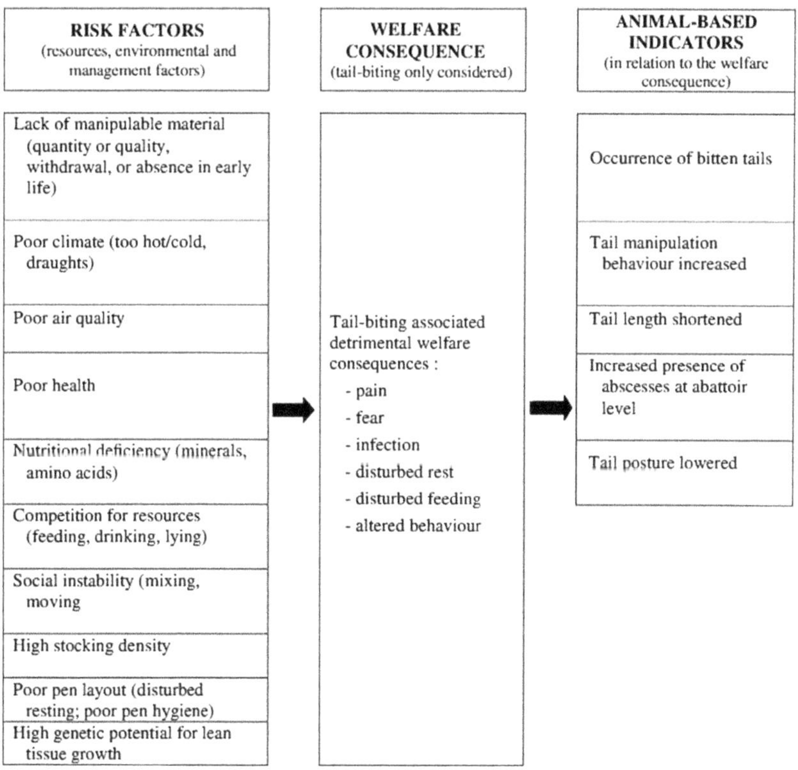

Figure 1 Associations between risk factors, welfare consequence and animal-based indicators of tail biting (risk factors to be controlled to avoid tail biting). Adapted from EFSA, 2014.

5.2 Alternatives to tail docking

Managing tail biting is complex and many factors have to be taken into consideration. All farms are different and one thing may work on one farm and not in another; therefore, adhering to what is effective by site may be one of the most effective ways to reduce tail biting.

One of the alternatives commonly suggested is the use of manipulable materials, such as straw hay. Straw hay has been reported to reduce tail biting by 50% (Hunter et al., 2001; Zonderland et al., 2008). Other enrichment objects have been studied but pigs seem to prefer objects that are ingestible, have chewable properties and destructibility (Van de Weerd et al., 2003).Pigs lose interest rather quickly when provided with toys compared to the use of manipulating straw, as they spent 2% of their time manipulating a toy compared to 20% of their time manipulating straw (Scott et al., 2006). Although straw is a good alternative because it allows rooting behaviours that distracts pigs, which shifts them away from tail biting, it is difficult to dispose of, especially with manure systems that cannot handle this type of material.

Another alternative is the removal of the tail biter. Removing the tail biter fixes the problem temporarily but in the experience of the author, it seems to not always be just one culprit, but several. Identifying the culprit(s) is time consuming and finding space for biters in barns can also be difficult in small to medium farms. Therefore, a combination of risk factors must be taken into consideration and the option that best works for each specific farm should be used.

The use of pig advisory services improves farm productivity through better housing, management and nutrition, and positively affects pig welfare by promoting detailed record keeping of financial and farm performance records (van Staaveren et al., 2016). The Teagasc Advisory Service provides farms with consultation visits, herd performance and financial monitoring, organizes discussion groups and provides education and training for the pig sector (van Staaveren et al., 2016). van Staaveren et al. (2016) reported that keeping of financial and performance-related records with the Teagasc eProfit Manager (ePM; which monitors over 150 parameters detailing technical and performance data of the herd) system was associated with a lower prevalence of moderate tail lesions, and further suggests that improving performance is likely associated with improvements in welfare.

5.3 Teeth clipping/resection

Very few piglets on US farms are now teeth clipped. While it was a common practice before in the 1990s, farms discovered that teeth clipping of all the piglets was not necessary. Today, skilled workers target teeth clipping to the piglets who scar other piglets or the sow's udder. We estimate that fewer than 10% of piglets on Modern US farms have their teeth clipped today.

Needle teeth, the deciduous canines and the corner incisors (Hay et al., 2004) are sharp and pointed, protrude out from the jaw at an angle and, therefore, can cause damage to other piglets and sow udders. Piglets display aggressive behaviours towards siblings to establish teat order and cause lesions to the faces of littermates (Marchant-Forde, 2009). Although resection has proven to reduce face injuries in piglets, it may also cause internal mouth lesions for piglets with resected teeth (Lewis et al., 2005a), resulting in pulpits (Hay et al., 2004), deep wounds to the tongue and lips (Burger, 1983), and splinters that can become embedded in the gums (Hunter et al., 1993). Additionally, the wounds seen with teeth resection may be the entry points for infection (Hunter et al., 1993) and could possibly lead to death. Hay et al. (2004) found that clipping teeth led to higher numbers of teeth with opening of the pulp cavity, teeth with fractures, haemorrhage, infiltration and abscesses compared to intact teeth, and resection was painful (with pain lasting up to the 15th day of life). Teeth-clipped pigs display irregular behaviours, such as chomping (Lewis et al., 2005a), but there are very few reports on the effect teeth clipping has on piglets. Chomping is attributed to irritation, pain or discomfort due to possible presence of teeth fragments and blood in the mouth (Batailee et al., 2002).

5.4 Alternatives to teeth clipping

One of the alternatives to replace teeth clipping is grinding of the teeth with a rotating grindstone. However, teeth grinding increases the mean call vocalizations, negatively affects body weight, growth rates tend to be lower, cortisol concentrations increase compared to clipping of the teeth one week post-treatment, and takes longer to perform than clipping of the teeth (Merchant-Forde et al., 2009). Additionally, teeth grinding may possibly cause additional handling stress for the pigs. Lewis et al. (2005a) suggested that grinding the teeth has some negative piglet welfare implications, but is better than the pain and discomfort associated with teeth clipping. However, they found that teeth clipping was more effective at reducing facial injuries in piglets than grinding. They concluded that grinding is recommended over clipping or leaving the teeth intact.

EU legislation permits both teeth clipping and grinding (Directive, 2001/93/EC), but only when there is evidence that there are injuries due to piglet aggression and injury to sow teats. Even when there is evidence, producers must have altered the environment and management practices in some way to prevent aggressive behaviours (Lewis et al., 2005a,b). This philosophy is largely followed in the United States.

The other alternative to teeth clipping is to leave teeth intact only in LBW piglets. There are two types of sibling competition: (1) direct competition – aggressive competition for teats, mainly in the first hours after birth, when ownership of a teat is established and where litter mates die or survive by suckling opportunistically (Hartsock and Graves, 1976; de Passille et al., 1988, 1989) – and (2) indirect competition – larger or more competitive piglets gain more weight than their litter mates, by possibly stimulating and draining their teats more efficiently and garnering to their respective teats more hormones and nutrients (Fraser et al., 1979; Thompson and Fraser, 1986). LBW piglets are disadvantaged in both types of competition (fail to secure their own teat and die during the first few days of life) and, therefore, by selectively tooth clipping only the larger piglets in the litter, LBW piglets have a decrease in mortality (as the presence of their needle teeth may allow them to defend their teat better) and weight gain is greater (Robert et al., 1995). Robert et al. (1995) found that in larger litters (12–14 piglets) LBW piglets that did not have their teeth clipped had a lower mortality rate than control litters which did have their teeth clipped (32% mortality compared to 39%, respectively), whereas HBW piglets had a trend in the opposite direction (14.4% mortality in the experimental groups and 13.2% in the controls). They concluded that selective tooth clipping works by assisting small litter mates (as intact teeth contribute to indirect competition) at the expense of the larger ones (only in large litters) contributes to uniform weaning weights and helps the most vulnerable pigs stay alive until fostering or another intervention, but does not improve overall growth and survival.

Leaving needle teeth intact in all piglets may not be a logical alternative since it can also lead to increased piglet mortality and damage sow udders. Farrowing crate restrictions prevent sows from expressing normal behaviours that would otherwise be expressed in the wild. In semi-natural conditions, sows remain in the nest with piglets for the first few days after birth (Lewis et al., 2005b), but after the first week they begin to leave the nest for long periods of time, returning only to suckle their piglets (Jensen and Rhebo, 1987). In farrowing crates, sows cannot reduce the amount of contact with their piglets and piglet behaviour directed towards the udder is very high (Cox and Cooper, 2001), and thus, sows begin to change postures more frequently (more ventral lying and dog sitting during lactation; Cronin et al., 1992, 1998). The change in postures, possibly due to frustration, leads to accidental crushing of piglets. Lewis et al. (2005b) concluded that intact teeth injured sows and caused discomfort and linked sow avoidance behaviour to increased piglet mortality. They further reported that teeth clipping was more beneficial to sow welfare than grinding or intact teeth due to the number of teat lesions observed on sows being less (at 11 days post-partum) when piglets were teeth clipped.

6 Transportation

Piglets are often transported to other facilities upon weaning, which causes additional stress to piglets. Currently in the United States, animals can be transported for 28 h without provision of water and feed or off-loading. It was previously believed that weaning stress was so great that transportation did not have an effect. Piglets are many times transported for many hours without feed and water. When piglets are transported for more than 24 h without feed and water, there is an apparent loss in body weight, negative effects on physiology and behaviour (Garcia et al., 2015). Garcia et al. (2016) reported that weaned pigs not provided with water during 32 h of transport lost significantly more weight than those that did have access to water, and further suggested that provision of water was more important than feed during transportation. Regardless of provision of feed and water, transportation is a complex stressor made up of many factors including fluctuating temperatures, stocking density, mixing with unfamiliar pigs and motion (Lambooij et al., 1993). Therefore, transportation has the potential to affect the health and welfare of pigs, especially in pigs already experiencing weaning stress; therefore, weaning stress and transport can cause additive negative effects. Transport of weaned pigs has many common factors with transport of finishing pigs, including mixing, fasting, temperature fluctuations, vibration, novelty and noise (Lewis et al., 2008). Similar to studies in finishing pigs, transport has been shown to cause behavioural and physiological changes indicative of stress in weaned pigs (Sutherland et al., 2014). Finding methods to reduce stress during transport could help reduce some of the negative effects on piglet performance after arrival to new facilities.

7 New technologies

Piglets have few biological products that are commonly used. These include supplemental iron, antibiotics and vaccines. Some nutritional supplements may have value. There are opportunities for innovative approaches to improve piglet health and welfare that are from entirely new classes of biological or electronic technologies.

7.1 Pheromones, interomones and semiochemicals

Pigs have a large number of olfactory receptors – four times as many as humans (Nguyen et al., 2012). Pigs are considered macro-osmotic while humans are micro-osmotic species. In addition, the pig's olfactory acuity is thousands-fold more sensitive than humans; that is, they can detect odour molecules at a much lower concentration. Impacting the olfactory environment may hold promise to improve pig health and welfare.

One can think of semiochemicals (pheromones, interomones and inter-species biologically relevant odours) as we do nutrients. When a nutrient is deficient, pigs grow slower and have health problems. When a pig is weaned from its mother, it is missing the maternal and some adult male odours from its environment. A lack of maternal-neonatal-rich olfactory environment contributes to the stressfulness of weaning. Alarm pheromones contribute to stress at other times. A lack of adult odours can encourage fighting among neonates. Unfamiliar neonates or sows must fight until they understand each other's olfactory signals and ultimately its place in the social order.

Pheromones are an important releaser of a variety of behaviours in the receiving animal, and can be found in faeces, urine, saliva, vaginal mucus and/or secreted from cutaneous scent glands (Rekwot et al., 2001; Mucignat-Caretta, 2014). They are classified as molecules released by individuals that elicit a behaviour or physiological change in the endocrine or reproductive system of other individuals of the same species (Tirindelli et al., 2009). Although pheromones typically act among individuals of the same species (intraspecific), they can be detected by different species (interspecific). An interomone is a pheromone in one species that changes the behaviour of another species without the requirements to benefit or harm the emitter or the receiver (McGlone et al., 2014).

The use of pheromones is not new to pork production. The boar salivary pheromone, androstenone, was reported to reduce aggressive behaviour in piglets in the 1980s (McGlone, 1985; McGlone and Morrow, 1988). Androstenone also reduces unwanted behaviours in other species of animals (acting as an interomone), such as barking and jumping in dogs (Stop That! Sentry, Omaha, NE) (McGlone et al., 2014).

Morrow and McGlone (1990) described sources of maternal odours in lactating sows. They showed that skin washings from sows are recognized by the piglet starting at a few hours of age. Patrick Pageat and his research group collected similar biological fluids and called them pig appeasing pheromones (Pageat, 2001; US patent 6,077,867). McGlone and Anderson (2002) showed that this mixture of fatty acids found on a sow's udder will stimulate weight gain, reduce fighting and increase feeding behaviours in weaned pigs.

A second maternal-neonatal odorous molecule was then discovered to be biologically active in the pig. The interomone 2-methyl-2-butenal (2M2B) is a rabbit maternal pheromone that elicits nursing behaviour in rabbit pups. Recently, 2M2B was shown to be beneficial in pigs by improving post-weaning performance (McGlone et al., 2016). Interestingly, 2M2B increased growth in newly weaned piglets (McGlone et al., 2016), but has also been marketed as a calming spray or collar for cats and dogs (Perrigo Animal Health). Animals spend the majority of their life communicating through olfactory signals. One way of improving animal experiences in a production setting is to incorporate the use of volatile compounds or semiochemicals such as pheromones into their environment. Animals, especially pigs 'see' their world through olfaction, due to the high number of olfactory receptors (Groenen et al., 2012). Therefore, manipulating the olfactory environment in pig production settings may provide numerous benefits.

We expect more semiochemicals will be discovered in pigs and other farm animal species. Modulation of the olfactory environment can improve reproduction, reduce stress and improve performance using what has been one approach that has been called 'clean, green and ethical' technology (Scaramuzzi and Martin, 2008).

7.2 Automated detection systems

Our pig barns are far from smart barns. While we may carry a phone that is more powerful than a mainframe computer of 40 years ago, the pig barn and pig pasture are simple systems. They have not evolved as have consumer, military or business electronics. However, the pig production unit will evolve as technology evolves.

A smart barn would have many automated features. Labour is hard to find in commercial agriculture. Automated systems could replace some people in production agriculture. Furthermore, automated systems would potentially do a better job at 24/7/365 surveillance of a pig barn or pasture. Scientists and engineers are developing sensors such as cough detectors, piglet scream sensors (for when a sow starts to crush a piglet), pig counting and body weight sensing devices, among others. Livestock and poultry automation in the future is as sure as tractors replacing humans and draft horses and milking machines replacing human milkers.

Health and welfare compromises in pigs have wide-ranged consequences. Automatic detection of compromised health and welfare can be accomplished with new technology that includes measures of health, growth and behaviour (Matthews et al., 2016). Audio, visual and other modalities that automatically monitor behaviours are certain to be of great benefit to the pig industry. We show some reported applications of this technology below (Table 1).

Audio sensors (automated processing techniques) can be used to detect and classify certain acoustic events such as coughing, sneezing, screaming, barking and the assessment of this audio data (through spectral analysis) can help classify sick pigs (Matthews et al., 2016). An example of an audio system is the STREMDO, a system that automatically measures the duration and intensity from vocalizations (Manteuffel and Schon, 2002; Schon et al., 2004). Audio sensing can also be used to measure external stimuli coming from the environment that may have behavioural impacts on the animals (Marx et al., 2003; Broucek, 2014).

Video is another automated system that can be used to measure general pig activity. Pixel differences between consecutive images (provide a measure of the group rather than an individual; Rushen et al., 2012) can detect fast pig movements that may be indicative of aggression, but can also be due to other behaviours such as playing and chasing (Viazzi et al., 2014). Non-invasive measurements (that do not require handling) of surface temperature in pigs from infrared video (McManus et al., 2016) have supported the evaluation of automated assessment of thermal comfort measured from thermal distribution of pigs (Cook et al., 2015).

The use of sensors to measure water consumption (Meiszberg et al., 2009), RFID at feeding and drinking (Andersen et al., 2014; Maselyne et al., 2015a,b), accelerometers combined with temperature sensors to detect abnormal body temperatures due to infection (Martinez-Aviles et al., 2015) and pressure mats that measure animal gaits for detection of lamenesses (Meijer et al., 2014) are all modalities that may aid in reducing the amount of labour and cost associated with labour, along with improving animal welfare.

Table 1 Current levels of automation for each behavioural category. Adapted from Matthews et al., 2016

	Automation categories		
Behavioural categories	Behaviour detection	Behaviour monitoring	Automated detection of behavioural change
Daily activity budget	Requires monitoring behaviour over time	Location-based (Andersen et al., 2014); Locomotor activity (Chung et al., 2014); Drinking (Madsen and Kristensen, 2005).	
Feeding, drinking, and elimination	Drinking (Meiszberg et al., 2009).	Feeding (Fernández et al., 2011; Andersen et al., 2014; Maselyne et al., 2015b); Commercial systems: I-BOX 360°, Farmex, Eliskool 2/Elister 2/Tristar systems, IVOG, FIRE; Elimination (Zhu et al., 2009).	Drinking (Madsen and KtifletKto.HK)
Posture and locomotion	Locomotion (Lind et al., 2005; Kongsro, 2013).	Spatial distribution (Cook et al., 2015; Nasirahmadi et al., 2015); gait (Starvrakakis et al., 2014); General activity (Leroy et al., 2006).	Activity (Martínez-Avilés et al., 2015).
Social behaviour	Aggression (Oczak et al., 2012, 2014; Viazzi et al., 2014; Lee et al., 2016); Clustening (Shao and Kin, 2008); Vocalisation (Maniteuffel and Schön, 2002; Schön et al., 2004).		
Disease-specific	Coughing (Chedad et al., 2001; Exadaktylos et al., 2008; Chung et al., 2013).	Coughing (Vandermeulen et al., 2013; Hemeryck and Berckmans, 2015; Hemeryck et al., 2015).	

7.3 Gene editing

Animal health challenges will be one of the biggest challenges for efficient livestock production in the future, due to resistance to existing drugs and the pressure to reduce antibiotic use in agriculture (Plastow, 2016). According to Plastow (2016), genome-wide analysis studies identify genetic factors influencing variation in how animals respond to diseases (genetic variation in the susceptibility to disease), which can be polygenic or even oligogenic (depending on the number of genes distributed across the genome). Boddicker et al. (2014) found that for porcine reproductive respiratory syndrome virus (PRRSV) one region on chromosome 4 was found to explain more than 10% of the variation in viraemia and growth after infection and reported the effect to be dominant so that only one copy of the allele is necessary to get the benefit. Further, Hess et al. (2014) confirmed the previous findings using a second strain of the virus and suggested that it could be possible to select for pigs that are less impacted by PRRSV when it is present. Scientists have developed genetically modified (GM) pigs intended for commercial production: One has a gene inserted to increase milk production and another has an enzyme added to break down plant phytase to make phosphorous more available (which has a large environmental advantage). Both GM pigs did not enter commercial herds largely due to perceived consumer negative reactions.

Gene editing is a newer technique that is being investigated (Perez-Pinera et al., 2012). In gene editing, the natural genome is changed to activate or inactivate genes. One viable example is a pig that had a PRRSV receptor protein inactivated. These pigs do not succumb to PRRSV infection. One can see the economic and welfare value to such a gene-edited pig. The consumer reaction to such a gene-edited pig is yet to be seen.

Gene editing may hold great potential as a way to improve animal welfare. The disease-resistant opportunities are as many as we have infectious diseases. However, beyond infectious diseases, one can imagine many other welfare advantages. Imagine a pig born with no needle teeth to clip, or male pigs that do not need to be castrated because the boar taint molecules are suppressed. For every animal welfare and animal health problem, there is a potential gene editing opportunity. But ethical thinking and consumer understanding are far behind in knowing how to handle these opportunities.

8 Conclusion

The commercial swine industry has changed both the production environment and the genetics of the pig significantly in the past half century. At the same time that pigs grow faster, are leaner and have larger litters, the consumers have adopted a variable view of production agriculture. A half century ago, low-cost food was the primary consideration for both producer and consumer. Today, we have a mixed bag of consumer desires. Some consumers still seek low-cost, high-quality pork. Others care about the environment, food safety, taste and animal welfare, among other concerns. Social media has accelerated the plethora of views among consumers. Bloggers and videographers have a powerful effect on consumer trends. At the same time, retail food companies (restaurants and grocery stores) have requirements for on-farm practices including animal welfare. All parties share the goal of improving animal welfare. The pork industry has opportunities to improve animal welfare and at the same time profitability and sustainability of the agricultural enterprise. The four major areas to improve piglet welfare are to reduce pre-weaning mortality, reduce weaning stress, prevent pain during standard practices and reduce stress associated with the transport of piglets. But in addition to making progress in these four long-investigated areas, we have opportunities to improve pig welfare in new areas such as pheromones, automated technologies and gene editing.

9 Where to look for further information

For further information about these topics or other animal well-being topics, please refer to Texas Tech University Laboratory of Animal Behavior, Physiology and Welfare (www.depts.ttu.edu/animalwelfare/).

10 References

Adell, N., Puig, P., Rojas-Olivares, A., Caja, G., Carné, S. and Salama, A. A. (2012). A bivariate model for retinal image identification in lambs. *Computers and Electronics in Agriculture*, 87, 108–112.

Allen, A., Golden, B., Taylor, M., Patterson, D., Henriksen, D. and Skuce, R. (2008). Evaluation of retinal imaging technology for the biometric identification of bovine animals in Northern Ireland. *Livestock Science*, 116(1), 42–52.

Andersen, H. L., Dybkjær, L. and Herskin, M. S. (2014). Growing pigs' drinking behaviour: number of visits, duration, water intake and diurnal variation. *Animal*, 8(11), 1881–8.

Babot, D., Hernández-Jover, M., Caja, G., Santamarina, C. and Ghirardi, J. J. (2006). Comparison of visual and electronic identification devices in pigs: On-farm performances. *Journal of Animal Science*, 84(9), 2575–81.

Bataillie, G., Rugraff, Y., Chevillon, P. and Meunier-Saluaum, M. (2002). Caudectomie et section des dents chez le porcelet: conséquences comportementales, zootechniques et sanitaires. *Techni Porc*, 25(1), 5–13.

Baxter, E. M., Rutherford, K. M.D., D'Eath, R. B., Arnott, G., Turner, S. P., Sandøe, P., Moustsen, V. A., Thorup, F., Edwards, S. A. and Lawrence, A. B. (2013). The welfare implications of large litter size in the domestic pig II: management factors. *Animal Welfare*, 22, 219–38.

Berkes, J., Viswanathan, V. K., Savkovic, S. D. and Hecht, G. (2003). Intestinal epithelial responses to enteric pathogens: effects on the tight junction barrier, ion transport, and inflammation. *Gut*, 52(3), 439–51.

Berkeveld, M., Langendijk, P., Bolhuis, J. E., Koets, A. P., Verheijden, J. H. and Taverne, M. A. (2007). Intermittent suckling during an extended lactation period: Effects on piglet behavior. *Journal of Animal Science*, 85, 3415–24.

Boddicker, N. J., Bjorkquist, A., Rowland, R. R., Lunney, J. K., Reecy, J. M. and Dekkers, J. C. (2014). Genome-wide association and genomic prediction for host response to porcine reproductive and respiratory syndrome virus infection. *Genetics Selection Evolution*, 46, 1.

Boudry, G., Péron, V., Le Huërou-Luron, I., Lallès, J. P. and Sève, B. (2004). Weaning induces both transient and long-lasting modifications of absorptive, secretory, and barrier properties of piglet intestine. *The Journal of Nutrition*, 134(9), 2256–62.

Bovey, K. E., Widowski, T. M., Dewey, C. E., Devillers, N., Farmer, C., Lessard, M. and Torrey, S. (2014). The effect of birth weight and age at tail docking and ear notching on the behavioral and physiological responses of piglets. *Journal of Animal Science*, 92(4), 1718–27.

Brouček, J. (2014). Effect of Noise on Performance, Stress, and Behaviour of Animals. *Slovak Journal of Animal Science*, 111.

Brown, E. G., Krebs, L. B., Boone, C. L. and Cauthen, T. (2016). Effects of Intermittent Suckling on Sow and Piglet Performance. *Texas Journal of Agriculture and Natural Resources*, 22, 55–60.

Burger, A. (1983). Consequences of clipping incisors in piglets (Doctoral dissertation, Thesis, Tierarztliche Fakultat Ladwig-Maximilians-Universitat, Munchen, Deutschland).

Caja, G., Hernandez-Jover, M., Conill, C., Garin, D., Alabern, X., Farriol, B. and Ghirardi, J. (2005). Use of ear tags and injectable transponders for the identification and traceability of pigs from birth to the end of the slaughter line. *Journal of Animal Science*, 83, 2215–24.

Campbell, J. M., Crenshaw, J. D. and Polo, J. (2013). The biological stress of early weaned piglets. *Journal of Animal Science and Biotechnology*, 4(1), 1.

Cook, N. J., Chabot, B., Lui, T., Bench, C. J. and Schaefer, A. L. (2015). Infrared thermography detects febrile and behavioural responses to vaccination of weaned piglets. *Animal*, 9(02), 339–46.

Corkery, G. P., Gonzales-Barron, U. A., Butler, F., Mc Donnell, K. and Ward, S. (2007). A preliminary investigation on face recognition as a biometric identifier of sheep. *Transactions of the ASABE*, 50(1), 313–20.

Cox, L. N., and Cooper, J. J. (2001). Observations on the pre-and post-weaning behaviour of piglets reared in commercial indoor and outdoor environments. *Animal Science*, 72(1), 75–86.

Cronin, G. M., Dunsmore, B. and Leeson, E. (1998). The effects of farrowing nest size and width on sow and piglet behaviour and piglet survival. *Applied Animal Behaviour Science*, 60(4), 331–45.

Cronin, G. M., Simpson, G. J. and Hemsworth, P. H. (1996). The effects of the gestation and farrowing environments on sow and piglet behaviour and piglet survival and growth in early lactation. *Applied Animal Behaviour Science*, 46(3), 175–92.

Daxenberger, A., Hageleit, M., Kraetzl, W. D., Lange, I. G., Claus, R., Le Bizec, B. and Meyer, H. H. (2001). Suppression of androstenone in entire male pigs by anabolic preparations. *Livestock Production Science*, 69(2), 139–44.

De Passille, A. M. B. and Rushen, J. (1989). Suckling and teat disputes by neonatal piglets. *Applied Animal Behaviour Science*, 22(1), 23–38.

De Passille, A. M. B., Rushen, J. and Hartsock, T. G. (1988). Ontogeny of teat fidelity in pigs and its relation to competition at suckling. *Canadian Journal of Animal Science*, 68(2), 325–38.

Deitch, E. A., Rutan, R. and Waymack, J. P. (1996). Trauma, shock, and gut translocation. *New Horizons*, 4(2), 289–99.

EFSA Panel on Animal Health and Welfare 2007. Scientific Opinion of the Panel on Animal Health and Welfare on a request from Commission on the risks associated with tail biting in pigs and possible means to reduce the need for tail docking considering the different housing and husbandry systems. *EFSA Journal* 611, 1–13.

Eide, I. (1963). Farrowing crate. USA patent 3,077,861. Publication date 19, Feb 1963.

Esbenshade, K. L. and Britt, J. H. (1985). Active immunization of gilts against gonadotropin-releasing hormone: effects on secretion of gonadotropins, reproductive function, and responses to agonists of gonadotropin-releasing hormone. *Biology of Reproduction* 33(3), 569–77.

Fahmy, M. H. and C. Bernard. (1971). Causes of mortality in Yorkshire pigs from birth to 20 weeks of age. *Canadian Journal of Animal Science*, 51, 351–9.

Fordyce, G., Beaman, N., Laing, A., Hodge, P., Campero, C. and Shepherd, R. K. (1989). An evaluation of calf castration by intra-testicular injection of a lactic acid solution. *Australian Veterinary Journal*, 66(9), 272–6.

Fraser, D. (2003). Assessing animal welfare at the farm and group level: the interplay of science and values. *Animal Welfare*, 12(4), 433–43.

Fraser, D., Thompson, B. K., Ferguson, D. K. and Darroch, R. L. (1979). The 'teat order' of suckling pigs: 3. Relation to competition within litters. *The Journal of Agricultural Science*, 92(02), 257–61.

Fredriksen, B., Johnsen, A. M. S. and Skuterud, E. (2011). Consumer attitudes towards castration of piglets and alternatives to surgical castration. *Research in Veterinary Science*, 90(2), 352–7.

Garcia, A., Pirner, G., Picinin, G., May, M., Guay, K., Backus, B., Sutherland, M. and McGlone, J. (2015) Effect of provision of feed and water during transport on the welfare of weaned pigs. *Animals*, 5(2), 407–25.

Garcia, A., Sutherland, M., Pirner, G., Picinin, G., May, M., Backus, B. and McGlone, J. (2016). Impact of Providing Feed and/or Water on Performance, Physiology, and Behavior of Weaned Pigs during a 32-h Transport. *Animals*, 6(5), 31.

Giri, S. C., Yadav, B. P. S. and Panda, S. K. (2002). Chemical castration in pigs. *Indian Journal of Animal Sciences*, 72(6), 451–3.

Groenen, M. A., Archibald, A. L., Uenishi, H., Tuggle, C. K., Takeuchi, Y., Rothschild, M. F., Rogel-Gaillard, C., Park, C., Milan, D., Megens, H. J. and Li, S. (2012). Analyses of pig genomes provide insight into porcine demography and evolution. *Nature*, 491, 393–8.

Hansson, M., Lundeheim, N., Nyman, G. and Johansson, G. (2011). Effect of local anaesthesia and/or analgesia on pain responses induced by piglet castration. *Acta Veterinaria Scandinavica*, 53(34), 1–9.

Harley, S., Boyle, L. A., O'Connell, N. E., More, S., Teixeira, D. and Hanlon, A. (2014). Docking the value of pig meat? Prevalence and financial implications of welfare lesions in Irish slaughter pigs. *Animal Welfare*, 23, 275–85.

Hartsock, T. G. and Graves, H. B. (1976). Neonatal behavior and nutrition-related mortality in domestic swine. *Journal of Animal Science*, 42(1), 235–41.

Hay, M., Rue, J., Sansac, C., Brunel, G. and Prunier, A. (2004). Long-term detrimental effects of tooth clipping or grinding in piglets: a histological approach. *Animal Welfare*, 13(1), 27–32.

Heim, G., Mellagi, A. P., Bierhals, T., de Souza, L. P., de Fries, H. C., Piuco, P., Seidel, E., Bernardi, M. L., Wentz, I., Bortolozzo, F. P. (2012). Effects of cross-fostering within 24h after birth on pre-weaning behaviour, growth performance and survival rate of biological and adopted piglets. *Livestock Science*, 31, 150(1), 121–7.

Herpin, P., Damon, M. and Le Dividich, J. (2002). Development of thermoregulation and neonatal survival in pigs. *Livestock Production Science*, 78(1), 25–45.

Herskin, M. S., Thodberg, K. and Jensen, H. E. (2015). Effects of tail docking and docking length on neuroanatomical changes in healed tail tips of pigs. *Animal*, 9(04), 677–81.

Hess, A., Boddicker, N., Rowland, R. R. R., Lunney, J. K., Plastow, G. S. and Dekkers, J. C. M. (2014, August). Genetic Parameters and Effects for a Major QTL of Piglets Experimentally Infected with a Second Porcine Reproductive and Respiratory Syndrome Virus. In Proceedings, 10th World Congress of Genetics Applied to Livestock Production.

Hewson, C. J., Dohoo, I. R., Lemke, K. A. and Barkema, H. W. (2007). Canadian veterinarians' use of analgesics in cattle, pigs, and horses in 2004 and 2005. *The Canadian Veterinary Journal*, 48(2), 155.

Hillier, S. G. (1994).Current concepts of the roles of follicle stimulating hormone and luteinizing hormone in folliculogenesis. *Human Reproduction* 9(2), 188–91.

Hunter, E. J., Jones, T. A., Guise, H. J., Penny, R. H. C. and Hoste, S. (2001). The relationship between tail biting in pigs, docking procedure and other management practices. *The Veterinary Journal*, 161(1), 72–9.

Ijaz, A., Abalkhail, A. A. and Khamas, W. A. H. (2000). Effect of intra testicular injection of formalin on seminiferous tubules in Awassi lambs. *Pakistan Veterinary Journal*, 20(3), 129–34.

Jensen, P. and Redbo, I. (1987). Behaviour during nest leaving in free-ranging domestic pigs. *Applied Animal Behaviour Science*, 18(3–4), 355–62.

Johnson, A. K., Julie, M.-T. and McGlone, J. J. (2001). Behavior and Performance of Lactating Sows and Piglets Reared Indoors or Outdoors. *Journal of Animal Science*, 79, 2588–96.

KilBride, A. L., Mendl, M., Statham, P., Held, S., Harris, M., Marchant-Forde, J. N., Booth, H., and Green, L. E. (2014). Risks associated with Preweaning mortality in 855 litters on 39 commercial outdoor pig farms in England. *Preventive Veterinary Medicine*, 117, 189–99.

King, J. W. B. and Gajić, Z. (1969) The repeatability of maternal performance in inbred, outbred and linecross large white sows. *Animal Science*, 11, 47–51.

Knauer, M., Stalder, K. J., Karriker, L., Baas, T. J., Johnson, C., Serenius, T., Layman, L. and McKean, J. D. A descriptive survey of lesions from cull sows harvested at two Midwestern US facilities. *Preventive Veterinary Medicine*, 14, Dec 2007, 82(3), 198–212.

Kuller, W. I., Soede, N. M., van Beers-Schreurs, H. M. G., Langendijk, P., Taverne, M. A. M., Verheijden, J. H. M. and Kemp, B. (2004). Intermittent suckling: Effects on piglet and sow performance before and after weaning. *Journal of Animal Science*, 82(2), 405–13.

Lambooij, W., van Putten. Transport of pigs. Livestock Handling and Transport. (1993). Grandlng ed. CAB International, Wallingford, UK, pp. 213–31.

Langendijk, P., Dieleman, S. J., Van Den Ham, C. M., Hazeleger, W., Soede, N. M. and Kemp, B. (2007). LH pulsatile release patterns, follicular growth and function during repetitive periods of suckling and non-suckling in sows. *Theriogenology*, 67(5), 1076–86.

Le Dividch, J. and Seve, B. (2000, September). Energy requirement of the young piglet. The weaner pig: Nutrition and management. In Proceedings of a British Society of Animal Science Occasional Meeting, University of Nottingham, UK.

Lewis, E., Boyle, L. A., Brophy, P., O'doherty, J. V. and Lynch, P. B. (2005). The effect of two teeth resection procedures on the welfare of sows in farrowing crates. Part 2. *Applied Animal Behaviour Science*, 90(3), 251–65.

Lewis, E., Boyle, L. A., Lynch, P. B., Brophy, P. and O'doherty, J. V. (2005). The effect of two teeth resection procedures on the welfare of piglets in farrowing crates. Part 1. *Applied Animal Behaviour Science*, 90(3), 233–49.

Lewis, N. Transport of early weaned pigs. (2008). *Behavioral Science*, 110, 128–35.

Livingston, D. H., Mosenthal, A. C. and Deitch, E. A. (1995). Sepsis and multiple organ dysfunction syndrome: a clinical-mechanistic overview. *New horizons* (Baltimore, Md.), 3(2), 257–66.

Lovell, J. S. (2016). Understanding farm animal abuse. *The Routledge International Handbook of Rural Criminology*, 137.

Lou, Z. and Hurnick, J. F. (1994). An ellipsoid farrowing crate: its ergonomical design and effects on pig productivity. *Journal of Animal Science*, 72, 2610–16.

Lumb, S. (2007) Towards a more 'humane' castration for piglets. *Pig Progress*, 23, 24–6

Main, R. G., Dritz, S. S., Tokach, M. D., Goodband, R. D. and Nelssen, J. L. (2004). Increasing weaning age improves pig performance in a multisite production system. *Journal of Animal Science*, 82(5), 1499–507.

Manteuffel, G. and Schön, P. C. (2002). Measuring pig welfare by automatic monitoring of stress calls. *Bornimer Agrartech. Ber*, 29, 110–18.

Marchant-Forde, J. N., Lay, D. C., McMunn, K. A., Cheng, H. W., Pajor, E. A. and Marchant-Forde, R. M. (2009). Postnatal piglet husbandry practices and well-being: the effects of alternative techniques delivered separately. *Journal of Animal Science*, 87(4), 1479–92.

Martínez-Avilés, M., Fernández-Carrión, E., López García-Baones, J. M. and Sánchez-Vizcaíno, J. M. (2015). Early detection of infection in pigs through an online monitoring system. *Transboundary and emerging diseases*, 64(2), 364–73.

Marx, G., Horn, T., Thielebein, J., Knubel, B. and Von Borell, E. (2003). Analysis of pain-related vocalization in young pigs. *Journal of Sound and Vibration*, 266(3), 687–98.

Maselyne, J., Adriaens, I., Huybrechts, T., De Ketelaere, B., Millet, S., Vangeyte, J., van Nuffel., A. and Saeys, W. (2015). 5.5. Assessing the drinking behaviour of individual pigs using RFID registrations. Ilan, H. (ed.), p. 209.

Maselyne, J., Saeys, W. and Van Nuffel, A. (2015). Review: Quantifying animal feeding behaviour with a focus on pigs. *Physiology and Behavior*, 138, 37–51.

Matthews, S. G., Miller, A. L., Clapp, J., Plötz, T. and Kyriazakis, I. (2016). Early detection of health and welfare compromises through automated detection of behavioural changes in pigs. *The Veterinary Journal*, 217, 43–51.

McCracken, B. A., Gaskins, H. R., Ruwe-Kaiser, P. J., Klasing, K. C. and Jewell, D. E. (1995). Diet-dependent and diet-independent metabolic responses underlie growth stasis of pigs at weaning. *Journal of Nutrition*, 125(11), 2838–45.

McGlone, J. J. (1985). Olfactory cues and pig agonistic behavior: evidence for a submissive pheromone. *Physiology and Behavior*, 34(2), 195–8.

McGlone, J. J. (2002). Housing options for farrowing: Considerations for Animal Welfare and Economics. Purdue Extension. PIH 01-01-02, 6p.

McGlone, J. J. and Blecha, F. (1987). An examination of behavioral, immunological and productive traits in four management systems for sows and piglets. *Applied Animal Behaviour Science*, 18, 269–86.

McGlone, J. J. and Anderson, D. L. (2002). Synthetic maternal pheromone stimulates feeding behavior and weight gain in weaned pigs. *Journal of Animal Science*, 80, 3179–83.

McGlone, J. J. and Hellman, J. M. (1988). Local and general anesthetic effects on behavior and performance of two-and seven-week-old castrated and uncastrated piglets. *Journal of Animal Science*, 66(12), 3049–58.

McGlone, J. J. and Morrow, J. L. (1988). Reduction of pig agonistic behavior by androstenone. *Journal of Animal Science*, 66(4), 880–4.

McGlone, J. J., Akins, C. K. and Green, R. D. 1991. Genetic variation of sitting frequency and duration in pigs. *Applied Animal Behavior*, 30, 319–22.

McGlone, J. J., Nicholson, R. I., Hellman, J. M. and Herzog, D. N. (1993). The development of pain in young pigs associated with castration and attempts to prevent castration-induced behavioral changes. *Journal of Animal Science*, 71(6), 1441–6.

McGlone, J. J., Thompson, W. G. and Guay, K. A. (2014). Case study: The pig pheromone androstenone, acting as an interomone, stops dogs from barking. *The Professional Animal Scientist*, 30(1), 105–8.

McGlone, J. J., Thompson, G., Devaraj, S. G. (2016). A natural interomone 2-methyl-2 butenal stimulates feed intake and weight gain in weaned pigs. *Animal*, 11 (2), 306–8.
McGlone, J. J. (2006). Comparison of sow welfare in the Swedish deep-bedded system and the US crated-sow system. *Journal of American Veterinary Medical Association*, 229, 1377–80.
McKean, J. D. (2001). The importance of traceability for public health and consumer protection. *Revue scientifique et technique* (International Office of Epizootics), 20(2), 363–71.
McManus, C., Tanure, C. B., Peripolli, V., Seixas, L., Fischer, V., Gabbi, A. M., Menegassi, S. R., Stumpf, M. T., Kolling, G. J., Dias, E. and Costa, J. B. G. (2016). Infrared thermography in animal production: An overview. *Computers and Electronics in Agriculture*, 123, 10–16.
Meijer., E., Bertholle, C. P., Oosterlinck, M., vander Staay, F., Back, W. and van Nes, A. (2014). Pressure mat analysis of the longitudinal development of pig locomotion in growing pigs after weaning pigs. *BMC Veterinary Research*, 10, 1–11.
Meiszberg, A. M., Johnson, A. K., Sadler, L. J., Carroll, J. A., Dailey, J. W. and Krebs, N. (2009). Drinking behavior in nursery pigs: Determining the accuracy between an automatic water meter versus human observers. *Journal of Animal Science*, 87(12), 4173–80.
Melin, L., Mattsson, S., Katouli, M. and Wallgren, P. (2004). Development of Post-weaning Diarrhoea in Piglets. Relation to Presence of Escherichia coli Strains and Rotavirus. *Journal of Veterinary Medicine*, Series B, 51(1), 12–22.
Merks, J. W. M. and Lambooy, E. (1990). Injectable electronic identification systems in pig production. *Pig News and Information*, 11(1), 35–6.
Moeser, A. J., Vander Klok, C., Ryan, K. A., Wooten, J. G., Little, D., Cook, V. L. and Blikslager, A. T. (2007). Stress signaling pathways activated by weaning mediate intestinal dysfunction in the pig. *American Journal of Physiology-Gastrointestinal and Liver Physiology*, 292(1), G173–81.
Morrow-Tesch, J. and McGlone, J. J. (1990). Sources of maternal odors and the development of odor preferences in baby pigs. *Journal of Animal Science*, 68, 3563–71.
Mucignat-Caretta, C. (ed.), (2014). *Neurobiology of Chemical Communication*. CRC Press.
Nguyen, D. T., Lee, K., Choi, H., Choi, M., Le, M. T., Song, N., Kim, J., Seo, H. G., Oh, J. W., Lee, K., Kim, T. and Park, C. (2012). The complete swine olfactory subgenome: expansion of the olfactory gene repertoire in the pig genome. *BMC Genomics*, 13, 584–96.
Niekamp, S. R., Sutherland, M. A., Dahl, G. E. and Salak-Johnson, J. L. (2007). Immune responses of piglets to weaning stress: Impacts of photoperiod. *Journal of Animal Science*, 85(1), 93–100.
Noonan, G. J., Rand, J. S., Priest, J., Ainscow, J. and Blackshaw, J. K. (1994). Behavioural observations of piglets undergoing tail docking, teeth clipping and ear notching. *Applied Animal Behaviour Science*, 39(3–4), 203–13.
Norman, A. W. and Litwack, G. (1997). *Hormones*, New York: Academic Press. New York.
Nuntapaitoon, M. and P. Tummaruk. (2015). Piglet Preweaning mortality in a commercial swine herd in Thailand. *Tropical Animal Health and Production*. 47, 1539–46.
Pageat, P. (2001). Pig appeasing pheromone to decrease stress, anxiety and aggressiveness. USA Patent 6,169,113. Publication date 2 January 2001.
Paoli, M. A., Lahrmann, H. P., Jensen, T. and D'Eath, R. B. (2016). Behavioural differences between weaner pigs with intact and docked tails. *Animal Welfare*, 25(2), 287–96.
Pederson, L. J., Jensen, H. and Thodberg, K. (2008). Cross fostering of piglets in farrowing pens. Proceedings of the 20th Nordic Symposium of the International Society of the International Society of Applied Ethology. Norway.
Perez-Pinera, P., Ousterout, D. G. and Gersbach, C. A. Advances in Targeted Genome Editing. (2012). *Current Opinion in Chemical Biology*, 16, 268–77.
Plastow, G. S. (2016). Genomics to benefit livestock production: improving animal health. *Revista Brasileira de Zootecnia*, 45(6), 349–54.
Price, E. O., Hutson, G. D., Price, M. I. and Borgwardt, R. (1994). Fostering in swine as affected by age of offspring. *Journal of Animal Science*, 72(7), 1697–701.
Prunier, A., Bonneau, M., Von Borell, E. H., Cinotti, S., Gunn, M., Fredriksen, B., Giersing, M., Morton, D. B., Tuyttens, F. A. and Velarde, A. (2006). A review of the welfare consequences of surgical

castration in piglets and the evaluation of non-surgical methods. *Animal Welfare-Potters Bar then Thampstead*. August 1, 15(3), 277.
Prunier, A., Mounier, A. M. and Hay, M. (2005). Effects of castration, tooth resection, or tail docking on plasma metabolites and stress hormones in young pigs. *Journal of Animal Science*, 83(1), 216–22.
Prunier, A., Mounier, A. M., Bregeon, A. and Hay, M. (2001). Influence of tail docking, tooth resection and castration on plasma cortisol, ACTH, glucose and lactate in piglets. In Proceedings of the 4th International Conference on Farm Animal Endocrinology, Parme, Italy, 7–10 October 2001.
Rault, J. L. and Lay, D. C. (2011). Nitrous oxide by itself is insufficient to relieve pain due to castration in piglets. *Journal of Animal Science*, 89, 3318–25.
Rault, J. L., Lay, D. C. and Marchant-Forde, J. N. (2011). Castration induced pain in pigs and other livestock. *Animal Behaviour Science*, 135(3), 214–25.
Rea, C. P. (1986). Universal ear notching system in swine. Agricultural Guide. Swine Management. University of Missouri-Colombia Extension. 2505.
Rekwot, P. I., Ogwu, D., Oyedipe, E. O. and Sekoni, V. O. (2001). The role of pheromones and biostimulation in animal reproduction. *Animal Reproduction Science*, 65(3), 157–70.
Robert, S. and Martineau, G. P. (2001). Effects of repeated cross-fosterings on preweaning behavior and growth performance of piglets and on maternal behavior of sows. *Journal of Animal Science*, 79(1), 88–93.
Robert, S., Thompson, B. K. and Fraser, D. (1995). Selective tooth clipping in the management of low-birth-weight piglets. *Canadian Journal of Animal Science*, 75(3), 285–9.
Robertson, J. B., Laird, R., Hall, J. K. S., Forsyth, R. J., Thomson, J. M. and Walker-Love, J. (1966) 'A comparison of two indoor farrowing systems for sows', *Animal Science*, 8, 171–8.
Rushen, J., Chapinal, N. and De Passille, A. M. (2012). Automated monitoring of behavioural-based animal welfare indicators. *Animal Welfare-The UFAW Journal*, 21(3), 339.
Rusk, C. P., Blomeke, C. R., Balschweid, M. A., Elliot, S. J. and Baker, D. (2006). An evaluation of retinal imaging technology for 4-H beef and sheep identification. *Journal of Extension*, 44(5), 1–33.
Sandercock, D. A., Smith, S. H., Di Giminiani, P. and Edwards, S. A. (2016). Histopathological Characterization of Tail Injury and Traumatic Neuroma Development after Tail Docking in Piglets. *Journal of Comparative Pathology*, 155, 276.
Scaramuzzi, R. J. and Martin, G. B. (2008). The importance of interactions among nutrition, seasonality and socio-sexual factors in the development of hormone-free methods for controlling fertility. *Reproduction in Domestic Animals*, 43, 129–36.
Schneider, F., Falkenberg, H., Kuhn, G., Nürnberg, K., Rehfeldt, C. and Kanitz, W. (1998). Effects of treating young boars with a GnRH depot formulation on endocrine functions, testis size, boar taint, carcass composition and muscular structure. *Animal Reproduction Science*, 50(1), 69–80.
Schön, P. C., Puppe, B. and Manteuffel, G. (2004). Automated recording of stress vocalisations as a tool to document impaired welfare in pigs. *Animal Welfare*, 13(2), 105–10.
Scott, K., Taylor, L., Gill, B. P. and Edwards, S. A. (2006). Influence of different types of environmental enrichment on the behaviour of finishing pigs in two different housing systems: 1. Hanging toy versus rootable substrate. *Applied Animal Behaviour Science*, 99(3), 222–9.
Sørensen, J. T., Rousing, T., Kudahl, A. B., Hansted, H. J. and Pedersen, L. J. (2016). Do nurse sows and foster litters have impaired animal welfare? Results from a cross-sectional study in sow herds. *Animal*, 10(04), 681–6.
Spencer, J. D., Boyd, R. D., Cabrera, R. and Allee, G. L. (2003). Early weaning to reduce tissue mobilization in lactating sows and milk supplementation to enhance pig weaning weight during extreme heat stress. *Journal of Animal Science*, 81(8), 2041–52.
Spreeuwenberg, M. A. M., Verdonk, J. M. A. J., Gaskins, H. R. and Verstegen, M. W. A. (2001). Small intestine epithelial barrier function is compromised in pigs with low feed intake at weaning. *The Journal of Nutrition*, 131(5), 1520–7.
Støier, S., Larsen, H. D., Aaslyng, M. D. and Lykke, L. (2016). Improved animal welfare, the right technology and increased business. *Meat Science*.
Sutherland, M. A. and Tucker, C. B. (2011). The long and short of it: A review of tail docking in farm animals. *Applied Animal Behaviour Science*, 135(3), 179–91.

Sutherland, M. A., Bryer, P. J., Krebs, N. and McGlone, J. J. (2008). Tail docking in pigs: acute physiological and behavioural responses.

Sutherland, M. A., Bryer, P. J., Krebs, N. and McGlone, J. J. (2009). The effect of method of tail docking on tail-biting behaviour and welfare of pigs. *Animal Welfare*, 18(4), 561–70.

Sutherland, M. A., Davis, B. L. and McGlone, J. J. (2011). The effect of local or general anesthesia on the physiology and behavior of tail docked pigs. *Animal*, 5(08), 1237–46.

Sutherland, M. A., Backus, B. L. and McGlone, J. J. (2014) Effects of transport at weaning on the behavior, physiology and performance of pigs. *Animal*, 4, 657–69.

Tao, X., Xu, Z. and Men, X. (2016). Transient effects of weaning on the health of newly weaning piglets. *Czech Journal of Animal Science*, 61(2), 82–90.

Thodberg, K., Jensen, K. H. and Jørgensen, E. (2010). The risk of tail-biting in relation to level of tail-docking. In Proceedings of the 44th Congress of the International Society for Applied Ethology (isae).

Thompson, B. K. and Fraser, D. (1986). Variation in piglet weights: development of within-litter variation over a 5-week lactation and effect of farrowing crate design. *Canadian Journal of Animal Science*, 66(2), 361–72.

Tirindelli, R., dibattista, M., Pifferi, S. and Menini, A. From pheromones to behavior. 2009. *Physiological Reviews*, 89, 921–56.

Tokach, M. D., Goodband, R. D., Nelssen, J. L., Kats, L. J., Goodband, B. and Tokach, M. (1992). Influence of weaning weight and growth during the first week postweaning on subsequent pig performance. In Kansas State University Swine Day 1992. Report of Progress 667. Kansas State University, pp. 19–21.

Turpin, D. L., Langendijk, P., Chen, T. Y., Lines, D. and Pluske, J. R. (2016). Intermittent Suckling Causes a Transient Increase in Cortisol That Does Not Appear to Compromise Selected Measures of Piglet Welfare and Stress. *Animal*, 6(3), 24.

Van de Weerd, H. A., Docking, C. M., Day, J. E., Avery, P. J. and Edwards, S. A. (2003). A systematic approach towards developing environmental enrichment for pigs. *Applied Animal Behaviour Science*, 84(2), 101–18.

van Staaveren, N., Teixeira, D. L., Hanlon, A. and Boyle, L. A. (2016). Pig carcass tail lesions: the influence of record keeping through an advisory service and the relationship with farm performance parameters. *Animal*, 1–7.

Viazzi, S., Ismayilova, G., Oczak, M., Sonoda, L. T., Fels, M., Guarino, M., Vranken, E., Hartung, J., Bahr, C. and Berckmans, D. (2014). Image feature extraction for classification of aggressive interactions among pigs. *Computers and Electronics in Agriculture*, 104, 57–62.

Von Borell, E., Baumgartner, J., Giersing, M., Jäggin, N., Prunier, A., Tuyttens, F. A. M. and Edwards, S. A. (2009). Animal welfare implications of surgical castration and its alternatives in pigs. *Animal*, 3(11), 1488–96.

Voulodimos, A. S., Patrikakis, C. Z., Sideridis, A. B., Ntafis, V. A. and Xylouri, E. M. (2010). A complete farm management system based on animal identification using RFID technology. *Computers and Electronics in Agriculture*, 70(2), 380–8.

Wayne, I. P. F., Wayne, F., Rusk, I. D. C. P. and Richert, B. T. Identification of Swine by Auricular Vein Patterns. https://pdfs.semanticscholar.org/5c2e/df3cadb5eb0cd139167992976eb1b6d4e1e7.pdf. Accessed April 17, 2017.

Wemelsfelder, F. and Putten, V. G. (1985). Behaviour as a possible indicator for pain in piglets. I.V.O.-report. Instituut voor Veeteeltkundig Onderzoek 'Schoonoord' no. B-260.

Westin, R., Holmgren, N., Hultgren, J., Ortman, K., Linder, A. and Algers, B. (2015). Post-mortem findings and piglet mortality in relation to strategic use of straw at farrowing. *Preventive Veterinary Medicine*, 119, 141–52.

White, R. G., DeShazer, J. A., Tressler, C. J., Borcher, G. M., Davey, S., Waninge, A., Parkhurst, A. M., Milanuk, M. J. and Clemens, E. T. (1995). Vocalization and physiological response of pigs during castration with or without a local anesthetic. *Journal of Animal Science*, 73, 381–6.

Widowski, T. and Torrey, S. (2002). Swine Fact Sheet. http://csm.pork.org/filelibrary/Factsheets/Well-Being/SWINE%20WELFAREFACTSHT-neona.pdf. Accessed April 17, 2017.

Zijlstra, R. T., Whang, K. Y., Easter, R. A. and Odle, J. (1996). Effect of feeding a milk replacer to early-weaned pigs on growth, body composition, and small intestinal morphology, compared with suckled littermates. *Journal of Animal Science*, 74(12), 2948–59.

Zonderland, J. J., Wolthuis-Fillerup, M., Van Reenen, C. G., Bracke, M. B., Kemp, B., Den Hartog, L. A. and Spoolder, H. A. (2008). Prevention and treatment of tail biting in weaned piglets. *Applied Animal Behaviour Science*, 110(3), 269–81.

Zurbrigg, K. (2006). Sow shoulder lesions: Risk factors and treatment effects on an Ontario farm. *Journal of Animal Science*, 84(9), 2509–14.

Chapter 10

Welfare of pigs during finishing

Jonathan Amory, Writtle University College, UK; and Nina Wainwright, British Pig Executive (BPEX), UK

1 Introduction

Finisher pigs, those of a post-weaning age kept for slaughter, make up the majority of the world pig population of approximately 1 billion (2014 – latest figures from FAOstat accessed on 19 October 2017). Although many of the countries in the world keep pigs in small groups that scavenge or are fed household waste, high-income countries (per capita GDP levels in excess of 30 000 USD) raise more than 95% of their pigs in intensive systems (Gilbert et al., 2015). In the European Union (EU), neonates (piglets) may be reared in extensive outdoor systems along with the sow, but keeping post-weaned pigs in systems other than indoor housing is rare (European Food Safety Authority, 2007). The intensive production system is characterised by higher animal density, larger farms, use of concentrated foods and control of the production environment, particularly temperature, humidity and lighting. For finishing pigs there are a range of production systems available, each with their own advantages and disadvantages; however, the majority will be reared in a partly or fully slatted floor system with restricted or no access to straw (Table 1).

The welfare of an animal has been defined by its ability to cope with its environment (Broom, 1986); a finishing pig in any production system is exposed to a range of challenges that threaten its well-being. These challenges or stressors can be either physiological or psychological in nature, and both the mental health and physical health of an animal are important to its welfare (Farm Animal Welfare Council, 2009). In addition, welfare scientists have more recently focused on the positive aspects of welfare that are considered important for a sentient animal. Particular risks to the welfare of finisher pigs relate to

http://dx.doi.org/10.19103/AS.2017.0013.25

Table 1 Housing systems for EU finishing pigs* (adapted from (Hendriks et al., 1999)

	Without/restricted access to straw			With straw	
System	Partly slatted	Fully slatted	Solid concrete	Solid concrete	Deep litter
%	47	44	3	4	2

* Note the total population in this report was stated to be approximately 69 million animals.

nutritional management, the physical environment and a lack of mental stimulation leading to such problems as tail-biting. This chapter discusses these concerns and current practical measures adopted in welfare assessment.

2 Nutrition management and welfare of finishing pigs

2.1 Gastric ulcers

A key objective of pig production is to let animals attain their slaughter weight at least cost as quickly as possible. Intensive systems often feed finely ground pelleted diets to increase the digestibility of nutrients and reduce feed wastage, but this can have important consequences on finisher pigs' welfare. Abattoir studies from major pig-producing countries have reported high prevalence of gastric ulcers varying from 19% to 79% (Robertson et al., 2002; van den Berg et al., 2005; Amory et al., 2006; Swaby and Gregory, 2012; Omotosho et al., 2016).

There are many risk factors associated with gastric ulceration and most commonly with intensified husbandry; however, the main risk factors are those concerned with the diet. Interruption in the diet or fasting are major factors inducing gastric ulcers in pigs (Davies et al., 1994; Lawrence et al., 1998), but causes may vary from disease (Dybkjær et al., 1998) to general environmental stressors, such as transport (Lawrence et al., 1998), large group size (Baeckstroem et al., 1988), low social ranking (Hessing et al., 1994), providing dam rather than river water (Robertson et al., 2002) and a lack of environmental enrichment (Ramis et al., 2005). A particularly important dietary risk factor is the small particle size often associated with feeding a pelleted diet.

A number of authors have reported increased gastric ulceration with reduced particle size in the ground cereal proportion of the diet given to pigs (Potkins et al., 1989; Wondra et al., 1995a,b; Millet et al., 2012a,b; Cappai et al., 2013); however, a recent study found that particle size or pelleting has no effect on stomach ulceration (Ball et al., 2015). Cappai et al. (2013) described a mechanism for the pathogenesis due to the increased fluidity of stomach contents, an associated increase in acidity in the proximal part of the stomach and the lack of secretory cells in the oesophageal mucosa predisposing that region to developing bleeding gastric ulcers. Infection due to *Helicobacter suis* may also have an important role in ulcer formation (Haesebrouck et al., 2009). It is important to note that because gastric ulcers cause increased mortality in finishing pigs and are considered detrimental to economic production, they are an unfortunate trade-off in the pursuit of increased feed efficiency. The increasing demand for grain for human and animal food can be met by searching for alternative feed ingredients, for example, high-fibre by-products from food and bio-fuel production that may increasingly replace single grain- and soybean-based diets, though finding suitable alternatives is not without challenges (Bakare et al., 2013; Zijlstra and Beltranena, 2013).

Crude fibre has been reported to have an additional dietary fibre that may break up the more fluid digesta from fine diets, preventing the erosion of the pars oesophageal region (Potkins et al., 1989). A number of reports have highlighted the protective effect straw bedding, presumably consumed as part of natural rooting activities, appears to have on gastric ulcer development (Potkins et al., 1989; Amory et al., 2006; Scott et al., 2006).

An additional benefit of a high-fibre diet may be the increased post-prandial satiety of pigs due to bulkiness of the fibre source or its ability to absorb water. Satiety, defined as an absence of hunger (Yeates and Main, 2008), is an important animal welfare concern and is an opportunity to establish positive well-being in pig production. Using pig behaviour to identify satiety is complex as gastric distension associated with a bulky food might reduce feeding behaviour through abdominal discomfort whilst the animal is still motivated to feed, due to nutrient restriction. Day et al. (1996) investigated this through use of a second-order reinforcement schedule where the pig was instructed to press a paddle to indicate feeding motivation thus removing the potential confounder of further ingestion of food. They found that dried citrus pulp, a fibre source of high water-holding capacity, reduced feeding motivation, for up to three hours post-feeding compared to only one hour by a low-fibre diet of an equivalent level of nutrients. There have been few studies of finishing pigs since, with most research carried out in sows where food deprivation is a well-recognised concern. Leeuw et al. (2008) provided a detailed review of both sows and finishing pigs to ascertain the welfare benefits of both bulky and fermentable fibres, concluding that it is fermentable dietary fibre that provides the longer term postprandial satiety effects, reducing physical activity and appetitive behaviour for many hours after a meal.

2.2 Nutritional deficiencies and pig behaviour

Gross inadequacies in nutrition provision will have obvious effects on pig behaviour and consequently welfare for example, lack of food leads to hunger (in itself a welfare concern) and can lead to an increase in foraging, and greater competition and aggressive acts around any remaining food or when food supply is renewed, resulting in fighting lesions and injuries (Day et al., 1995). Similar effects will also be seen, usually at lower level, when there are less obvious issues for example, acute or chronic nutrient imbalance; pigs have shown that they can accurately select between different diets in order to obtain the correct balance of specific amino acids (Kyriazakis et al., 1991) – this could be reflected in time spent foraging if these are not correctly balanced in their provided diet.

Whilst there may be intrinsically apparent biological pathways between nutritional deficits and behaviour (e.g. hunger → food searching), individual nutrients can have less obvious effects – for example, specific amino acids play a role in neurotransmission through the blood–brain barrier and can therefore affect the animal's behaviour at this precise chemical/physiological level, for example, Popova (2006). Edwards (2006) stated that a changed protein metabolism may pathologically affect neurotransmitter balances in the central nervous system. It has been suggested that individual pigs performing tail-biting at obsessively high levels may have reduced protein uptake – this has several similarities with metabolic disorders found in humans. Deficiencies in metabolism of phenylalanine (an amino acid) in humans lead to the rare condition of phenylketonuria (PKU) (Moyle et al., 2007), a symptom of disruptive behaviours such as self-harm or tantrums (Brumm et al., 2010); individuals with PKU often have low birth weight (Lee et al., 2005).

In several species, it is suggested that gut discomfort leads to increased chewing behaviour which stimulates saliva production – the additional saliva is thought to have a role in reducing gut discomfort. Kubo et al. (2015) also suggested that a repetitive chewing action in itself may calm pigs. Early stages of crib-biting in horses can be alleviated by providing them with antacids (Mills and Macleod, 2002); chewing behaviour, such as tail or ear-biting, in pigs may similarly be directed at objects within the pen, but may also be directed at pen-mates. Pigs with conditions such as ileitis (porcine proliferative enteropathy, PPE) or gastric ulcers may also therefore be more likely to tail-bite. Almond and Bilkei (2006) showed that groups of pigs which had been vaccinated against the bacterium *Lawsonia intracellularis* (associated with ileitis) had lower rates of cannibalism-related culling (an alternative description of severe tail-biting) and mortality.

2.3 Salt as a welfare concern

Salt is mainly considered as a welfare concern because pigs are susceptible to salt poisoning when dietary levels are too high (0.4–0.5%) and pigs are not ingesting sufficient water, (The Pig Site, n.d.). In many species, animals will seek out additional salt when stressed (Denton et al., 1999), and this can be recreated by administration of ACTH (adrenocorticotrophic hormone – a hormone produced in response to stress). It has therefore been suggested that stress and salt-seeking behaviour in pigs may be observed as increased foraging and chewing behaviour, potentially directed at pen-mates' tails; however, ACTH injection in the pig does not trigger such a response (Jankevicius and Widowski, 2003), but this does not rule out that stress does cause this behaviour, only that ACTH may not be the appropriate model. Salt-deficient pigs do show increased rooting behaviour (Beattie and O'Connell, 2002). Anecdotally, increasing salt levels at critical tail-biting periods on individual farms has been reported as being successful in reducing tail-biting behaviour, suggesting that on these farms, at these key periods, salt-seeking behaviour may be occurring in response to stressors; therefore, increasing the amount of salt in gradual pace is recommended, rather than feeding pigs initially with higher salt content.

3 Physical and social environment and welfare of finishing pigs

3.1 Group size, stocking density and competition for resources

Aggression is a natural behaviour in pigs with a function of improving social position (Camerlink et al., 2014) and possibly chasing away strangers (Puppe, 1998). However, it has an impact on their welfare in a number of ways: increasing skin lesions with their associated pain and risk of infection (Turner et al., 2006), stress physiology (Escribano et al., 2015) and probably becoming a part of a wider pattern of social tension in finishing pigs (Camerlink et al., 2014).

Various aspects of pen management affect levels of aggression in pigs. Pigs are regrouped for various management reasons, such as reducing stocking density or making efficient use of pen spaces, and can be grouped into a finisher category right up to the few weeks before being slaughtered (Stookey and Gonyou, 1994). When previously unacquainted finisher pigs are put together in a common pen, they show agonistic behaviour (aggressive and submissive actions) that varies in intensity over a day and that increases in frequency and fight time compared with newly weaned pigs (Stukenborg et al., 2011). Larger group sizes are associated with poor growth post-weaning; however, this association is not found in pigs above a live weight of 69 kg (Turner et al., 2003). There is also little evidence of a relationship between group size and aggression (Turner et al., 2001; O'Connell et al., 2004; Schmolke et al., 2004) although pigs from large-size groups were less aggressive in encountering those from other groups (Turner et al., 2001).

Reducing space allowance below a critical value (Box 1) results in a decline in growth rate, indicating reduced well-being (Gonyou et al., 2006), and may still be more economical (Edwards et al., 1988; Vermeer et al., 2014) epitomising the tension between production and welfare. An interdisciplinary Dutch welfare initiative experimented with a 'ComfortClass' pen design, with separated rooting and dunging areas and a space allowance of up to 2.4 m^2 per pig based on the perceived space required to meet the needs of the animals (de Greef et al., 2011). This study concluded that this level resulted in a good quality of life for these finishing pigs, based on the absence of signs of poor welfare. A follow-up study of the ComfortClass pen found detrimental effects of increased stocking density, even at elevated levels, and increased group sizes in terms of welfare and growth (Vermeer et al., 2014); however, the performance benefits of more space/lower group sizes did not offset the increased housing costs.

Box 1 Space allowance and pig welfare

Floor allowance for pig production is often calculated by the following formula:

A=k*BW(0.667)

where A = floor space allowance (m^2/BW), BW = body weight (kg) and k = space allowance coefficient

Gonyou et al. (2006) investigated the effects of stocking density on production efficiency from a range of published articles and concluded that crowding was detrimental to pig production (measured as average daily gain) in fully and partially slatted systems when the value of k was less than 0.035.

Competition for access to pen resources can be important for welfare. Increased competition for feeding spaces is associated with increased aggression (Spoolder et al., 1999) with lightweight animals particularly at risk of welfare impairment (Rasmussen et al., 2006). Competition for enrichment devices such as a tyre swing or block of straw can also increase fighting in finisher pigs (Schaefer et al., 1990; Bulens et al., 2016). Pigs seem to be less competitive to access drinking facility (Turner et al., 1999).

An article by Bernard Peet, on the well-known The PigSite, describes the importance of pen design in managing lying and dunging behaviour in order to reduce social tension through competition for resources or disturbance of lying pigs (Peet, n.d.). Managing ventilation to provide a cooler dunging area, organising drinkers to be away from dunging areas, ensuring the lying area is as undisturbed as possible through correct positioning of feeders (Fig. 1) and ensuring pigs are not forced to lie in 'undesirable' areas will reduce aggression and possible tail-biting, as discussed in the following section.

3.2 Mounting behaviour

With surgical castration to reduce boar taint considered to be a major welfare concern, many major pig-producing countries in the EU have signed a common European declaration on alternatives. A 2014 progress report on this initiative describes one alternative as the rearing of entire males, commonly carried out in the United Kingdom, Spain and Portugal (Backus et al., 2014). However, this practice is associated with sexual behaviour manifesting as mounting of other animals that can result in skin lesions and leg injuries (Rydhmer et al., 2006; Fredriksen et al., 2008) that is a particular welfare concern to finisher pigs. Mounting is carried out mainly by male pigs mostly, but the behaviour is equally displayed towards male and female animals in mixed groups (Hintze et al., 2013). Immunocastration is not straightforward at present, but may provide a solution (Rydhmer et al., 2010); however, mounting behaviour can occur well before the onset of puberty (Clark and D'Eath, 2013) and is a trait that varies from pig to pig (Hintze et al., 2013). The underlying biological mechanisms and the approaches to reduce this behaviour within the industry are not clear at present.

3.3 Flooring and lameness

Lameness in pigs can be caused by a wide range of factors for example, injury or infection in the foot or joint, or longer term skeletal and joint problems such as osteochondrosis (Stavrakakis et al., 2014). Injury or infection (e.g. bacterial infection via abrasions) is likely to be due to suboptimal flooring (rough-edged flooring or inappropriate slot size or slat shape, permanently damp flooring) either during the finisher stage or carried through from earlier housing; it can also be caused or exacerbated by behaviours such as running, mounting, fighting or fleeing (which can also have genetic components). Steps within a pen can contribute to mechanical injury. Deep strawed areas can increase the risk of lameness due to entanglement; outdoor pigs can also risk damaging limbs while trying to move through deep mud or slipping on ice.

Lameness and flooring are unquestionably linked to pig production, and the main flooring/substrate types all have their inherent risks. The material used for making a floor

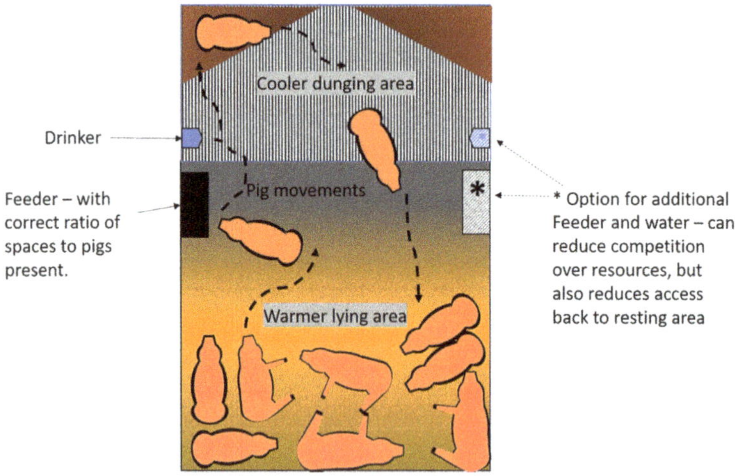

Figure 1 Optimum pen design to allow free movement and undisturbed lying behaviour (adapted from (Peet, n.d.)).

needs to be nonabrasive, non-slippery and preferably elastic (to some extent) to cushion the pigs while moving or resting. Slatted flooring is more complex, with the construction, profile and size of slots (gaps) being important in protecting the feet from injury (European Food Safety Authority (EFSA), 2007). EFSA (2005) concluded that width of a gap should not exceed half the width of the contact area between the foot and the floor and the solid area between the gaps should be sufficient enough to support the foot; they also noted that whilst several studies produced recommendations of slat width based on specific weights of pig, these may be confounded by slat type, shape and material. Construction profile is critical: sharp edges may cause cut injuries as well as a compressive stress (Webb and Nilsson, 1983; Webb, 1984). Foot and leg injuries tend to be higher in pigs on slatted systems than those on solid floors (European Food Safety Authority, 2005), whereas pigs housed in unbedded systems have a greater severity of pressure and bursae than those in bedded systems. Some of these may not feel the pain because they have no lesion or infection, but all have a potential to have their locomotion and social interactions affected and hence welfare.

As noted in EFSA (2005) and in the development of Real Welfare protocols, observation of lesions and bursae on pigs' limbs can be particularly difficult in typical housing – this may mean that suitability of housing is suboptimal but goes unnoticed until lameness develops. It is difficult to get to eye level with pigs' legs within a pen, as observers are likely to see either inquisitive heads and backs, or the back of rear legs if the pigs are flighty. Flighty animals are at increased risk of leg injuries and lameness (Stavrakakis, 2014).

There is a genetic component to lameness in pigs, with leg weakness and osteochondrosis found to be genetically correlated with lean meat content, and this points to a negative effect of selecting for lean meat content in the carcase (European Food Safety Authority, 2007). Stavrakakis et al. (2014) used automated gait tracking of pigs in order to identify issues, with the aim of aiding farmers in selecting breeding animals.

Lameness is considered to be a key outcome measure or indicator of the welfare of pigs in individual, group and herd levels (alongside tail-biting and other measures); a high level of lameness can be due to a range of factors. Identifying and addressing these factors can therefore improve the welfare of current and future batches of pigs on the farm.

3.4 Outdoor production

Whilst outdoor housing of pigs is often seen by consumers as achieving the highest welfare standards, it has inherent welfare risks to manage, which are rarely found in indoor systems. Exposure to a wide range of wildlife increases the pigs' exposure to parasites (European Food Safety Authority, 2007); lack of access to wallows or shaded areas in hot temperatures can result in sunburn (NADIS, 2003). Excess rain can make the fields impassable to pigs, and frozen water supply and frozen ground are also some of the challenges to pigs.

Indoor systems (often termed 'intensive' but covering a wide range of housing and system types, population densities etc.) were developed commercially from the 1930s (Woods, 2012) in order to improve pig production by taking away stressors and production challenges associated with outdoor rearing. Pigs housed indoors can be kept at an optimal temperature range, with ready access to dry bedding, clean food and water, and pen layouts enabling the stockperson to observe the animals for problems more frequently and more accurately (mud coverage on outdoor animals, especially in summer months can make observation difficult). Indoor housing does take away the range of environmental stimulation that the pig receives, for example, temperature range, weather and light levels. This less varied environment can potentially lead animals that are less able to cope with changes in daily routine to activities such as movement around the unit or loading.

4 Environmental enrichment and tail-biting

4.1 Environmental enrichment

Effective environmental enrichment for animals is achieved by providing them with what meets their behavioural or psychological needs in a situation where food, water and social and physical comfort are already provided. For most animals in captivity this relates to the appetitive stage of food (or other resource) attainment because food presentation and availability are predictable (Jensen, 2002). Pigs have evolved to root and investigate their environment with their snout and specialised rostral disc to obtain a balanced and nutritional and satiating diet (down to specific amino acids); this behaviour can account for 10–20% of their active time (Stolba and Wood-Gush, 1989). Suitable environmental enrichment for pigs is therefore one that they can root, chew and ideally ingest with some nutritional benefits; where such substrates or objects are provided, pigs show less unwanted agonistic behaviours such as tail-biting and aggression/fighting. Finding the ideal environmental enrichment for pigs has been the subject of many research papers as this is an area of commercial and welfare significance (Fig. 2).

Whilst papers often focus on comparing a small number of different substrates or objects to find pigs' preference and effects on both desirable and undesirable behaviours, a different approach was followed by van de Weerd and Day (2009) who described objects based on their properties, rather than by size, substance, etc. This system makes comparison of enrichment types far more useful in a practical environment, as well as comparisons of research papers, and helps to avoid creating a prescriptive list of recommended objects or substrates. Pigs were shown to have long-term (five-day) preference for objects that were edible and destructible. Additional work suggests that nutritional objects are also preferred (Day et al., 1996). Studnitz et al. (2007) confirmed this with additional detail, finding that the best materials for maintaining pigs' investigatory behaviour are complex, changeable, destructible, manipulable and contain sparsely distributed edible parts. Grandin (1989) showed that pigs prefer clean objects; this can be an issue where objects have become detached from their suspending chain or rope and are rooted into the dunging area of the pen, if objects are suspended at pig-height near the dunging area of the pen, or in a crowded pen with no specific dunging area. Longevity of use of enrichment is a key factor in commercial units where objects cannot be replaced daily.

Commercial pigs experience a wide range of enrichment types, both intended and incidental. In practice, this means that pigs may be finding objects around the pen that they choose to root and chew on, but which have not been intentionally provided as enrichment, for example, wooden gates; pen fittings such as hinges, rubber curtains across kennel entrances; baler twine; wellington boots; etc. In outdoor systems turf, earth, wildlife, etc. might all be considered by the pigs as suitable rootable objects/substrates and be a

Figure 2 'Suitable rootables' – cartoon produced for the Tail Biting Project (2006–09): implementation of existing knowledge on tail-biting by the development and evaluation of prevention and outbreak husbandry advisory tools. Research partners: University of Bristol, University of Newcastle. Industrial partners: RSPCA, BPEX, pig producers.

satisfactory outlet for rooting and chewing behaviours. This distinction between intended and incidental enrichment only becomes an issue on-farm where pigs are damaging pen fittings, etc. resulting in increased farm maintenance time and costs, or where pen fittings are chewed into a dangerous state (revealing wires, sharp edges etc., or allowing pigs to mix). Whilst car tyres have historically been recommended as enrichment for pigs, their destructive capabilities meant that there is a welfare risk from the exposed wire core of the tyres, resulting in entanglement, distress and lesions (NADIS, 2002). In academic exercises or where enrichment-related welfare issues are being addressed, provision needs to be made to recognise all objects or substrates that the pigs are choosing to root or chew on, whether intentional or not. An alternative interpretation of the EU wording on provision of enrichment has been to state that enrichment is appropriate (i.e. sufficient quality and quantity) *only* when pigs are not tail-biting (Compassion in World Farming, 2011), but this overlooks the importance of many other factors in tail-biting – good enrichment can ameliorate some, but not all, stressors attributable to suboptimal husbandry (Edwards, 2006). Typical enrichment provision for different systems is shown in Table 2.

Six per cent of pigs in Europe are housed in straw pens, (the United Kingdom and Switzerland are atypical with 25% and 20%, respectively, of finisher pigs housed on straw) (European Food Safety Authority, 2007). Apart from fulfilling the key properties of enrichment, straw offers additional benefits of providing thermal and physical comfort (Fraser et al., 1991; Arey and Bruce, 1993; Lyons et al., 1995; Schrøder-Petersen and Simonsen, 2001) and gastric comfort: pigs that had straw bedding or received rolled cereals had nearly no gastric lesions (Nielsen and Ingvartsen, 2000) which may reduce tail-biting risk by reducing gut discomfort and drive to chew. Straw is also generally provided in much larger quantities than other forms of enrichment, increasing pigs' opportunity to interact with it and reducing competition among pigs to access it compared with objects (see Box 2 for key points).

Box 2 Straw – the key points

Barley straw preferred to wheat straw

Clean, dry, bright straw (rather than damp and mouldy or old, dry and dusty) (EFSA Panel on Animal Health and Welfare, 2014)

Long straw, rather than chopped (Pedersen et al., 2005)

Regular replenishment required, for example, daily (EFSA Panel on Animal Health and Welfare, 2014)

Also effective when provided solely as enrichment (dispenser) rather than bedding (EFSA Panel on Animal Health and Welfare, 2014)

Deep straw systems – lying area needs to be dry (DEFRA, 2003)

Table 2 Typical enrichment provision for different systems of pig production

System	Enrichment provided	Enrichment properties	Notes
Indoor, slatted	Suspended objects, generally plastics or wood (branch segments) suspended from chain or rope. Floor-based objects such as large plastic containers, logs. Large stones or rounded pebbles. Specific objects such as acetyl pipe helicopters, commercially available objects such as Bite-Rite. Occasionally cardboard boxes	Wide range of enrichment properties, for example, large solid metal or plastic objects limited malleable or deformable properties; wood good difficult to keep clean; limited nutritional properties	Can be limited by cost per pen (large number of pens per farm) and replacement rate. Objects and broken objects must not risk injury to pigs (or farm staff) or disruption to mechanised slurry systems
Indoor, solid (or part solid) floor	As above plus opportunity for bedding substrate, generally straw, occasionally sawdust or shavings		Where slats are present, bedding provision is limited due to risk of blockage to slurry system
Solid floor	Straw provision in some or all of pen; depending on pen size, entire bales may be provided		Where straw is present as bedding, additional enrichment objects are provided less often
Deep straw system	Straw		
Outdoor-housing + paddock	Strawed lying area		Parasite risk from external contamination

4.2 Removing straw

In many units pigs move between housing types as they mature, for example, from weaner housing to grower pens then finisher housing. Each housing type may have different enrichment (or straw) opportunities, for example, fully slatted weaner decks → strawed grower yards → slatted finisher pens. Pigs may also be moved between units at the different stages; some pigs may therefore experience both strawed and non-strawed pens. Day et al., (2002) suggested that a change in straw provision may be more important than the straw *per se* in altering pigs' behaviour for example, increasing tail-biting, as suggested by other authors (Ruiterkamp, 1985; Statham et al., 2011) and has also been found to be the case for enrichment removal (Munsterhjelm et al., 2009).

4.3 Tail-biting

Tail-biting is a multifactorial behaviour in pigs and can result in 'mild' lesions (e.g. underlying bruising but no visible surface lesions) through to severe damage (also described as cannibalism), and can affect a significant percentage of pigs on a unit. Tail-biting has both welfare and financial impacts on a unit, with severe outbreaks potentially affecting all or a majority of pigs on a farm. Harley et al. (2012) found that 58.1% of finisher pigs reaching slaughter had detectable tail lesions, with an additional 1.03% having severe lesions (based on 36963 pigs, 99% of which were tail docked). As a consequence, 42% of profit margin was eroded at slaughter due to tail-biting with losses due to reduced growth and pig meat lost to carcase condemnation totalling €1.70 per pig (Harley et al., 2012). Despite routine tail docking (a measure used to reduce tail-biting risk) the overall cost of tail-biting to the Dutch pig industry was calculated to be €8 million each year (EFSA Panel on Animal Health and Welfare, 2014). Whilst tail-biting research has previously been considered in terms of the outcome of the behaviour and its obvious production and welfare impact that is, bitten tails, abscesses, potential pyaemia in the pig, on-farm deaths, partial and whole carcase condemnations, research has more recently focused on the welfare of the pigs driven to perform this behaviour, and therefore the potential to help prevent the behaviour, by either identifying individuals predisposed to tail-biting (Brunberg et al., 2013) or adjusting the environment to better adapt the pigs to. Pigs which tail-bite are demonstrating an abnormal behaviour (not reported in wild boar or feral pig populations) indicating an inability to cope with their environment.

Whilst tail-biting can occur at any stage of pig production, it has more obvious production and welfare impacts at finisher stages of production due to the size of lesion inflicted and the value of the animals at this stage.

Taylor et al. (2010) described three types of tail-biting based on the pigs' underlying motivation. The most commonly described tail-biting is boredom tail-biting, attributed to the pig being unable to find suitable rooting and chewing objects within its environment, resulting in rooting and chewing directed towards the tail (and/or ears, especially if the tail is docked to a short length); this behaviour is thought to start with innocuous manipulation of the tail in the mouth (described as 'tail in mouth' behaviour by Schrøder-Petersen and Simonsen, 2001), and to develop biting and chewing in some cases into stronger biting and chewing, resulting in lesions. A second type of tail-biting is frustration tail-biting, where a pig bites the tail of another pig that is blocking its access to resources such as feed, transit through an area of the pen, or to optimal lying areas; the biter pig may learn that this behaviour is rewarded with access to what it wants, and so the pattern is reinforced. A third category of tail-biting is generally observed in fanatic pigs, often the smallest in a group, which perform tail-biting at unusual (and sometimes extremely high) levels; the underlying cause of this behaviour in these individuals is unknown, but it may be linked to their slower growth for example, due to protein imbalance (and the diet provided to pigs in their group may not cater to their individual needs). By identifying the underlying cause of tail-biting, it is expected that tail-biting can be prevented or reduced by addressing the specific problem for that group, building or unit. This approach does not rule out the possibility of other tail-biting ontogenies such as tail suckling in weaners. Once tail-biting occurs within a group, other pigs of the same group may become attracted to the scent of blood (and potentially the protein or salt content of the blood) and can result in outbreak behaviour where significant numbers of pigs become involved, both as biters and victims (see Box 3 for Genetics and Tail-biting).

Box 3 Genetics and tail-biting

Not all pigs perform tail-biting under the same environment and stressors as their pen-mates; therefore, internal variation in propensity is also a factor in tail-biting. This genetic component is also the subject of study (e.g. Breuer et al., 2005) as identifying and removing high-risk factors from the commercial pig population would help with pig management and improve the welfare of the commercial pig herd (provided tail-biting propensity is not linked to beneficial behavioural or disease-susceptible traits). There are correlations among tail-biting, lean tissue growth rate and back fat thickness (Beattie et al., 2005); lower back fat thickness also emerged as a risk factor (Moinard et al., 2003). However, lean meat is perceived as a desirable product by consumers, and Breuer et al., (2005) suggested that selection for a leaner carcase in modern pig production has resulted in pigs that are more prone to tail-biting.

Boredom tail-biting is likely to be best addressed by providing optimal environmental enrichment such as straw or other suitable substrates or objects meeting the criteria established by Van de Weerd et al. (2003). Frustration tail-biting should be reduced by identifying the critical resource or resources and by increasing access to them, for example, by adding feed spaces, removing cinch points in traffic around the pen and altering pen layout to stop traffic across intended lying areas. Fanatic pigs are best dealt with by identifying the individual pigs that are tail-biting and removing them from the pen. Housing pigs with a range of tail lengths in the same group is a significant factor in tail-biting for example, housing short and long docked pigs together, undocked and half-docked pigs (reviewed by Taylor et al., 2012). Whilst variation in appearance of tail lengths is likely to be a factor in higher tail-biting due to novel tail shape being of more interest, a range of background factors may also be important (Box 4).

Box 4 Variation in tail lengths

Mixed groups – presence of two tail lengths within a group can indicate mixing of two or more litters (handled by two or more stockpeople when tail docked) – mixing of pigs is generally avoided due to associated aggression and stressors.

'Sunday pigs' – piglets born at the weekend – may not receive all procedures that other litters receive as well as tail docking; this could include procedures such as iron injections, with subsequent effects on health.

Gilts – female pigs identified as potential future breeding animals – may be left long-tailed to help identify them at sorting (short docked tails are avoided for breeding gilts as the projecting stump is considered to make mounting uncomfortable for the boar). Gilts present in a finisher group have therefore been remixed with mainstream finishers, with attendant mixing stressors.

Runts – may not be tail docked in some instances (to avoid additional stress to a weak piglet) – when present in mainstream groups may therefore be targeted as weaker individuals, but are also potentially more likely to become tail-biters.

Variation in tail lengths can be avoided by

1. Setting a standard tail length on the unit (based on length of tail remaining, not length removed)
2. Streamlining pig movements on the farm so that pig groups are not mixed later in the system (e.g. permanent gilt groups or 'smalls'/runt groups, or do not regroup different tail length groups if stockpeople cannot agree on a set length to dock to)
3. Enabling pigs to be trained and be familiar with docking equipment to reduce variation by individual stockpeople (and review of all other training to ensure no other aspects of pig keeping are being overlooked)

5 Practical welfare assessment of finisher pigs

5.1 Outcomes approach to welfare assessment

Welfare assessments have generally been based on physically measurable features of the animals' environment ('inputs') for example, pen size, temperature, water flow rate, etc. but this approach has been challenged in recent years by the 'Outcomes' approach. Outcome measures (or welfare outcomes) are based on observations of the animals for example, their behaviour, presence and location of lesions, lameness, etc. Outcome measures have been developed, validated and refined for pigs by programmes such as Welfare Quality (European-wide development, initiated in 2004), Real Welfare (initial work by S. Mullan (Mullan, 2009)) currently applied to all Red Tractor finisher units – UK) and AssureWel (RSPCA/Soil Association – applied to their certified finisher units).

Outcome schemes collect information from a representative sample of animals per unit (or per scheme) depending on the specificity of information needed, and what will be done with the information for example, a small sample from a large number of units to give a scheme-wide or national overview, or a large enough sample per unit to be able to benchmark and compare specific welfare measures accurately between different units and measure changes over time at individual unit level (Mullan et al., 2009). The sample pigs are examined by the assessor using specific criteria for each outcome measure, and presence or absence of the welfare measure is recorded (or levels of severity depending on the measure and scheme – see additional material). A key factor in valid use of outcome measures is adhering to the definitions to make objective rather than subjective assessments, and ensuring that assessors have been standardised (Mullan et al., 2011) – see Box 5.

Box 5 Definitions of lameness in welfare assessment

All three major outcome-based welfare assessment schemes recognise the importance of lameness as an indicator of welfare and management on a unit. The definition as used in Real Welfare has been agreed by vets to describe a level of lameness of significant welfare concern with affected pigs experiencing pain and/or discomfort:

Record as lame: any pig that when standing will not bear full weight on the affected limb and/or appears to be standing on its toes. When moving, there is a shortened stride with minimum or no weight-bearing on the affected limb and a swagger of the hind quarters. The pig may still be able to trot and gallop.

Additionally, consider whether pig should also be recorded under hospital pig category. This protocol does not include pigs that are showing stiffness or uneven gait as this is more subjective and difficult to interpret (AHDB Pork).

Identifying that an outcome such as lameness is 'an issue' (i.e. abnormally high) on a unit is just the starting point – the aim of these programmes is not just to record the prevalence of each measure, but to investigate and address likely causes of the problem.

5.2 Play as a positive welfare indicator

In addition to looking for indicators of negative or poor welfare in pigs, it is possible to look for positive behaviours such as play as well (Lawrence, 1987). Play in pigs can be locomotor or object based, and can be social or isolated (Newberry et al., 1988). Careful observation is needed to separate object-based play from investigatory behaviour and social/locomotor behaviour from fighting. Play behaviour can include throwing objects in the air and holding an object whilst being chased, rather than nudging, rooting and biting which would be more investigatory. Social play in pigs can also be difficult to distinguish from fighting, for example, behaviours such as 'shove' may be difficult to classify in all situations (Newberry et al., 1988), and it is likely that behaviour which began as play may turn into more agonistic behaviour for example, if playing pigs disturb resting animals. Play behaviour in pigs can be accompanied by a range of identifiable vocalisations which help to characterise the behaviours (Reimert et al., 2013) – where observations do not include sound it will be extremely difficult to identify all occasions of play. Social play in pigs can be instigated by head bobs, similar to play bows in dogs (Newberry et al., 1988). Studies on play in pigs by Wright (2012) concluded that the absence of play behaviour in indoor weaner and grower/finisher pigs could be evidence of poor welfare, (and presence of play behaviour in adult sows could be viewed as positive welfare) due to the prevalence of play behaviours in the different age groups (juvenile pigs in indoor housing would be expected to be observed playing in a functional observation window). As often noted, the presence of an observer at the pen will affect pigs' behaviour (Martin and Bateson, 2007). Projects comparing welfare outcome measures come up across several obstacles when trying to incorporate play as a welfare indicator, and it is generally not included in final protocols. One difficulty is in forming a clear definition of the behaviour that can be reliably recorded by numerous observers, given the subtle differences in between play and non-play, as well as practical difficulties in observing pigs within a commercial environment. The potentially short duration of the behaviour in finishers also meant that observation could be hit and miss. Presence of a novel observer could also affect pigs' likelihood of playing differentially in different environments.

5.3 Vocalisations as a welfare indicator

Vocalisation by pigs is now being recognised as a useful indicator of their behaviour and welfare, with automated systems being developed (e.g. Schön et al., 2004; Vandermeulen et al., 2015). High pitched vocalisations can be an indication of pain, distress or fear, for example, pigs being squashed due to huddling behaviour in low temperatures (Hillmann et al., 2004), competition and aggression when/if feed becomes limited or tail-biting incidence (Vandermeulen et al., 2015). Pig buildings without any stressors are likely to be characterised by low-pitched contentment grunts and few/no squeals. Friel et al. (2016) demonstrated that pigs from more stressed environments had lower rates of grunting than those from more enriched environments. Grandin (2010) used pig squeals when handled (at abattoir) as an indication of welfare, with plants passing if less than 5% of pigs squealed during the handling process (a series of potentially stressful events). Barking vocalisations are usually the result of pigs being startled ('alarm calls' – Chan et al., 2011) – a commercial environment should aim to reduce extreme startling events, for example, with handlers making pigs aware that they have entered a building, following set routines and introducing novel objects with care.

These different vocalisations can be readily picked up with experience, but can be difficult to characterise for use by unfamiliar observers. McLeman et al. (2008) showed that juvenile pigs (growers/finishers) could distinguish between conspecifics (both familiar and unfamiliar) by use of auditory cues alone. Provision of an acoustic environment which infringes on pigs' ability to hear and communicate can be assumed to reduce their welfare. The DEFRA code of regulations for pigs (DEFRA, 2003) states that 'Pigs shall not be exposed to constant or sudden noise. Noise levels above 85 dBA shall be avoided in that part of any building where pigs are kept'.

6 Future trends

Tear staining (chromodacryorrhea) is the occurrence of dark staining below the inner corner of the eye. Whilst this has been considered a reliable indicator of stress in rats (Harkness and Ridgway, 1980) and considered as a general indicator of welfare in pigs relating to air quality and clinical disease (Gruber, 2002; Whay et al., 2007), it has only recently been investigated as a specific measure of stress in pigs. Tear staining in pigs has now been shown to correlate with stress in pigs under controlled conditions, and more recently in pigs under commercial conditions: tear staining scores correlated with tail and ear damage, and lower tear staining scores were found in groups housed with environmental enrichment objects rather than control groups (Telkänranta et al., 2016). Presence of high degree of tear staining in a population could therefore be a useful starting point to investigate potential stressors.

7 Conclusion

There are particular challenges to the welfare of finishing pigs due to their significant physiological and behavioural needs combined with the economic pressures that drive intensive production. Global issues, particularly population growth combined with a growing demand for meat products from an increasingly urban population (Satterthwaite et al., 2010), are a potential threat to animal welfare with the associated drive towards higher levels of intensive production. Extensive reports of these issues have resulted in the inclusion of the welfare of finisher pigs in major programmes, such as Welfare Quality and the EU PiG innovation group (part of the EU Horizon 2020 programme). The World Organisation for Animal Health (OIE) first published a work on international standards on animal welfare in 2005, with 2016 seeing the 25th edition of the OIE Terrestrial Animal Health Code that makes general recommendations for a range of livestock species (OIE, 2016a) – a specific chapter on pig welfare is currently in development (OIE, 2016b). Recent international advances have been reported in regions such as Asia, the Far East (Murray et al., 2014) including the first Compassion in World Farming Good Pig Production awards launched in China in 2014 (Compassion in World Farming, n.d.).

Researchers working with industries have increasingly focused on practical welfare assessment and animal-based outcomes, as exemplified by Welfare Quality and the Real Welfare and AssureWel schemes in the United Kingdom. However, perhaps the greatest potential for further improvement comes from technological advancement. Precision Livestock Farming (PLF) is the automated continuous monitoring of animals. PLF approaches may be used in future to monitor a range of indicators pertinent to pig welfare including weight gain (Lu et al., 2016), drinking behaviour (Kashiha et al., 2013), aggression (Oczak et al., 2014), lying patterns (Nasirahmadi et al., 2015), lameness (Stavrakakis et al., 2015) and mounting behaviour (Nasirahmadi et al., 2016). Improvements in genetic technologies may provide solutions to particular finisher pig welfare solutions including breeding animals that are less aggressive (D'Eath et al., 2009), have better leg conformation to reduce lameness (Lu et al., 2016) and are less susceptible to heat stress (Rauw et al., 2017).

With an historic emphasis on suffering negative welfare indicators are well established, but increasing attention is being given to the concept of positive welfare (Mendl and Paul, 2004; Boissy et al., 2007; Mellor, 2012). In an influential 2009 report, the Farm Animal Welfare Council suggested that farm animals should have a 'Life Worth Living' and that the 'balance of an animal's experiences must be positive over its lifetime, (Farm Animal Welfare Council, 2009). Whilst this might be arguable (can negative experiences really be mitigated in this way?), clearly robust measures of both negative and positive welfare

are needed to determine whether an animal does indeed have a life worth living. Future welfare codes are likely to include positive affective states as requirements alongside the minimisation of factors causing a reduction in well-being. Behaviours such as play and social interaction may provide useful indicators, but perhaps validation of positive indicators provides the main challenge in future pig welfare research.

8 Where to look for further information

Flooring and lameness:
http://pork.ahdb.org.uk/media/2113/Action-37-Flooring.pdf4

Outcomes approach to welfare assessment:
http://pork.ahdb.org.uk/health-welfare/welfare/real-welfare/
http://www.welfarequality.net/network/45848/7/0/40
http://www.assurewel.org/pigs

9 References

Almond, P. K. and G. Bilkei. 2006. Effects of oral vaccination against Lawsonia intracellularis on growing-finishing pig's performance in a pig production unit with endemic porcine proliferative enteropathy (PPE). *DTW Dtsch. Tierarztl. Wochenschr.* 113:232–5.

Amory, J. R., A. M. Mackenzie and G. P. Pearce. 2006. Factors in the housing environment of finisher pigs associated with the development of gastric ulcers. *Vet. Rec.* 158:260–4.

Arey, D. S. and J. M. Bruce. 1993. A note on the behaviour and performance of growing pigs provided with straw in a novel housing system. *Anim. Prod.* 56:269–72.

Backus, G., S. Støier, M. Courat, M. Bonneau and M. Higuera. 2014. First progress report from the European declaration on alternatives to surgical castration of pigs. Available from: https://ec.europa.eu/food/sites/food/files/animals/docs/aw_prac_farm_pigs_cast-alt_declaration_progress-report_20141028.pdf.

Baeckstroem, L., U. Wisconsin, T. Crenshaw and D. Shenkman. 1988. Gastric ulcers in swine: Effects of dietary fiber, corn particle size and preslaughter stress. In: *Rapport - Sveriges Lantbruksuniversitet, Veterinaermedicinska Fakulteten.* Institutionen foer Husdjurshygien, Sweden. Available from: http://agris.fao.org/agris-search/search.do?recordID=SE8811273.

Bakare, A. G., S. P. Ndou, and M. Chimonyo. 2013. Influence of physicochemical properties of fibrous diets on behavioural reactions of individually housed pigs. *Livest. Sci.* 157:527–34.

Ball, M. E. E., E. Magowan, K. J. McCracken, V. E. Beattie, R. Bradford, A. Thompson and F. J. Gordon. 2015. An investigation into the effect of dietary particle size and pelleting of diets for finishing pigs. *Livest. Sci.* 173:48–54.

Beattie, V. E. and N. E. O'Connell. 2002. Relationship between rooting behaviour and foraging in growing pigs. *Anim. Welf.* 11:295–303.

Beattie, V. E., K. Breuer, N. E. O'Connell, I. A. Sneddon, J. T. Mercer, K. A. Rance, M. E. M. Sutcliffe and S. A. Edwards. 2005. Factors identifying pigs predisposed to tail biting. *Anim. Sci.* 80:307–12.

Boissy, A., G. Manteuffel, M. B. Jensen, R. O. Moe, B. Spruijt, L. J. Keeling, C. Winckler, B. Forkman, I. Dimitrov, J. Langbein, M. Bakken, I. Veissier and A. Aubert. 2007. Assessment of positive emotions in animals to improve their welfare. *Physiol. Behav.* 92:375–97.

Breuer, K., M. E. M. Sutcliffe, J. T. Mercer, K. A. Rance, N. E. O'Connell, I. A. Sneddon and S. A. Edwards. 2005. Heritability of clinical tail-biting and its relation to performance traits. *Livest. Prod. Sci.* 93:87–94.

Broom, D. M. 1986. Indicators of poor welfare. *Br. Vet. J.* 142:524–6.

Brumm, V. L., D. Bilder and S. E. Waisbren. 2010. Psychiatric symptoms and disorders in phenylketonuria. *Mol. Genet. Metab.* 99(Suppl. 1):S59–63.

Brunberg, E., P. Jensen, A. Isaksson and L. J. Keeling. 2013. Behavioural and brain gene expression profiling in pigs during tail biting outbreaks – Evidence of a tail biting resistant phenotype. *PLoS ONE* 8:e66513.

Bulens, A., S. Van Beirendonck, J. Van Thielen, N. Buys and B. Driessen. 2016. Long-term effects of straw blocks in pens with finishing pigs and the interaction with boar type. *Appl. Anim. Behav. Sci.* 176:6–11.

Camerlink, I., S. P. Turner, W. W. Ursinus, I. Reimert and J. E. Bolhuis. 2014. Aggression and Affiliation during Social Conflict in Pigs. *PLoS ONE* 9:e113502.

Cappai, M. G., M. Picciau and W. Pinna. 2013. Ulcerogenic risk assessment of diets for pigs in relation to gastric lesion prevalence. *BMC Vet. Res.* 9:36. doi:10.1186/1746-6148-9-36.

Chan, W. Y., S. Cloutier and R. C. Newberry. 2011. Barking pigs: Differences in acoustic morphology predict juvenile responses to alarm calls. *Anim. Behav.* 82:767–74.

Clark, C. C. A. and R. B. D'Eath. 2013. Age over experience: Consistency of aggression and mounting behaviour in male and female pigs. *Appl. Anim. Behav. Sci.* 147:81–93.

Compassion in World Farming. 2011. Providing enrichment for pigs. Available from: https://www.ciwf.org.uk/media/3818880/providing-enrichment-for-pigs.pdf.

Compassion in World Farming. nd. China Awards | Compassion in Food Business. Available from: https://www.compassioninfoodbusiness.com/awards/china-awards/.

Davies, P. R., J. J. Grass, and W. E. (University of M. (USA) C. of V. M. Marsh. 1994. Time of slaughter affects prevalence of lesions of the pars oesophagea of pigs. Available from: http://agris.fao.org/agris-search/search.do?recordID=TH2000001484.

Day, J. E. L., I. Kyriazakis and A. B. Lawrence. 1995. The effect of food deprivation on the expression of foraging and exploratory behaviour in the growing pig. *Appl. Anim. Behav. Sci.* 42:193–206.

Day, J. E. L., I. Kyriazakis and A. B. Lawrence. 1996. The use of a second-order schedule to assess the effect of food bulk on the feeding motivation of growing pigs. *Anim. Sci.* 63:447–55.

Day, J. E. L., A. Burfoot, C. M. Docking, X. Whittaker, H. A. M. Spoolder and S. A. Edwards. 2002. The effects of prior experience of straw and the level of straw provision on the behaviour of growing pigs. *Appl. Anim. Behav. Sci.* 76:189–202.

D'Eath, R. B., R. Roehe, S. P. Turner, S. H. Ison, M. Farish, M. C. Jack and A. B. Lawrence. 2009. Genetics of animal temperament: Aggressive behaviour at mixing is genetically associated with the response to handling in pigs. *Animal* 3:1544–54.

DEFRA. 2003. *Code of Recommendations for the Welfare of Livestock: Pigs.* DEFRA Publications, London.

Denton, D. A., J. R. Blair-West, M. I. McBurnie, J. A. P. Miller, R. S. Weisinger and R. M. Williams. 1999. Effect of adrenocorticotrophic hormone on sodium appetite in mice. *Am. J. Physiol. - Regul. Integr. Comp. Physiol.* 277:R1033–40.

de Greef, K. H., H. M. Vermeer, H. W. J. Houwers and A. P. Bos. 2011. Proof of Principle of the Comfort Class concept in pigs: Experimenting in the midst of a stakeholder process on pig welfare. *Livest. Sci.* 139:172–85.

Dybkjær, L., L. G. Paisley, L. Vraa-Andersen and G. Christensen. 1998. Associations between behavioural indicators of `stress' in weaner pigs and respiratory lesions at slaughter. *Prev. Vet. Med.* 34:175–90.

Edwards, S. A. 2006. Tail biting in pigs: Understanding the intractable problem. *Vet. J.* 171:198–9.

Edwards, S. A., A. W. Armsby and H. H. Spechter. 1988. Effects of floor area allowance on performance of growing pigs kept on fully slatted floors. *Anim. Sci.* 46:453–9.

EFSA Panel on Animal Health and Welfare. 2014. Scientific Opinion concerning a Multifactorial approach on the use of animal and non-animal-based measures to assess the welfare of pigs. *EFSA J.* 12:3702.

Escribano, D., A. M. Gutiérrez, F. Tecles and J. J. Cerón. 2015. Changes in saliva biomarkers of stress and immunity in domestic pigs exposed to a psychosocial stressor. *Res. Vet. Sci.* 102:38–44.

European Food Safety Authority. 2007. Opinion of the Scientific Panel on Animal Health and Welfare on a request from the Commission related to animal health and welfare in fattening pigs in relation to housing and husbandry. *EFSA J.* 5:564.

European Food Safety Authority (EFSA). 2005. Opinion of the Scientific Panel on Animal Health and Welfare (AHAW) on a request from the Commission related to welfare of weaners and rearing pigs: Effects of different space allowances and floor. *EFSA J.* 3:585.

Farm Animal Welfare Council. 2009. Farm animal welfare in Great Britain: Past, present and future. Available from: https://www.gov.uk/government/uploads/system/uploads/attachment_data/file/319292/Farm_Animal_Welfare_in_Great_Britain_-_Past__Present_and_Future.pdf.

Fraser, D., P. A. Phillips, B. K. Thompson and T. Tennessen. 1991. Effect of straw on the behaviour of growing pigs. *Appl. Anim. Behav. Sci.* 30:307–18.

Fredriksen, B., B. M. Lium, C. H. Marka, B. Mosveen and O. Nafstad. 2008. Entire male pigs in farrow-to-finish pens–Effects on animal welfare. *Appl. Anim. Behav. Sci.* 110:258–68.

Friel, M., H. P. Kunc, K. Griffin, L. Asher and L. M. Collins. 2016. Acoustic signalling reflects personality in a social mammal. *R. Soc. Open Sci.* 3:160178.

Gilbert, M., G. Conchedda, T. P. V. Boeckel, G. Cinardi, C. Linard, G. Nicolas, W. Thanapongtharm, L. D'Aietti, W. Wint, S. H. Newman and T. P. Robinson. 2015. Income disparities and the global distribution of intensively farmed chicken and pigs. *PLoS ONE* 10:e0133381.

Gonyou, H. W., M. C. Brumm, E. Bush, J. Deen, S. A. Edwards, T. Fangman, J. J. McGlone, M. Meunier-Salaun, R. B. Morrison, H. Spoolder, P. L. Sundberg and A. K. Johnson. 2006. Application of broken-line analysis to assess floor space requirements of nursery and grower-finisher pigs expressed on an allometric basis. *J. Anim. Sci.* 84:229–35.

Grandin, T. 1989. Effect of rearing environment and environmental enrichment on behavior and neural development in young pigs. Available from: http://agris.fao.org/agris-search/search.do?recordID=US201300166011.

Grandin, T. 2010. Auditing animal welfare at slaughter plants. *Meat Sci.* 86:56–65.

Gruber, T. 2002. Aufstallung, Fütterung, Hygiene, Gesundheit und Management von Mastschweinen in biologisch bewirtschafteten Betrieben. PhD thesis, University of Veterinary Medicine, Vienna.

Haesebrouck, F., F. Pasmans, B. Flahou, K. Chiers, M. Baele, T. Meyns, A. Decostere and R. Ducatelle. 2009. Gastric helicobacters in domestic animals and nonhuman primates and their significance for human health. *Clin. Microbiol. Rev.* 22:202–23.

Harkness, J. and M. Ridgway. 1980. Chromodacryorrhea in laboratory rats (Rattus norvegicus): Etiologic considerations. *Lab. Anim. Sci.* 30:841–4.

Harley, S., S. J. More, N. E. O'Connell, A. Hanlon, D. Teixeira and L. Boyle. 2012. Evaluating the prevalence of tail biting and carcase condemnations in slaughter pigs in the Republic and Northern Ireland, and the potential of abattoir meat inspection as a welfare surveillance tool. *Vet. Rec.* 171:621.

Hendriks, H. J. M., A. M. Van de Weerdhof and N. InfoMil. 1999. Dutch notes on BAT for pig and poultry intensive livestock farming. Minist. Hous. Spat. Plan. Environ. MANMF. Available from: http://files.gamta.lt/aaa/Tipk/tipk/4_kiti%20GPGB/37.pdf.

Hessing, M. J. C., A. M. Hagelsø, W. G. P. Schouten, P. R. Wiepkema and J. A. M. Van Beek. 1994. Individual behavioral and physiological strategies in pigs. *Physiol. Behav.* 55:39–46.

Hillmann, E., C. Mayer, P.-C. Schön, B. Puppe and L. Schrader. 2004. Vocalisation of domestic pigs (Sus scrofa domestica) as an indicator for their adaptation towards ambient temperatures. *ResearchGate* 89:195–206.

Hintze, S., D. Scott, S. Turner, S. L. Meddle and R. B. D'Eath. 2013. Mounting behaviour in finishing pigs: Stable individual differences are not due to dominance or stage of sexual development. *Appl. Anim. Behav. Sci.* 147:69–80.

Jankevicius, M. L. and T. M. Widowski. 2003. Exogenous adrenocorticotrophic hormone does not elicit a salt appetite in growing pigs. *Physiol. Behav.* 78:277–84. doi:10.1016/S0031-9384(02)00970-8.

Jensen, P. 2002. *The Ethology of Domestic Animals: An Introductory Text.* CABI Pub, Wallingford.

Kashiha, M., C. Bahr, S. A. Haredasht, S. Ott, C. P. H. Moons, T. A. Niewold, F. O. Ödberg and D. Berckmans. 2013. The automatic monitoring of pigs water use by cameras. *Comput. Electron. Agric.* 90:164–9.

Kubo, K., M. Iinuma and H. Chen. 2015. Mastication as a stress-coping behavior. *BioMed Res. Int.* 2015:11. doi:10.1155/2015/876409. Available from: http://www.ncbi.nlm.nih.gov/pmc/articles/PMC4450283/.

Kyriazakis, I., G. C. Emmans and C. T. Whittemore. 1991. The ability of pigs to control their protein intake when fed in three different ways. *Physiol. Behav.* 50:1197–203.

Lawrence, A. 1987. Consumer demand theory and the assessment of animal welfare. *ResearchGate* 35:293–5.

Lawrence, B. V., D. B. Anderson, O. Adeola and T. R. Cline. 1998. Changes in pars esophageal tissue appearance of the porcine stomach in response to transportation, feed deprivation, and diet composition. *J. Anim. Sci.* 76:788–95.

Lee, P. J., D. Ridout, J. H. Walter and F. Cockburn. 2005. Maternal phenylketonuria: Report from the United Kingdom Registry 1978–97. *Arch. Dis. Child.* 90:143–6.

Leeuw, J. A. de, J. E. Bolhuis, G. Bosch and W. J. J. Gerrits. 2008. Effects of dietary fibre on behaviour and satiety in pigs: Symposium on 'Behavioural nutrition and energy balance in the young'. *Proc. Nutr. Soc.* 67:334–42.

Lu, M., Y. Xiong, K. Li, L. Liu, L. Yan, Y. Ding, X. Lin, X. Yang and M. Shen. 2016. An automatic splitting method for the adhesive piglets' gray scale image based on the ellipse shape feature. *Comput. Electron. Agric.* 120:53–62.

Lyons, C. A. P., J. M. Bruce, V. R. Fowler and P. R. English. 1995. A comparison of productivity and welfare of growing pigs in four intensive systems. *Livest. Prod. Sci.* 43:265–74.

Martin, P. and P. Bateson. 2007. *Measuring Behaviour: An Introductory Guide.* Updated edition edition. Cambridge University Press, Cambridge and New York.

McLeman, M. A., M. T. Mendl, R. B. Jones and C. M. Wathes. 2008. Social discrimination of familiar conspecifics by juvenile pigs, Sus scrofa: Development of a non-invasive method to study the transmission of unimodal and bimodal cues between live stimuli. *Appl. Anim. Behav. Sci.* 115:123–37.

Mellor, D. J. 2012. Animal emotions, behaviour and the promotion of positive welfare states. *N. Z. Vet. J.* 60:1–8.

Mendl, M. and E. Paul. 2004. Consciousness, emotion and animal welfare: Insights from cognitive science. *Anim. Welf.* 13:17–25.

Millet, S., S. Kumar, J. De Boever, R. Ducatelle and D. De Brabander. 2012a. Effect of feed processing on growth performance and gastric mucosa integrity in pigs from weaning until slaughter. *Anim. Feed Sci. Technol.* 175:175–81.

Millet, S., S. Kumar, J. De Boever, T. Meyns, M. Aluwé, D. De Brabander and R. Ducatelle. 2012b. Effect of particle size distribution and dietary crude fibre content on growth performance and gastric mucosa integrity of growing–finishing pigs. *Vet. J.* 192:316–21.

Mills, D. S. and C. A. Macleod. 2002. The response of crib-biting and windsucking in horses to dietary supplementation with an antacid mixture. *Ippologia* 13:33–41.

Moinard, C., M. Mendl, C. J. Nicol and L. E. Green. 2003. A case control study of on-farm risk factors for tail biting in pigs. *Appl. Anim. Behav. Sci.* 81:333–55. doi:10.1016/S0168-1591(02)00276-9.

Moyle, J. J., A. M. Fox, M. Arthur, M. Bynevelt and J. R. Burnett. 2007. Meta-analysis of neuropsychological symptoms of adolescents and adults with PKU. *Neuropsychol. Rev.* 17:91–101.

Mullan, S. 2009. An evaluation of including some welfare outcome measures within UK pig farm assurance schemes [Ph.D.]. University of Bristol. Available from: http://ethos.bl.uk/OrderDetails.do?uin=uk.bl.ethos.503854.

Mullan, S., W. J. Browne, S. A. Edwards, A. Butterworth, H. R. Whay and D. C. J. Main. 2009. The effect of sampling strategy on the estimated prevalence of welfare outcome measures on finishing pig farms. *Appl. Anim. Behav. Sci.* 119:39–48.

Mullan, S., S. A. Edwards, A. Butterworth, H. R. Whay and D. C. J. Main. 2011. Inter-observer reliability testing of pig welfare outcome measures proposed for inclusion within farm assurance schemes. *Vet. J.* 190:e100–9.

Munsterhjelm, C., O. A. T. Peltoniemi, M. Heinonen, O. Hälli, M. Karhapää and A. Valros. 2009. Experience of moderate bedding affects behaviour of growing pigs. *Appl. Anim. Behav. Sci.* 118:42–53.

Murray, G., K. Ashley and R. Kolesar. 2014. Drivers for animal welfare policies in Asia, the Far East and Oceania. *Rev. Sci. Tech. Int. Off. Epizoot.* 33:77–83.

NADIS. 2002. Pig Health - Tail Biting. Available from: http://www.nadis.org.uk/bulletins/tail-biting.aspx.

NADIS. 2003. Pig Health - Sunburn and Heatstroke/Heatstress. Available from: http://www.nadis.org.uk/bulletins/sunburn-and-heatstrokeheatstress.aspx?altTemplate=PDF.

Nasirahmadi, A., U. Richter, O. Hensel, S. Edwards and B. Sturm. 2015. Using machine vision for investigation of changes in pig group lying patterns. *Comput. Electron. Agric.* 119:184–90.

Nasirahmadi, A., O. Hensel, S. A. Edwards and B. Sturm. 2016. Automatic detection of mounting behaviours among pigs using image analysis. *Comput. Electron. Agric.* 124:295–302.

Newberry, R. C., D. G. M. Wood-Gush and J. W. Hall. 1988. Playful behaviour of piglets. *Behav. Processes.* 17:205–16.

Nielsen, E. K. and K. L. Ingvartsen. 2000. Effect of cereal type, disintegration method and pelleting on stomach content, weight and ulcers and performance in growing pigs. *Livest. Prod. Sci.* 66:271–82.

O'Connell, N. E., V. E. Beattie and R. N. Weatherup. 2004. Influence of group size during the post-weaning period on the performance and behaviour of pigs. *Livest. Prod. Sci.* 86:225–32.

Oczak, M., S. Viazzi, G. Ismayilova, L. T. Sonoda, N. Roulston, M. Fels, C. Bahr, J. Hartung, M. Guarino, D. Berckmans and E. Vranken. 2014. Classification of aggressive behaviour in pigs by activity index and multilayer feed forward neural network. *Biosyst. Eng.* 119:89–97.

OIE. 2016a. *Terrestrial Animal Health Code.* World Organization for Animal Health (OIE), Paris. Available from: http://www.oie.int/en/international-standard-setting/terrestrial-code/access-online/

OIE. 2016b. Report of the OIE ad hoc group on animal welfare and pig production systems. Available from: http://www.oie.int/fileadmin/SST/adhocreports/Animal%20welfare%20and%20pig%20production%20systems/AN/A_AW_Pig_Production_Systems_March_2016.pdf.

Omotosho, O. O., B. O. Emikpe, O. T. Lasisi, and T. A. Jarikre. 2016. Prevalence, distribution and pattern of gastric lesions in slaughtered pigs in south-western Nigeria. *Onderstepoort J. Vet. Res.* 83:6.

Pedersen, L. J., L. Holm, M. B. Jensen and E. Jørgensen. 2005. The strength of pigs' preferences for different rooting materials measured using concurrent schedules of reinforcement. *Appl. Anim. Behav. Sci.* 94:31–48.

Peet, B. nd. Using pig behaviour to optimize pen design. *Pig Site.* Available from: http://www.thepigsite.com/articles/928/using-pig-behaviour-to-optimize-pen-design/ (Accessed at 1 August 2017)

Popova, N. K. 2006. From genes to aggressive behavior: The role of serotonergic system. *BioEssays.* 28:495–503.

Potkins, Z. V., T. L. Lawrence and J. R. Thomlinson. 1989. Oesophagogastric parakeratosis in the growing pig: Effects of the physical form of barley-based diets and added fibre. *Res. Vet. Sci.* 47:60–7.

Puppe, B. 1998. Effects of familiarity and relatedness on agonistic pair relationships in newly mixed domestic pigs. *Appl. Anim. Behav. Sci.* 58:233–9.

Ramis, G., S. Gómez, F. Pallarés and A. Muñoz. 2005. Comparison of the severity of esophagogastric, lung and limb lesions at slaughter in pigs reared under standard and enriched conditions. *Anim. Welf.* 14:27–34.

Rasmussen, D. K., R. Weber and B. Wechsler. 2006. Effects of animal/feeding-place ratio on the behaviour and performance of fattening pigs fed via sensor-controlled liquid feeding. *Appl. Anim. Behav. Sci.* 98:45–53.

Rauw, W. M., E. J. Mayorga, S. Lei, J. C. M. Dekkers, J. F. Patience, N. K. Gabler, S. M. Lonergan and L. H. Baumgard. 2017. Effects of genetics on thermal regulatory responses to repeated heat stress exposure in pigs. *J. Anim. Sci.* 95:4–4.

Reimert, I., J. E. Bolhuis, B. Kemp and T. B. Rodenburg. 2013. Indicators of positive and negative emotions and emotional contagion in pigs. *Physiol. Behav.* 109:42–50.

Robertson, I. D., J. M. Accioly, K. M. Moore, S. J. Driesen, D. W. Pethick and D. J. Hampson. 2002. Risk factors for gastric ulcers in Australian pigs at slaughter. *Prev. Vet. Med.* 53:293–303.

Ruiterkamp, W. A. 1985. Het gedrag van mestvarkens in relatie tot huisvesting. Available from: https://bib.vetmed.fu-berlin.de/ResourceList/details/201962

Rydhmer, L., G. Zamaratskaia, H. K. Andersson, B. Algers, R. Guillemet and K. Lundström. 2006. Aggressive and sexual behaviour of growing and finishing pigs reared in groups, without castration. *Acta Agric. Scand. Sect. — Anim. Sci.* 56:109–19.

Rydhmer, L., K. Lundström and K. Andersson. 2010. Immunocastration reduces aggressive and sexual behaviour in male pigs. *Animal* 4:965–72.

Satterthwaite, D., G. McGranahan and C. Tacoli. 2010. Urbanization and its implications for food and farming. *Philos. Trans. R. Soc. B Biol. Sci.* 365:2809–20.

Schaefer, A. L., M. O. Salomons, A. K. W. Tong, A. P. Sather, and P. Lepage. 1990. The effect of environment enrichment on aggression in newly weaned pigs. *Appl. Anim. Behav. Sci.* 27:41–52.

Schmolke, S. A., Y. Z. Li and H. W. Gonyou. 2004. Effects of group size on social behavior following regrouping of growing–finishing pigs. *Appl. Anim. Behav. Sci.* 88:27–38.

Schön, P., B. Puppe and G. Manteuffel. 2004. Automated recording of stress vocalisations as a tool to document impaired welfare in pigs. *Anim. Welf.* 13:105–10.

Schrøder-Petersen, D. L., and H. B. Simonsen. 2001. Tail Biting in Pigs. *Vet. J.* 162:196–210.

Scott, K., D. J. Chennells, F. M. Campbell, B. Hunt, D. Armstrong, L. Taylor, B. P. Gill and S. A. Edwards. 2006. The welfare of finishing pigs in two contrasting housing systems: Fully-slatted versus straw-bedded accommodation. *Livest. Sci.* 103:104–15.

Spoolder, H. a. M., S. A. Edwards and S. Corning. 1999. Effects of group size and feeder space allowance on welfare in finishing pigs. *Anim. Sci.* 69:481–9.

Statham, P., L. Green and M. Mendl. 2011. A longitudinal study of the effects of providing straw at different stages of life on tail-biting and other behaviour in commercially housed pigs. *Appl. Anim. Behav. Sci.* 134:100–8.

Stavrakakis, S. 2014. *Biomechanical Studies of Locomotion in Pigs.* Newcastle University, Newcastle. Available from: https://theses.ncl.ac.uk/dspace/handle/10443/2510.

Stavrakakis, S., J. H. Guy, O. M. E. Warlow, G. R. Johnson and S. A. Edwards. 2014. Walking kinematics of growing pigs associated with differences in musculoskeletal conformation, subjective gait score and osteochondrosis. *Livest. Sci.* 165:104–13.

Stavrakakis, S., W. Li, J. H. Guy, G. Morgan, G. Ushaw, G. R. Johnson and S. A. Edwards. 2015. Validity of the Microsoft Kinect sensor for assessment of normal walking patterns in pigs. *Comput. Electron. Agric.* 117:1–7.

Stolba, A. and D. G. M. Wood-Gush. 1989. The behaviour of pigs in a semi-natural environment. *Anim. Sci.* 48:419–25.

Stookey, J. M. and H. W. Gonyou. 1994. The effects of regrouping on behavioral and production parameters in finishing swine. *J. Anim. Sci.* 72:2804–11.

Studnitz, M., M. B. Jensen, and L. J. Pedersen. 2007. Why do pigs root and in what will they root?: A review on the exploratory behaviour of pigs in relation to environmental enrichment. *Appl. Anim. Behav. Sci.* 107:183–97.

Stukenborg, A., I. Traulsen, B. Puppe, U. Presuhn and J. Krieter. 2011. Agonistic behaviour after mixing in pigs under commercial farm conditions. *Appl. Anim. Behav. Sci.* 129:28–35.

Swaby, H. and N. G. Gregory. 2012. A note on the frequency of gastric ulcers detected during post-mortem examination at a pig abattoir. *Meat Sci.* 90:269–71. doi:10.1016/j.meatsci.2011.06.015.

Taylor, N. R., D. C. J. Main, M. Mendl and S. A. Edwards. 2010. Tail-biting: A new perspective. *Vet. J.* 186:137–47.

Taylor, N. R., R. M. A. Parker, M. Mendl, S. A. Edwards and D. C. J. Main. 2012. Prevalence of risk factors for tail biting on commercial farms and intervention strategies. *Vet. J.* 194:77–83.

Telkänranta, H., J. N. Marchant-Forde and A. E. Valros. 2016. Tear staining in pigs: A potential tool for welfare assessment on commercial farms. *Anim. Int. J. Anim. Biosci.* 10:318–25.

The Pig Site. nd. Salt poisoning, water deprivation. Pig Site. Available from: http://www.thepigsite.com/pighealth/article/525/salt-poisoning-water-deprivation/ (Accessed at 1 August 2017).

Turner, S. P., D. J. Allcroft and S. A. Edwards. 2003. Housing pigs in large social groups: A review of implications for performance and other economic traits. *Livest. Prod. Sci.* 82:39–51.

Turner, S. P., S. A. Edwards and V. C. Bland. 1999. The influence of drinker allocation and group size on the drinking behaviour, welfare and production of growing pigs. *Anim. Sci.* 68:617–24.

Turner, S. P., M. J. Farnworth, I. M. S. White, S. Brotherstone, M. Mendl, P. Knap, P. Penny and A. B. Lawrence. 2006. The accumulation of skin lesions and their use as a predictor of individual aggressiveness in pigs. *Appl. Anim. Behav. Sci.* 96:245–59.

Turner, S. P., G. W. Horgan and S. A. Edwards. 2001. Effect of social group size on aggressive behaviour between unacquainted domestic pigs. *Appl. Anim. Behav. Sci.* 74:203–15.

Van de Weerd, H. A., C. M. Docking, J. E. L. Day, P. J. Avery and S. A. Edwards. 2003. A systematic approach towards developing environmental enrichment for pigs. *Appl. Anim. Behav. Sci.* 84:101–18.

van de Weerd, H. A. and J. E. L. Day. 2009. A review of environmental enrichment for pigs housed in intensive housing systems. *Appl. Anim. Behav. Sci.* 116:1–20.

van den Berg, A., F. Brülisauer and G. Regula. 2005. Prävalenz von Veränderungen der kutanen Magenschleimhaut bei Schlachtschweinen in der Schweiz. *Schweiz. Arch. Für Tierheilkd.* 147:297–303.

Vandermeulen, J., C. Bahr, E. Tullo, I. Fontana, S. Ott, M. Kashiha, M. Guarino, C. P. H. Moons, F. a. M. Tuyttens, T. A. Niewold and D. Berckmans. 2015. Discerning pig screams in production environments. *PLoS ONE.* 10:e0123111.

Vermeer, H. M., K. H. de Greef and H. W. J. Houwers. 2014. Space allowance and pen size affect welfare indicators and performance of growing pigs under Comfort Class conditions. *Livest. Sci.* 159:79–86.

Webb, N. G. 1984. Compressive stress on, and the strength of, the inner and outer digits of pigs' feet, and the implications for injury and floor design. *J. Agric. Eng. Res.* 30:71–80. Available from: http://agris.fao.org/agris-search/search.do?recordID=US201302564478.

Webb, N. G. and C. Nilsson. 1983. Flooring and injury – an overview. In: Baxter, S. H., Baxter, M. R. and McCormick, J. A.C. (Eds), *Farm Animal Housing and Welfare.* Nijhoff, The Hague, pp. 226–59.

Whay, H., C. Leeb, D. Main, L. Green and A. Webster. 2007. Preliminary assessment of finishing pig welfare using animal-based measurements. *Anim. Welf.* 16:209–11.

Wondra, K. J., J. D. Hancock, K. C. Behnke and R. H. Hines. 1995a. Effects of dietary buffers on growth performance, nutrient digestibility, and stomach morphology in finishing pigs. *J. Anim. Sci.* 73:414–20.

Wondra, K. J., J. D. Hancock, K. C. Behnke, R. H. Hines and C. R. Stark. 1995b. Effects of particle size and pelleting on growth performance, nutrient digestibility, and stomach morphology in finishing pigs. *J. Anim. Sci.* 73:757–63.

Woods, A. 2012. Rethinking the History of Modern Agriculture: British Pig Production, c.1910–65. *Twent. Century Br. Hist.* 23:165–91.
Wright, A. J. 2012. *Animal Welfare Assessment in Veterinary Education: Its Theory and Practical Application to Domestic Pigs (Sus scrofa domestica)*. University of London, London.
Yeates, J. W. and D. C. J. Main. 2008. Assessment of positive welfare: A review. *Vet. J.* 175:293–300.
Zijlstra, R. T. and E. Beltranena. 2013. Swine convert co-products from food and biofuel industries into animal protein for food. *Anim. Front.* 3:48–53.

Chapter 11

Transport and lairage of pigs

Jennifer M. Young, North Dakota State University, USA

1 Introduction

Pre-slaughter events of movement from finishing pen, loading, transportation, unloading and lairage are all opportunities to introduce stress to a pig. These events have all been identified as high-stress events for hogs (Meisinger, 2002), with transportation having been identified as the most stressful based on salivary cortisol concentrations (Geverink et al., 1998). Bradshaw et al. (1996) also reported elevated levels of cortisol, measured in the blood instead of saliva, as a result of loading and transportation.

While pre-slaughter treatment is important from an animal welfare standpoint, it can also be influential on pork quality. The reduction in pork quality due to pre-slaughter stress has been shown to be mostly influenced by the alteration in the rate and extent of pH decline post-mortem, which is related to muscle glycolytic potential (Hambrecht et al., 2005). In 1992, the incidence of pale, soft and exudative (PSE) pork at the packing plant was estimated at 10.2% (Cannon et al., 1996). By 2003, this estimate had increased to 15.5% (Person et al., 2005) and PSE pork and other major quality defects, primarily meat exudation and unacceptable colour, have been estimated to cost the pork industry over US$200 million annually (Faucitano, 2010). Edwards et al. (2010) showed increased serum lactate concentrations at moments of high pre-slaughter stress which have been shown to be correlated to increased drip loss percentage (Hambrecht et al., 2004, 2005).

Stress during pre-slaughter events has also been shown to be associated with dead-on-arrival (DOA) and non-ambulatory pigs (Schwartzkopf-Genswein et al., 2012) which have been estimated to cost the US pork industry US$50 to 100 million annually

http://dx.doi.org/10.19103/AS.2017.0013.26

(Fitzgerald et al., 2009) and the Canadian pork industry more than US$3 million annually. The design of loading facilities and method of handling at the farm; trailer design and microclimate; stocking density; the duration and distance of transportation; duration and design of lairage; and season can all affect the amount of stress to which a pig is exposed before slaughter. As a result, extensive research has been done on the impact of on-farm handling, transportation and lairage on animal welfare and pork quality.

2 On-farm loading facilities and handling

The design of the loading facilities and the method of handling at individual farms can vary greatly and have differing effects on animal welfare and pork quality. Variation in animal losses and stress response between differing farms of origin have been attributed in part to differences in barn design and in the handling of pigs as they are moved from the finishing pen to the truck (Ritter et al., 2008; Fitzgerald et al., 2009; Correa et al., 2010; Garcia and McGlone, 2015). Fitzgerald et al. (2009) reported a 0.93% spread in total losses (DOA and non-ambulatory) between the poorest and best performing farms. Sort within a barn played a significant role in the percentage of DOA or non-ambulatory pigs with pigs from the first sort averaging 0.51% and 0.78%, respectively, fewer losses than the second and third sorts (Fitzgerald et al., 2009).

Ritter et al. (2008) found a significant effect of distance (<24 m vs. 47–67 m) pigs were moved during loading on the percentage of pigs displaying signs of fatigue (open-mouth breathing, skin discolouration or muscle tremors) with pigs travelling a greater distance having a larger percentage (17.7 vs. 7.6%). This confirmed previous research where pigs travelling a long distance (61.0–91.4 m) when compared to pigs travelling a short distance (0–30.5 m) had a greater percentage of fatigued pigs (28.8 vs. 13.7%). Ritter et al. (2007, 2008) also found that pigs moved a greater distance while being loaded had a tendency to have a greater percentage of non-ambulatory pigs due to injury at the plant (2007: 0.24 vs. 0.04%; 2008: 0.13 vs. 0.05%). However, Ritter et al. (2007, 2008) found no differences in non-ambulatory due to fatigue or DOA between distance treatments. The angle of the ramp and bedding used on the ramp also impacted animal welfare at loading (Garcia and McGlone, 2015). An increase in ramp slope resulted in longer loading times and increased heart rates (Garcia and McGlone, 2015). There was a season effect on bedding type; however, feed was among the shortest loading times and straw was among the longest loading times during both summer and winter while other bedding types (sand, wood shavings and none) had varying effects on loading times (Garcia and McGlone, 2015).

The method of handling pigs during movement has been shown to impact animal welfare and pork quality (Carr et al., 2008; Correa et al., 2010; D'Souza et al., 1998). Negative handling (using the electric prod) prior to loading for transportation to market was shown to increase the incidence of PSE pork (D'Souza et al., 1998). Carr et al. (2008) showed that those pigs loaded using conventional methods of loading (extensive handler vocalization, extensive use of board/paddle and goading once with electric prod) had greater pH at both 1 and 30 hours post-mortem, greater firmness scores, were darker and more red than pigs loaded more passively (no handler vocalization, minimum use of board/paddle and no use of electric prod). Correa et al. (2010) compared the use of a board in combination with either an electric prod or a paddle. Use of the electric prod resulted in a quicker loading time; however, pigs slipped, fell, and overlapped more often and had more and longer vocalizations and greater heart rates than those pigs loaded using the board and a paddle (Correa et al., 2010). Pigs loaded using the electric prod had greater serum lactate concentrations, semimembranosus muscles that were darker with a greater pH and an increase in blood-splashed hams than those loaded using the paddle (Correa et al., 2010). However, greater blood lactate concentrations after loading or greater change in blood lactate concentrations during loading were associated with increased 24 h pH, decreased L* and decreased drip loss which suggests that improvements in animal handling during loading would not necessarily translate to better pork quality (Edwards et al., 2010).

3 Transport of pigs

3.1 Season and temperature

Season and temperature have been found to have significant effects on animal welfare and pork quality in previous studies. Haley et al. (2008b) found that DOA percentage was expected to increase 1.26 times for each 1°C increase in temperature. The percentage of DOA was expected to increase threefold when moving from a temperature range of 8.6°C–23.3°C to a range of 23.4°C–26.1°C and increase an additional twofold when the temperature range increased from 26.2°C–28.9°C to a range of 29.0°C–30.5°C (Haley et al., 2008b). Sutherland et al. (2009) also found an increase in percentage of DOA when the temperatures rose above 20°C. No differences were found for temperature on the percentage of pigs injured on the trailer while temperatures below 5°C resulted in an increase in non-ambulatory pigs (Sutherland et al., 2009). When temperatures were greater than 17°C, open-mouth breathing and skin discolouration were more prevalent (Kephart et al., 2010).

Sutherland et al. (2009) found that the percentage of DOA was greatest in winter months than in the summer months despite an increase in DOA as the temperature increased. This is supported by Fitzgerald et al. (2009) who also found that losses were greater in November and December compared to June and July. Fitzgerald et al. (2009) explained this by citing that pig market weights tend to increase during winter months as well as the total number of pigs marketed which means the seasonal differences in death losses may be explained by seasonal differences in trailer densities. Fitzgerald et al. (2009) supported this by showing that the mean weekly death loss percentage per trailer followed the same curve as the mean weekly average pig body weight per trailer. Winter months had more non-ambulatory pigs than summer months while there was a slight trend for summer months to have an increased number of injured during transport than the winter months (Sutherland et al., 2009).

Different physiological and behavioural measures have been used to draw conclusions regarding stress. Correa et al. (2014) found that lactate levels were greater in pigs harvested in the winter than in pigs harvested in the summer. Their explanation for this was that pigs stood more in lairage in the winter and, therefore, could not recover from the stress of loading and transport as readily as those harvested in the summer. When comparing pigs marketed in February, August and November, Carr et al. (2008) found that pigs marketed in February had lower cortisol levels than those marketed in August or November. In pigs lairaged for only 45 minutes, no differences were found between month marketed for cortisol level; however, for pigs lairaged for 3 h, pigs marketed in August had the greatest serum cortisol levels while pigs marketed in February had the lowest serum cortisol levels (Carr et al., 2008). Pigs marketed in summer had the greatest percentage of pigs exhibiting open-mouth breathing and skin discolouration compared to pigs marketed in the winter (Ritter et al., 2008).

Variations in pork quality were found between seasons, particularly with defects in quality. Correa et al. (2014) found that pigs harvested in the summer had a greater percentage of carcasses with skin damage and fighting-type bruises than those pigs harvested in the winter with there being no seasonal effect on mounting or other types of bruises. Arduini et al. (2014) found that the percentage of hams with haematomas decreased significantly by season from winter (greatest percentage) to autumn (lowest percentage). The percentage of hams with lacerations was greatest in autumn with no difference between other seasons while the percentage of hams with haemorrhages was lowest in the summer and veining was lowest in summer and autumn (Arduini et al., 2014). Gajana et al. (2013) found that the percentage of carcasses classified as PSE was greatest in autumn and smallest in the winter and spring while the percentage of carcasses classified as DFD was greatest in winter and spring.

3.2 Trailer design and microclimate

Trailers are used to move pigs from the farm to slaughter and come in a variety of styles, ranging from a small single-deck trailer to large three-deck trailers (often referred to as pot-belly trailers, see Fig. 1). Semi-trailers are commonly used for transporting pigs as they can accommodate a large number of animals on a single journey. However, there are multiple microclimates within a pot-belly trailer that contribute to different loading and transport conditions for pigs and can have significant effects on animal welfare.

Pot-belly trailers take longer to load and unload than straight-deck trailers (Ritter et al., 2008; Kephart et al., 2010; Torrey et al., 2013a). There is also an increased use of electric prods when loading and unloading a pot-belly trailer compared to a straight-deck trailer. Kephart et al. (2010) also found an increase in the percentage of pigs exhibiting open-mouth breathing on pot-belly trailers. These observations have also been shown to correspond to an increase in number of DOA and non-ambulatory pigs transported on a pot-belly trailer compared to a straight-deck trailer (Kephart et al., 2010; Correa et al., 2009; Correa, 2011). Correspondingly, Correa et al. (2009) found an increase in blood

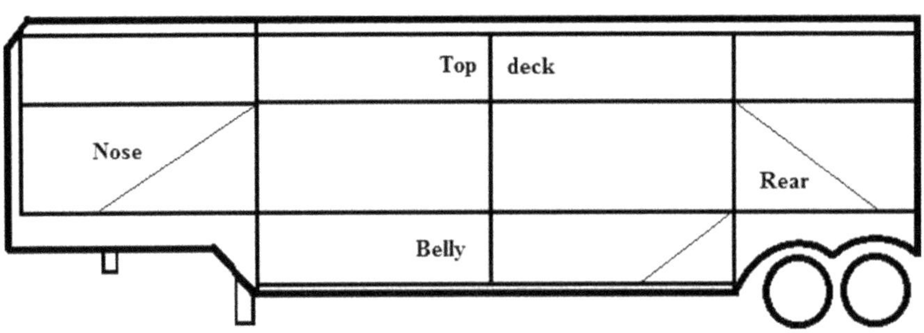

Figure 1 Example of semi-trailer.

creatine phosphokinase and lactate levels in those pigs transported on a pot-belly trailer compared to a straight-deck trailer. Serum cortisol concentration was also found to be greater in pigs transported on a pot-belly trailer compared to a straight-deck trailer (Weschenfelder et al., 2012).

Not only does the style of the trailer affect welfare but the location within the trailer. Kephart et al. (2010) found that most deaths occurred on the bottom deck regardless of trailer type (17 on bottom vs. 6 on top). However, this was in contrast to Ritter et al. (2006) who found no difference in DOA or non-ambulatory pigs between decks. Pigs which were in compartments where they had to navigate tight turns or ramps took longer to unload than those pigs in compartments where they could walk right off (Torrey et al., 2013a,b). Pigs loaded into the top rear compartment or the belly were found to have a greater heart rate than those pigs loaded into either the top or bottom front compartments (Goumon et al., 2013). When comparing the heart rate of pigs in the top deck (top four compartments), middle deck (rear two compartments on the middle level), belly (two compartments), and bottom nose compartment, Correa et al. (2014) found that pigs in the bottom nose compartment had the lowest heart rate compared to the other pigs when waiting at loading time during the summer with no significant differences in heart rate found at other times during transport. In the winter, there was no difference found in heart rate between compartments until the second half of the transport phase when pigs transported in the middle deck had the lowest heart rates (Correa et al., 2014). Lactate levels differed between pigs in each of the four subgroups studied with pigs in the bottom nose compartment having the greatest lactate levels and those in the middle deck having the lowest lactate levels (Correa et al., 2014).

Few studies have evaluated the effect of trailer type or location on pork quality. Correa et al. (2009) found a numerically, but not significant, increase in DFD loins when pigs were transported on a pot-belly compared to a straight-deck trailer when transported for 2 h. This was supported by results from Weschenfelder et al. (2012) who found no detrimental effects on pork quality when pigs were transported 7 h in a pot-belly trailer compared to a straight-deck trailer. In a study examining the effects of location within trailer, transportation duration and lairage duration, Newman et al. (2014) found no effect of location on ultimate pH or colour of the *longissimus dorsi* muscle. While Newman et al. (2014) found no difference in ultimate pH in the *longissimus dorsi*, Correa et al. (2014) reported differences in the *longissimus thoracis* and *semimembranosus* muscles. For the *longissimus thoracis* muscle, ultimate pH of carcasses from pigs transported on the upper deck or in the bottom nose compartment was lower than that of carcasses from pigs transported in the middle deck or in the belly while pH was greatest for carcasses from pigs transported in the bottom nose compartment with other compartments having similar pH for the *semimembranosus*. While drip loss percentages were similar between pigs transported on the top and bottom decks when marketed in spring and summer, pigs transported on the top deck tended to have lower drip loss percentages in winter and greater drip loss percentages in fall when compared to those transported on the bottom deck (Newman et al., 2014). Correa et al. (2014) found that drip loss percentage was greatest for those pigs that were transported on the upper deck supporting the fall data from Newman et al. (2014). Drip loss percentages were least for those pigs transported in the bottom nose with no differences being found between other compartments (Correa et al., 2014). These data show that the effects of trailer type and location within trailer vary greatly and are most likely dependent on a number of other factors such as genetics and season.

3.3 Stocking density

Several studies have examined the effect of stocking density on animal welfare and pork quality. As expected, Ritter et al. (2007) found no effect of floor space/pig on physical indicators of stress (open-mouth breathing, skin discolouration or muscle tremors) at loading. However, the number of pigs with skin discolouration at unloading decreased as floor space/pig increased, although the differences were relatively small (Ritter et al., 2007). Increasing floor space during transport from 0.39 to 0.48 m^2/pig corresponded to a decrease in the number of total non-ambulatory pigs by reducing the number of non-ambulatory, non-injured pigs with no differences found for number of non-ambulatory, injured pigs (Ritter et al., 2006). A tendency was observed for a decrease in floor space to correspond with an increase in percentage of dead pigs at the plant (Ritter et al., 2006). This is supported by findings of Fitzgerald et al. (2009) which showed that an increased trailer density resulted in greater death loss during transportation and lairage with a decrease in number of hogs per m^2, average body weight or both resulting in a decreased death loss.

Studies evaluating the effect of stocking density on pork quality have focused mainly on the incidence rates of PSE or DFD pork. Guàrdia et al. (2004) found that the risk of PSE pork increased when m^2 per 100 kg pig decreased when transportation time was greater than 3 h. In trips that lasted less than 3 h, the risk of PSE pork decreased when m^2 per 100 kg pig decreased (Guàrdia et al., 2004). Additionally, when m^2 per 100 kg pig increased, the risk of DFD pork increased (Guàrdia et al., 2005). Gajana et al. (2013) evaluated stocking densities of hogs (0.25–0.40 m^2/100 kg pig) and found that increased and decreased stocking densities resulted in a higher incidence of PSE meat with the lowest risk for PSE occurring when stocking density was 0.35 m^2/100 kg pig. Gajana et al. (2013) found that pH increased as space allotted increased.

3.4 Transport duration

The length of time hogs spend on the trailer between the farm and the slaughter facility impacts both animal welfare and pork quality. Pigs have been shown to experience travel sickness which may be associated with increased lysine vasopressin serum concentrations (Bradshaw et al., 1996). There appears to be an intermediate optimum for transport duration. Fitzgerald et al. (2009) found that hogs transported greater than 80 km had greater death loss than those transported fewer than 80 km. Those hogs transported between 40 and 80 km had the lowest percentage of death loss during transport or lairage (Fitzgerald et al., 2009). Pigs transported a longer time (6 h) compared to those transported a short duration (3 h) had greater serum cortisol concentrations (Newman et al., 2014). Additionally, pigs which were transported a short distance (<30 min) were found to be more difficult to handle at the plant, resulting in poorer quality pork (greater incidence of PSE), than pigs transported longer distances (Grandin, 1994). However, Gajana et al. (2013) found that pigs transported longer distances (>1.5 h) had an increased risk for PSE meat. The findings by Gajana et al. (2013) were supported by a decline in pH as transportation time increased. Arduini et al. (2014) evaluated defects (haematomas, laceration, haemorrhages and veining) in hams for four different transportation lengths (1 = 11–37 km; 2 = 38–86 km; 3 = 89–170 km; and 4 = 199–276 km). Hams from pigs transported in length 3 had the fewest number of defects in all categories while hams from pigs transported in length 4 had the greatest number of defects compared to hams from the other transportation lengths (Arduini et al., 2014). While Gajana et al. (2013) found that increased transportation duration resulted in an increased incidence of PSE and a decline in pH. Leheska et al. (2002) found that greater transportation durations resulted in an increased incidence of DFD pork. This is supported by Hambrecht et al. (2005) who found that pork from pigs in the long transport treatment had a greater ultimate pH than pork from pigs in the short transport treatment. Hambrecht et al. (2005) also found that longer transport times resulted in pork which was less red and less yellow than pork from pigs transported a shorter time. The increased ultimate pH could be related to the increased glycolytic potential due to increased lactate production with no difference in residual glycogen levels (Hambrecht et al., 2005). Pork from pigs transported longer also had decreased water-holding capacity as indicated by increased electrical conductivity values (Hambrecht et al., 2005). However, Newman et al. (2014) found an increased 24-h drip loss percentage in *longissimus dorsi* from pigs transported 3 h compared to those from pigs transported 3 h in the winter and fall but the opposite effect in the spring.

4 Pigs at the slaughter facility

4.1 At-plant handling

Increased stress at slaughter can result in poorer meat quality. High stress at slaughter resulted in increased serum concentrations of lactate, creatine phosphokinase, cortisol, epinephrine and norepinephrine, and decreased serum concentrations of glycogen (Hambrecht et al., 2004, 2005; Warriss et al., 1994). Greater serum lactate concentration levels post unloading at the plant were associated with an increase in 24 h pH and a decrease in drip loss (Edwards et al., 2010). Warriss et al. (1994) found that an increase in serum lactate concentrations as a result of a high-stress system corresponded to poorer meat quality and increased incidences of PSE and DFD pork. D'Souza et al. (1998) also found an increase in the incidence of PSE pork in a high stress situation. While ultimate pH was greater in pigs moved with electric prods (Hambrecht et al., 2004, 2005), 30 min pH was lower in pigs moved without the use of electric prods (Hambrecht et al., 2004). Pork from pigs moved with electric prods had poorer water-holding capacity as evidenced by greater electrical conductivity, greater filter-paper-measured moisture content and greater 24 h and 48 h drip loss percentages (Hambrecht et al., 2004, 2005). Longissimus muscles from pigs in the electric prod treatment were lighter (greater L*) than those from pigs in the calm treatment (Hambrecht et al., 2004). The rapid decline of pH, decreased water-holding capacity and lighter colour shown by Hambrecht et al. (2004) in pigs moved with the electric prods would be indicative of a greater incidence of PSE pork.

4.2 Lairage duration

The effect of lairage duration on animal welfare and pork quality varied and was dependent on season (Carr et al., 2008; Hambrecht et al., 2005; Newman et al., 2014). Decreasing the duration of lairage from 3 h to 45 min resulted in decreased serum cortisol levels in pigs marketed in August and November but resulted in increased serum cortisol levels in pigs marketed in February (Carr et al., 2008). In pigs marketed in the fall, serum cortisol concentrations were greater in pigs given 3 h of lairage compared to 6 h, while the opposite was true for pigs marketed in the summer (Newman et al., 2014). Decreasing the duration of lairage from 3 h to less than 45 min resulted in pork with decreased glycolytic potential and residual glycogen levels which was darker, less red and less yellow (Hambrecht et al., 2005). However, Newman et al. (2014) found no differences in colour during the fall, winter and spring but did find differences in the summer with pigs lairaged for 6 h having greater a*, b*, saturation (C*) and hue angle values than those lairaged for 3 h. Ultimate pH of the *longissimus dorsi* was greater in pigs given 3 h of lairage compared to 6 h in the fall while the opposite was true in the summer and no differences were found in spring and winter (Newman et al., 2014).

5 Implications for industry practices

Regulations for transportation and slaughter have been present in the United States for decades, starting with the Humane Methods of Slaughter Act of 1958, which focused on livestock slaughtered for sale to the government. In 1978, the Humane Methods of Slaughter Act was expanded to cover all livestock slaughtered in federally inspected meat plants. In 1991, the American Meat Institute Foundation (AMIF) first published recommendations for animal welfare guidelines for meat packing operations with multiple revisions since then (Grandin, 2013). Packer compliance with AMIF's guidelines has become part of many customer purchase agreements since 1999 (Grandin, 2013). The AMIF guidelines provide recommendations and audit forms for transportation and handling animals at the packing plants (Grandin, 2013).

While the AMIF guidelines are for all livestock with sections addressing specific species, the United States National Pork Board (NPB) created two programmes to guarantee the humane handling of swine. While the Pork Quality Assurance Plus (PQA) programme is for the entire lifetime of a pig including the relationship of the producer with a veterinarian, an entire section of the handbook (National Pork Board, 2015a) is dedicated to providing proper care when handling and transporting pig (Good Production Practice #6). While PQA addresses all stages of swine production, it does make note that improper handling leads to stressed pigs which can lead to several negative consequences such as physical injury to the pig or handler, increases in the incidence of non-ambulatory pigs and increased time to load/unload pigs. Additionally, improper handling and transport significantly contributes to carcass shrink, trim loss and poor meat quality resulting in it being one of the largest profit-reducing issues facing the pork industry (National Pork Board, 2015a). In addition to PQA, the NPB also launched the Transport Quality Assurance (TQA) programme in 2002. The focus of TQA is to provide the most current and science-based information to help pig producers, handlers and transporters define the best practices for handling, moving and transporting pigs and the potential impacts those actions can have on animal welfare and pork quality (National Pork Board, 2015b).

6 Conclusion

In conclusion, there are many factors during loading, transportation and lairage that impact animal welfare and pork quality. Stress before slaughter has been shown to have severe, negative impacts on pork quality characteristics, so care must be taken to reduce the amount of stress an animal is exposed to prior to slaughter. By reducing losses due to DOA/non-ambulatory pigs, bruising and other carcass defects, and PSE pork and other quality defects, the swine industry can increase pork production on the same number of pigs and work towards producing a more consistent quality product for the consumers.

7 Future trends

With increasing consumer interest in the source of food and concern for animal welfare, more research into the impact of transportation and lairage on animal welfare will need to be conducted. Additionally, as the pork industry continues to seek improvement in quality, the impact of pre-slaughter events on pork quality will continue to be investigated in order to ensure that pigs are transported and handled in such a way that maximizes good animal welfare and pork quality. Previous research has tended to focus on one specific set of genetics for each study. This may change in the future to determine if certain genetics are less susceptible to the stress of transportation and lairage, resulting in improved animal welfare and pork quality.

8 Where to look for further information

There are many good literature reviews on transportation (Ritter et al., 2009; Schwartzkopf-Genswein et al., 2012). There are handbooks that provide guidelines for practice (Grandin, 2013; NPB, 2015a,b). In addition, the majority of work on the effect of transportation on animal welfare and pork quality comes from a few key research groups as evidenced by the list of references. Any of these research groups would be great resources for more information and possible collaboration.

9 References

Arduini, A., Redaelli, V., Luzi, F., Sall'Olio, S., Pace, V. and Nanni Costa, L. (2014), 'Effect of transport distance and season on some defects of fresh hams destined for DPO production', *Animals*, 4(3), 524–34.

Bradshaw, R. H., Parrott, R. F., Forsling, M. L., Goode, J. A., Llloyd, D. M., Rodway, R. G. and Broom, D. M. (1996), 'Stress and travel sickness in pigs: effects of road transport on plasma concentrations of cortisol, beta-endorphin and lysine vasopressin', *Anim. Sci.*, 63(3), 507–16.

Cannon, J. E., Morgan, J. B., McKeith, F. K., Smith, G. C., Sonka, S., Heavner, J. and Meeker, D. L. (1996), 'Pork chain quality audit survey: quantification of pork quality characteristics', *J. Muscle Foods*, 7(1), 29–44.

Carr, C. C., Newman, D. J., Rentfrow, G. K., Keisler, D. H. and Berg, E. P. (2008), 'Effects of slaughter date, on-farm handling, transport stocking density, and time in lairage on digestive tract temperature, serum cortisol concentrations, and pork lean quality of market hogs', *Prof. Anim. Sci.*, 24(3), 208–18.

Correa, J. A., Gonyou, H., Widowski, T., Bergeron, R., Lewis, N., Crowe, T., Torrey, S., Tamminga, E. and Faucitano, L. (2009), 'Effects of vehicle type on transport losses, blood stress indicators and pork quality in pigs', *Can. J. Anim. Sci.*, 89(1), 151.

Correa, J. A., Gonyou, H., Torrey, S., Widowski, T., Bergeron, R., Crowe, T., Laforest, J.-P. and Faucitano, L. (2010), 'Effects of different moving devices at loading on stress response and meat quality in pigs', *J. Anim. Sci.*, 88(12), 4086–93.

Correa, J. A. (2011), 'Effects of farm handling and transport on physiological response, losses and meat quality of commercial pigs', *Adv. Pork Prod., Proc. 2011 Banff Pork Seminar*, 22, 249–56.

Correa, J. A., Gonyou, H., Torrey, S., Widowski, T., Bergeron, R., Crowe, T., Laforest, J.-P. and Faucitano, L. (2014), 'Welfare of pigs being transported over long distances using a pot-belly trailer during winter and summer', *Animals*, 4(2), 200–13.

D'Souza, D. N., Warner, R. D., Dunshea, F. R. and Leury, B. J. (1998), 'Effect of on-farm and pre-slaughter handling on meat quality', *Aust. J. Agric. Res.*, 49(6), 1021–5.

Edwards, L. N., Grandin, T., Engle, T. E., Ritter, M. J., Sosnicki, A. A., Carlson, B. A. and Anderson, D. B. (2010), 'The effects of pre-slaughter pig management from the farm to the processing plant on pork quality', *Meat Sci.*, 86(4), 938–44.

Fitzgerald, R. F., Stalder, K. J., Matthews, J. O., Schultz Kaster, C. M. and Johnson, A. K. (2009), 'Factors associated with fatigued, injured, and dead pig frequency during transport and lairage at a commercial abattoir', *J. Anim. Sci.*, 87(3), 1156–66.

Gajana, C. S., Nkukwana, T. T., Marume, U. and Muchenje. (2013), 'Effects of transportation time, distance, stocking density, temperature and lairage time on incidences of pale soft exudative (PSE) and the physico-chemical characteristics of pork', *Meat Sci.*, 95(3), 520–5.

Garcia, A. and McGlone, J. J. (2015), 'Loading and unloading finishing pigs: effects of bedding types, ramp angle, and bedding moisture', *Animals*, 5(1), 13–26.

Geverink, N. A., Kappers, A., van de Burgwal, J. A., Lambooij, E., Blokhuis, H. J. and Wiegant, V. M. (1998), 'Effects of regular moving and handling on the behavioral and physiological responses of pigs to preslaughter treatment and consequences for subsequent meat quality', *J. Anim. Sci.*, 76(8), 2080–5.

Goumon, S., Brown, J. A., Faucitano, L., Bergeon, R., Widowski, T. M., Crowe, T., Connor, M. L. and Gonyou, H. W. (2013) 'Effects of transport duration on maintenance behavior, heart rate and gastrointestinal tract temperature of market-weight pigs in 2 seasons', *J. Anim. Sci.*, 91(10), 4925–35.

Grandin, T. (1994), 'Methods to reduce PSE and bloodsplash', *Proc. Allen D. Leman Swine Conf.*, 21, 206–9.

Grandin, T. (2013), 'Recommended animal handling guidelines & audit guide: a systematic approach to animal welfare', *American Meat Institute Foundation*, Washington, D.C.

Guàrdia, M. D., Estany, J., Balasch, S., Oliver, M. A., Gispert, M. and Diestre, A. (2004), 'Risk assessment of PSE condition due to pre-slaughter conditions and RYR1 gene in pigs', *Meat Sci.*, 67(3), 471–8.

Guàrdia, M. D., Estany, J., Balasch, S., Oliver, M. A., Gispert, M. and Diestre, A. (2005), 'Risk assessment of DFD meat due to pre-slaughter conditions in pigs', *Meat Sci.*, 70(4), 709–16.

Haley, C., Dewey, C. E., Widowski, T. and Friendship, R. (2008), 'Association between in-transit loss, internal trailer temperature, and distance traveled by Ontario market hogs', *Can. J. Vet. Res.*, 72(5), 385–9.

Hambrecht, E., Eissen, J. J., Nooijen, R. I. J., Ducro, B. J., Smits, C. H. M., den Hartog, L. A. and Verstegen, M. W. A. (2004), 'Preslaughter stress and muscle energy largely determine pork quality at two commercial processing plants', *J. Anim. Sci.*, 82(5), 1401–9.

Hambrecht, E., Eissen, J. J., Newman, D. J., Smits, C. H. M., Verstegen, M. W. A. and den Hartog, L. A. (2005), 'Preslaughter handling effects on pork quality and glycolytic potential in two muscles differing in fiber type composition', *J. Anim. Sci.*, 83(4), 900–7.

Kephart, K. B., Harper, M. T. and Raines, C. R. (2010) 'Observations of market pigs following transport to a packing plant', *J. Anim. Sci.*, 88(6), 2199–2203.

Leheska, J. M., Wulf, D. M. and Maddock, R. J. (2003), 'Effects of fasting and transportation on pork quality development and extent of postmortem metabolism', *J. Anim. Sci.*, 81(12), 3194–3202.

Meisinger, D. J. (2002), 'A system for ensuring pork quality', *National Pork Board Fact Sheet*, National Pork Board: Des Moines, IA, USA.

National Pork Board. (2015a), 'PQA Plus Education Handbook', *National Pork Board*, Des Moines, IA.

National Pork Board. (2015b), 'Transport Quality Assurance Handbook', *National Pork Board*, Des Moines, IA.

Newman, D., Young, J., Carr, C., Ryan, M. and Berg, E. (2014), 'Effect of season, transport length, deck location, and lairage length on pork quality and blood cortisol concentrations of market hogs', *Animals*, 4(4), 627–42.

Person, R. C., McKenna, D. R., Ellebracht, J. W., Griffin, D. B., McKeith, F. K., Scanga, J. A., Belk, K. E., Smith, G. C. and Savell, J. W. (2005), 'Benchmarking value in the pork supply chain: Processing and consumer characteristics of hams manufactured from different quality raw materials', *Meat Sci.*, 70(1), 91–7.

Ritter, M. J., Ellis, M., Brinkmann, J., DeDecker, J. M., Keffaber, K. K., Kocher, M. E., Peterson, B. A., Schlipf, J. M. and Wolter, B. F. (2006), 'Effect of floor space during transport of market-weight pigs on the incidence of transport losses at the packing plant and the relationships between transport conditions and losses', *J. Anim. Sci.*, 84(10), 2856–64.

Ritter, M. J., Ellis, M., Bertelsen, C. R., Bowman, R., Brinkmann, J., DeDecker, J. M., Keffaber, K. K., Murphy, C. M., Peterson, B. A., Schlipf, J. M. and Wolter, B. F. (2007), 'Effects of distance moved during loading and floor space on the trailer during transport on losses of market weight pigs on arrival at the packing plant', *J. Anim. Sci.*, 85(12), 3454–61.

Ritter, M. J., Ellis, M., Bowman, R., Brinkmann, J., Curtis, S. E., DeDecker, J. M., Mendoza, O., Murphy, C. M., Orellana, D. G., Peterson, B. A., Rojo, A., Schlipf, J. M. and Wolter, B. F. (2008), 'Effects of season and distance moved during loading on transport losses of market-weight pigs in two commercially available types of trailer', *J. Anim. Sci.*, 86(11), 3137–45.

Schwartzkopf-Genswein, K. S., Faucitano, L., Dadgar, S., Shand, P., González, L. A. and Crowe, T. G. (2012), 'Road transport of cattle, swine and poultry in North America and its impact on animal welfare, carcass and meat quality: a review', *Meat Sci.*, 92(3), 227–43.

Sutherland, M. A., McDonald, A. and McGlone, J. J. (2009), 'Effects of variation in the environment, length of journey and type of trailer on the mortality and morbidity of pigs being transported to slaughter', *Vet. Rec.*, 165(1), 13–18.

Torrey, S., Bergeron, R., Widowski, T., Lewis, N., Crowe, T., Correa, J. A., Brown, J., Gonyou, H. W. and Faucitano, L. (2013a), 'Transportation of market-weight pigs: I. Effect of season, truck type, and location within truck on behavior with a two-hour transport', *J. Anim. Sci.*, 91(6), 2863–71.

Torrey, S., Bergeron, R., Faucitano, L., Widowski, T., Lewis, N., Crowe, T., Correa, J. A., Brown, J., Hayne, S. and Gonyou, H. W. (2013b), 'Transportation of market-weight pigs: II. Effect of season and location within truck on behavior with an eight-hour transport', *J. Anim. Sci.*, 91(6), 2872–8.

Warriss, P. D., Brown, S. N. and Adams, S. J. M. (1994), 'Relationships between subjective and objective assessments of stress at slaughter and meat quality in pigs', *Meat Sci.*, 38(2), 329–40.

Weschenfelder, A. V., Torrey, S., Devillers, N., Crowe, T., Bassols, A., Saco, Y., Piñeiro, Saucier, L. and Faucitano, L. (2012), 'Effects of trailer design on animal welfare parameters and carcass and meat quality of three Pietrain crosses being transported over a long distance', *J. Anim. Sci.*, 90(9), 3220–31.

Chapter 12

Humane slaughter techniques for pigs

Susanne Støier, Leif Lykke and Lars O. Blaabjerg, Danish Meat Research Institute – Danish Technological Institute, Denmark

1 Introduction

This chapter deals with slaughter techniques for pigs. The key challenges addressed are:

- Maintaining a high level of animal welfare: This requires well managed production systems and technologies
- Carcass value: Humane pre-slaughter handling reduces losses due to injury and poorer meat quality
- Efficiency: The need for a high slaughter capacity and a fast line speed
- Working environment: Acceptable working conditions for operators at the slaughterhouse
- Legislation: The need to fulfil legal requirements, e.g. EU Regulation No. 1099/2009

Authorities, NGOs, markets and the public in general have a number of concerns and requirements for animal welfare. The relevant authorities and certain implemented regulations like EU Regulation No. 1099/2009 have increased awareness of animal welfare issues and have contributed to significant improvements in the handling of animals on the day of slaughter. Due to EU legislation, slaughterhouses must appoint an animal welfare officer, establish standard operating procedures and entrust the handling of animals to properly trained staff with a certificate of competence. The market has additional requirements, such as the need for meat to be of a uniformly high quality.

http://dx.doi.org/10.19103/AS.2016.0013.27

Fierce competition in the food industry and the drive for sustainability mean that production efficiency must increase. However, there is also a need for increased focus on the working conditions of operators. Production systems and technologies that improve animal welfare, product quality, efficiency and working conditions are therefore required for the future development of the meat industry.

In order to maintain a cost-effective production system, including an efficient and fast process flow, while also improving animal welfare, efficient and targeted technology is needed. Although operators are becoming more educated and aware of animal welfare issues, systems with less human involvement will be advantageous in slaughterhouse operation, since they will minimize the impact of unavoidable differences between operators. Through the development and introduction of new technology which takes animal behaviour into consideration, it is possible to achieve efficient production with a high level of animal welfare and at the same time to maintain a high level of quality and yield.

It is important to look at the production chain as a whole, since the chain is no stronger than its weakest link. The different process stages must be aligned, as the optimal performance of one stage often depends on the functioning of the rest of the stages in the chain. Stages of the production chain which require particular attention is driving of pigs, stunning, sticking and debleeding including

- The influence of animal handling on the day of slaughter (from lairage to slaughtering) on both animal welfare and product quality and value
- The importance of efficient systems in improving animal welfare, product quality, yield and operator working conditions
- Systems to monitor the handling of pigs and their level of welfare at the abattoir

The development of new systems and procedures for pre-slaughter handling needs to be based on accurate knowledge of animal behaviour in order to create efficient systems which cause less harm and do not require physical force during animal handling. Therefore, thorough insight in animal behaviour is a necessary qualification for the development and implementation of new technology. Furthermore, operators must be educated so that they possess understanding of animal behaviour and the signs of animal distress, as well as understanding of the operation and performance of the systems via which animals are handled.

This chapter will show how the application of important insights from research to the development of new procedures, methods and technologies has helped to improve levels of animal welfare during pig slaughter. Insight into the physiological effect of CO_2 has led to optimization of the gas stunning process, knowledge of pig behaviour has resulted in the group-based handling system, and increased understanding of the causes of injuries to animals and carcasses has provided the starting point for developing operational procedures to lower the incidence of such injuries.

In Section 2, we address the slaughter of pigs, including approved stunning methods and correct sticking procedure. Section 3 covers the advantages of group-based handling, and Section 4 reviews methods for the surveillance and documentation of animal welfare on the day of slaughter. Section 5 offers a brief review of how the suggested ways of improving animal welfare during pig slaughter can also have economic benefits for slaughterhouses.

2 Slaughtering: stunning, shackling and sticking

2.1 Moving pigs to the stunning area

After lairage, pigs are moved to the stunning area. If a system with automatic push-hoist gates (described below in Section 3) is not available, this has to be done manually, thus requiring human–animal interaction. When emptying pens, it is recommended that the operator move along the wall and behind the pigs in order to move them forward to the driveway. No more than 15 pigs should be handled at one time. Pigs that cannot walk should be evacuated from the pen and placed in an inspection pen, or a captive bolt pistol should be used on the spot instead of a conscious pig being dragged to a sick pen.

In some systems, pigs are moved into a single-file chute immediately prior to stunning. Since pigs are herd animals, this goes against their natural instincts and, depending on the system, they may need to be forced forward, for example by using electric goads. A limited use of electric goads is currently allowed on adult animals just before stunning (Council Regulation (EC) No.1099/2009). Pigs that are forced forward are known to vocalize. In a study, pigs that had been forced with electric goads to move back and forth four times prior to stunning were compared with pigs handled without the use of electric goads, and which had walked directly to the stunning area (Hambrecht et al., 2004). The use of electric goads resulted in significantly increased cortisol and lactate exsanguination concentrations and a more rapid decrease in pH measured 30 min after slaughter, thus indicating that the use of electric goads increases the stress level.

2.2 Stunning

The relevant regulatory authorities require that pigs be stunned before sticking (see, e.g., Council Regulation (EC) No. 1099/2009). The purpose of this procedure is to induce insensibility to pain until the animal is dead (McKinstry and Anil, 2004), and the stunning procedure must therefore ensure that no animal regains consciousness before, during or after sticking. However, suboptimal stunning can induce pain, and the handling prior to stunning can also involve fear and pain. Therefore, optimal handling prior to stunning combined with the correct stunning procedure is crucial to the welfare of the animals during this stage of the day of slaughter. For pigs, the most common stunning methods are electrical stunning and gas stunning.

2.2.1 Electrical Stunning

Electrical stunning requires the animal to be restrained, which is a potential stressor. However, it induces immediate unconsciousness (Troeger, 2008), which can be advantageous in terms of animal welfare.

It is crucial that the electrodes be positioned correctly to achieve the intended result (Velarde et al., 2000a). Improper placement of electrodes can cause incomplete stunning – immobilization might occur instead of stunning – and thus result in painful electric shocks and poor animal welfare. According to one survey in the United Kingdom, 15.6% of pigs were incorrectly stunned on the first attempt, and the stunning procedure had to be repeated (McKinstry and Anil, 2004). This must be regarded as being detrimental to the welfare of the animals. The failure of the first application could be due to short stun duration, poor tong position or insufficient current.

Additionally, some operators repeat the stunning process in order to suppress spontaneous kicking, aid the shackling and hoisting process, and thus to help reduce the stun-to-stick interval. An investigation has shown that the use of repeat application of electrical stunning following an initial stun, to ensure that pigs do not regain consciousness, is acceptable as an emergency procedure. However, ideally stunning should only be applied once to meet legislative requirements, and should be followed by prompt, efficient sticking (McKinstry and Anil, 2004).

Head-only electrical stunning is often used, but head-to-chest stunning is also a possibility. The latter method reduces the risk of return to sensibility (consciousness) before sticking and exsanguination because it produces cardiac arrest and thus death. Furthermore, head-to-chest stunning prevents carcass convulsions (Anil et al., 2000). The recommendation for head-only electrical stunning is to apply a minimum current of 1.3 A across the brain for at least one second to induce immediate loss of consciousness. Sticking should then be performed within 15 seconds of the end of the stun. For head-to-body electrical stunning a minimum current of 1.3 A using 50 Hz sine wave AC should be applied for at least one second to induce immediate loss of consciousness and cardiac ventricular fibrillation (EFSA, 2004). However, the technical reference data supporting the effectiveness of a 1.3 A current needs to be verified under commercial conditions. An automatic system for electrical stunning is available from MPS (see http://www.mps-group.nl/en/).

The Midas electric stunning system has three electrodes situated on the head and heart, respectively. The supporting restrainer belt allows an automatic transfer of the pigs to a conveyor belt. By the use of an optical camera, the presence and position of the pig are determined and in combination with the resistance data of the particular pig, the individual key stunning parameters can be decided. An optimal stunning current is thus obtained for the individual pig and a lower stunning current can be used than that used in traditional systems (see http://www.mps-group.nl/en/).

A drawback of electrical stunning is an increased number of haemorrhages in the meat and a reduced pH. However, these characteristics are not relevant to animal welfare if the pigs are successfully rendered unconscious by the procedure (Velarde et al., 2001, 2000b).

2.2.2 CO_2 stunning

CO_2 stunning was introduced to improve meat quality (since it leads to fewer haemorrhages and improved water holding capacity) and to avoid the need to restrain the pigs. Furthermore, CO_2 stunning allows pigs to be handled in small groups. However, CO_2 stunning is the subject of some discussion, since pigs show signs of aversion to the gas a few seconds before unconsciousness occurs (Dalmau et al., 2010; Llonch et al., 2012). Inert gases such as argon have been suggested as alternatives, but since these render pigs unconscious for less time they risk compromising animal welfare unless the stun-to-stick interval can be made very short (EFSA, 2004).

In order to achieve efficient CO_2 stunning while reducing negative welfare implications, it is essential to carefully control both CO_2 concentration and the duration of the stunning procedure. Investigation has shown that stunning in a deep lift system with 80% CO_2 for at least 100 seconds can be regarded as acceptable in terms of animal welfare, assessed based on clinical signs of consciousness (Nowak et al., 2007). However, depending on the facilities and their capacity, the necessary CO_2 concentration and the time required for stunning can vary. Thus, some methods to assess and maintain the quality of the stunning process are required. To investigate and improve the stunning process, a method for video surveillance of the entire CO_2 stunning process during commercial conditions has been developed. However, so far this remains primarily a tool for research purposes, and it is not commonly used during production (Larsen et al., 2015).

2.3 Shackling and sticking

After stunning, pigs are shackled and then killed. Pigs must be shackled in the order in which they were stunned, the chain must be well secured on the hind leg, and hoisting should be done as soon as possible after stunning.

After shackling, the sticking is carried out. The two carotid arteries or the vessels from which these arise should be systematically severed. Inadequate sticking constitutes a potential welfare issue when it causes bleeding to occur too slowly, thus creating a risk that the pigs may show signs of recovery during exsanguination. The length of the sticking wound has been shown to affect the rate of blood loss significantly, with a long sticking wound resulting in faster exsanguination (Anil et al., 2000). The study used head-only electrical stunning, and one of the challenges reported was convulsions (Anil et al., 2000) that may render a proper sticking wound difficult to perform. A German survey reported that, of 2707 animals examined, 1.1% showed signs of regaining consciousness and sensibility three minutes after sticking (Troeger, Moje and Schurr, 2005). If this is not detected, and the pig is then subjected to scalding, it might be scalded alive. The assumption must be that this is extremely painful for the pigs, and so that it severely compromises the animals' welfare. Ensuring the quality of the sticking process is therefore crucial to ensuring the welfare of the animal. Furthermore, incorrect sticking may lead to haemorrhages in the fore-end muscles and thereby decrease the value of the carcass.

To optimize the use of resources, collection of the blood is recommended, and systems including hollow knife blood collection systems and the corresponding identification and processing systems are available. The blood collection systems have to fulfil all present EU and USDA veterinary and hygienic demands. Generally, about three litres of blood per pig can be hygienically collected (see http://www.butina.eu/products/blood_collection/).

3 Group-based handling of pigs on the day of slaughter

To obtain a high level of animal welfare, a procedure for pre-slaughter handling of pigs has been developed in which pigs are kept in the same groups during transport, lairage and stunning (Gade and Christensen, 1999). The idea behind this procedure is to reduce levels of aggression and fighting by keeping unfamiliar pigs separate. Thus, the occurrence of skin damage and bruises is reduced and meat quality, especially water holding capacity, is improved (Støier et al., 2001). The procedure covers group treatment of the pigs from loading onto the transport lorry, where mixing of unfamiliar pigs is allowed, to unloading, lairage and stunning at the abattoir.

Pigs in groups of 15 are easier to move at unloading, during transfer to holding pens and during transfer to the stunner. In the lairage area, a series of flap gates and one

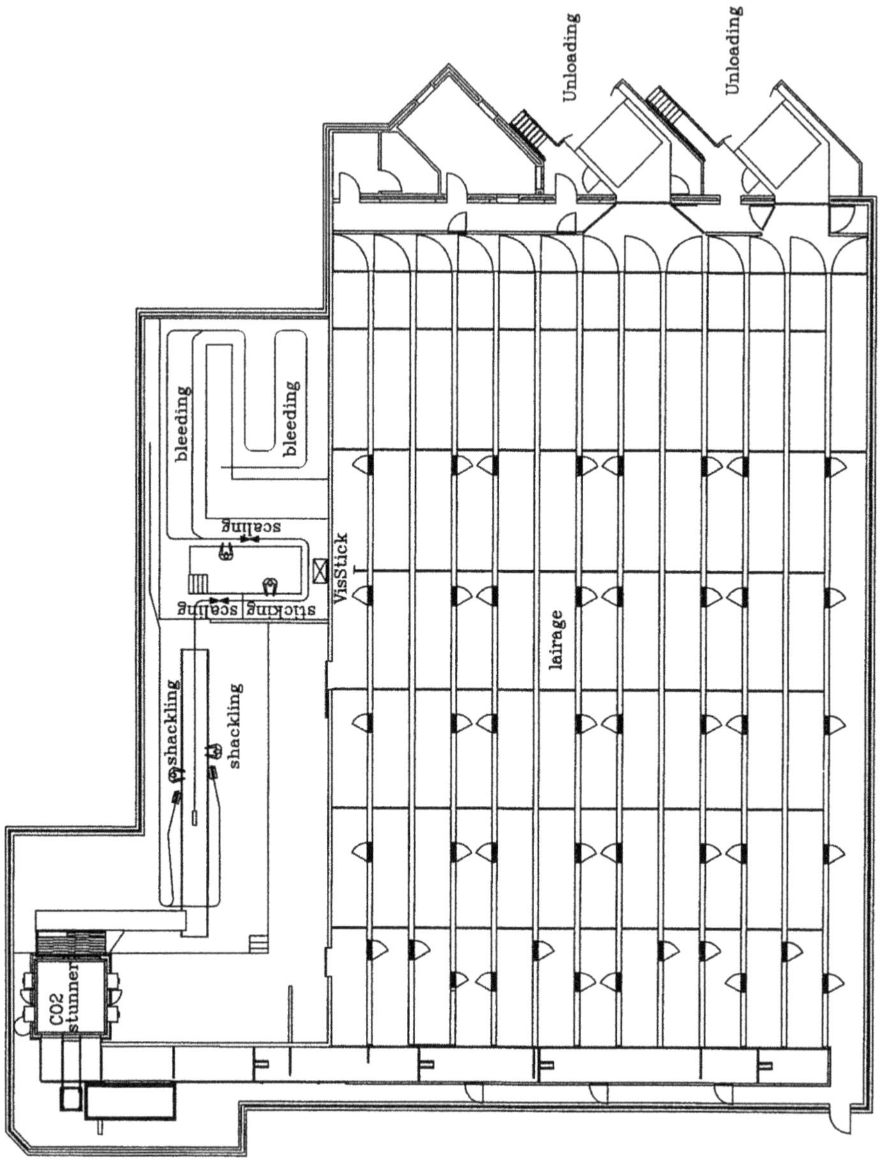

Figure 1 Procedure for handling pigs in groups at the slaughterhouse.

push-hoist gate divide the holding pen into an area suitable for 15 pigs, and the pigs move calmly into the lairage pen. In practice, the systems can be more or less automated. In a comparison test, the case of pigs kept in groups of 15 was compared to a case where pigs were kept in groups of 45 during transport and lairage. In the former case with the smaller group size, pigs showed less aggression towards each other and less skin damage in their legs and shoulders (Gade and Christensen, 1999).

Furthermore, pigs are moved using a driving board only. The procedure's design uses fact that animals have a tendency to move from a darker area towards a brighter area. It also ensures that pens, passageways and races are designed and constructed to allow the animals to move freely (Velarde et al., 2015). The principle therefore satisfies Council Regulation (EC) No.1099/2009, which says that pens, passageways and races should be designed and constructed to allow the animals to move freely in the required direction using their behavioural characteristics without distraction.

The final part of the procedure requires an automatic system to transfer the pigs from the holding pens to the stunning area. Fifteen pigs are moved forward using a series of push-hoist gates, and then sliding division gates subdivide the pigs into two or three groups depending on the capacity of the stunner. The smaller groups are automatically transferred to the CO_2 stunning equipment by a sliding division gate. The CO_2 equipment consists of a number of stunning boxes. The push-hoist gate system can be used to ensure compliance with the line speed, although it is important that the gates are finely adjusted so that they stop and do not drag an immobile pig (Gade, 2004). Sensors are integrated into the gates to register the gate pressure, to ensure that the operating gates do not drag a pig. All gates are operated mechanically.

The group-based CO_2 stunning procedure was patented by DMRI, and systems based on it have been installed in many slaughterhouses around the world. In such slaughterhouses, lairage systems encourage pigs to move in the required direction by having the pigs to move from darker to brighter areas, and/or through the use of floors at small angles of elevation (Kristensen et al., 2014). The development of this principle for slaughtering pigs in groups is therefore an example of how greater understanding of animal behaviour combined with technical insight and the use of technology makes it possible to implement an efficient system that works in practice at the slaughterhouse. No electric goads are needed to make the pigs move and the pigs are not restrained when gas stunned. Due to the more gentle handling of the animals, the incidence of skin damage and bruises is lowered, water holding capacity is improved (Støier et al., 2001) and the noise level in the production area is reduced (Gade and Christensen, 1999). Furthermore, the working conditions of operators also improve, due to lower workload and a lower noise level.

To complete the group-based procedure, DMRI has developed and implemented a system whereby the required traceability can be maintained even when pigs have not been marked by the farm supplier. Traceability is achieved by keeping the pigs in the same group from collection at the farm until they are identified using RFID tags on the slaughter line (Andersen et al., 2012). As the present tattoo marking methods used on pigs for slaughter are associated with a certain level of stress for the pigs, this new system contributes to the improvement of animal welfare (Nielsen et al., 2014). The system consists of IT (information technology) solutions combined with slaughterhouse inspection, trained operators and management focus.

4 Surveillance and documentation of animal welfare levels on the day of slaughter

Documentation of animal welfare through the entire production chain is increasingly demanded by both the market and the relevant regulatory authorities. Furthermore, surveillance of animal welfare could be an effective tool to identify and register problems in the handling of the animals on the day of slaughter. Monitoring animal welfare parameters provides the possibility of changing any inappropriate procedures and operations. The day of slaughter includes a complex of potentially stressful elements, the effects of which could be expressed by several different biological outcome measurements relevant for animal welfare (Gade, 2004). Animal welfare can be monitored and documented by different animal-based measurements but also more indirectly by resource-based measurements as reviewed by Brandt and Aaslyng (2015). In the EU project Welfare Quality®, indicators for animal welfare in primary production and at the slaughterhouse were identified (Welfare Quality®, 2009). Grandin (2010) proposed protocols based on resource and animal-based measurements, and suggested assessing animal welfare at abattoirs using an animal-based scoring system including stunning efficiency, percentage rendered insensible, falls, vocalization and the use of electric goads. These five measurements are easy to implement at slaughterhouses and highly repeatable (Grandin, 2010). However, it is extremely time consuming to register all of the criteria suggested by Welfare Quality and Grandin, and the criteria are therefore not useful on a routine basis. For this reason, there is a need for simple methods to monitor and document animal welfare on the day of slaughter. Automatic registration of welfare indicators as documentation of the level of animal welfare could be a useful tool for slaughterhouses and meat producers.

In principle, surveillance of animal welfare can take place in two different ways:

- An online method that measures and provides a result immediately and sounds an alarm if the parameter is out of the defined acceptable range. In this way, the conditions can be changed immediately.
- A retrospective documentation of animal welfare at a given stage at a given time. If there is an increase or a deviating pattern in the negative incidences, the registrations can be used as an underlying basis for subsequent changes in the conditions at the slaughterhouse or during transport.

Pre-slaughter handling associated with loading and slaughter as well as transport of pigs leads to increased levels of lactate, glucose and ear temperature (Becerril-Herrera et al., 2010; Mota-Rojas et al., 2012; Warriss et al., 1994); increased creatine kinase (CK); increased heart rate (Correa et al., 2010, 2013); increased incidence of skin damage (Mota-Rojas et al., 2006) as well as lower pH 30 minutes after slaughter (Van de Perre et al., 2010). However, these studies have investigated the effect of a single or a few potential stressors on the day of slaughter. Therefore, a study was performed with the aim of assessing the accumulated effects of the different potential stressors encountered from the pick-up facilities at the farm until killing at the slaughterhouse. The observational study conducted under Danish commercial conditions indicated that plasma concentration of lactate, glucose and CK in the blood at sticking might be relevant indicators of the welfare of pigs at the slaughterhouse (Brandt et al., 2013). According to Brandt and Aaslyng (2015), the development of automatic blood sample collection and automatic analysis of, for example, lactate or glucose may be beneficial in the continuous monitoring and documentation of animal welfare on the day of slaughter. However, further studies with more variable pre-slaughter conditions and a greater number of animals are needed to validate the potential in these plasma parameters.

Video recordings can be used to survey pre-slaughter handling and the behaviour of animals on arrival at the abattoir and in the lairage area. However, efficient use of these recordings would require an automated image analysis. It is possible to develop an algorithm by which the movement pattern of the animals can be analysed, i.e. if the animals move faster than the average or if they stand still (Gronskyte et al., 2015, 2016). Deviating patterns of movements such as stops, turns and falls can be determined, but so far it has not been possible to analyse the specific movements separately. More development is needed to set up a functional and operational system for video surveillance.

An example of a developed and implemented documentation system is the VisStick®. In accordance with regulations, slaughter pigs have to be stunned before killing (Council Regulation (EC) No. 1099/2009). Although pigs are anaesthetized before slaughter, it is the sticking and subsequent loss of blood that causes death. Afterwards, the animals are moved on to scalding and dehairing. At this point, a pig will occasionally show signs of life. With manual sticking, there is a certain risk that a pig will not be stuck properly. To minimize this risk, a vision system has been developed to check that pigs are in fact stuck after stunning (Borggaard et al., 2011; Lykke et al., 2010). A vision-based system – VisStick® – monitors the pigs after sticking and gives an alarm if no blood dripping from the pig's snout is registered. In approval tests of five typical installations, a detection range of 98 to 100% of unstuck pigs was achieved. False positive results occurred in a range from 0 to 0.64‰ (Borggaard et al., 2011).

The VisStick system has been implemented at most slaughter lines in Denmark and at several units in other Nordic countries. Furthermore, spraying hot water (60°C) on the faces of the pigs has been used to confirm death combined with behavioural and clinical observations (Arnold et al., 2014; Troeger and Meiler, 2006). For at least 5 seconds, water is applied to the head and front legs of the pigs. Short video sequences are recorded for each pig and analysed in real time. Pigs with movements following the hot water spraying can thus be detected and killed properly before further processing. Tests have shown that high intensity movements following hot water spraying are reliably differentiated from passive movements of the suspended pigs. However, not all movement types can be automatically detected yet, and therefore real-time software is being further developed (Parotat et al., 2015). If a robust vision system for detection can be developed, the method can probably be automated.

5 Improved value of meat products

It is well known that slaughter animals can show stress responses during the day of slaughter, which may adversely affect both the welfare of the animals and the meat quality. The driving and stunning system according to which pigs are kept in small groups has improved animal welfare, meat quality, productivity and working conditions. Similarly, optimized transport equipment has made more humane animal transport possible. The technical development of equipment combined with insight into animal behaviour and consideration of animal welfare has therefore improved the handling of pigs.

Furthermore, experience from improvement of pre-slaughter handling at commercial slaughterhouses, including optimization of the stunning systems such as the change from electrical stunning to CO_2 stunning, shows that it is possible to reduce the amount of PSE (pale, soft, exudative) meat, drip loss and the incidences of haemorrhages as well. Due to the increased value of the cuts and more products being acceptable for high price markets, the annual value of the quality improvements was estimated at $ 4 600 000 for a slaughterhouse killing 4 000 000 pigs annually.

6 Summary and future trends

Methods to make the production of meat more efficient and to increase line speed can be combined with improvement in the pre-slaughter handling of pigs. Keeping pigs in smaller groups during transport, lairage and gas stunning seems to be an almost optimal solution for handling pigs on the day of slaughter. However, the right gas composition during gas stunning of pigs, with the aim of eliminating aversive reactions, is still an interesting subject for investigation. Efficient online methods for surveillance of animal welfare would be useful tools to document and steadily improve pre-slaughter handling. The development of automatic blood sample collection and automatic analysis of, for example, lactate or glucose may be beneficial in the continuous monitoring and documentation of animal welfare on the day of slaughter. However, all measurable indicators of animal welfare must be combined with a decision about what constitute the limits of good, acceptable and unacceptable levels of welfare.

An increasing demand for documentation of food quality, including information about animal welfare during production, means that there is a need for effective systems for the surveillance of animal welfare levels on the day of slaughter. Video surveillance is one possible solution, but to be feasible this will require the development of automated image analysis of video recordings. Improvements to stunning methods, including alternative gas compositions, are still a relevant research issue. Furthermore, technical and flexible overall solutions from farm to slaughter, which fulfil the requirements for production at both small and large scales, are still needed.

7 Where to look for further information

An introduction to the subject is W. Klinth Jensen (ed.) (2004), *Encyclopedia of Meat Sciences*, Oxford: Elsevier (see P. B. Gade, Pre-slaughter handling, pp. 1012–20) and M. Dikeman and C. Devine (ed.) (2014), *Encyclopedia of Meat Sciences*, 2nd Edition, Elsevier (see vol. 3).

The European Food Safety Authority (EFSA) has published assessments of welfare aspects related to stunning and killing:

- Scientific Opinion on monitoring procedures at slaughterhouses for pigs (2013), http://www.efsa.europa.eu/en/efsajournal/pub/3523
- Opinion of the Scientific Panel on Animal Health and Welfare (AHAW) on a request from the Commission related to welfare aspects of the main systems of stunning and killing the main commercial species of animals (2004), http://www.efsa.europa.eu/en/efsajournal/pub/45

Furthermore, the references below and in a newly published paper:

- Støier, S., Larsen, H. D., Aaslyng, M. D. and Lykke, L. (2016). Improved animal welfare, the right technology and increased business. Meat Science. doi:10.1016/j.meatsci.2016.04.010.

8 References

Andersen, P. B., Sørensen, R., Steenberg, B., Lykke, L. and Madsen, N. T. (2012). Traceability system for slaughter of unmarked pigs. *58th International Congress of Meat Science and Technology*, 12–17 August 2012, Montreal, Canada.

Anil, M. H., Whittington, P. E. and McKinstry, J. L. (2000). The effect of the sticking method on the welfare of slaughter pigs. *Meat Science*, 55, 315–19.

Arnold, S., Parotat, S., Moje, M., Machtolf, M., von Wenzlawowicz, M., Troeger, K. and Lücker, E. (2014). Ascertaining death in pig slaughter using hot water. *60th International Congress of Meat Science and Technology*, 17–22 August, Punta del Este, Uruguay.

Becerril-Herrera, M., Alonso-Spilsbury, M., Ortega, M. E., Guerrero-Legarreta, I., Ramírez-Necoechea, R., Roldan-Santiago, P., Pérez-Sato, M., Soni-Guillermo, E. and Mota-Rojas, D. (2010). Changes in blood constituents of swine transported for 8 or 16 h to an abattoir. *Meat Science*, 86, 945–8.

Borggaard, C., Claudi-Magnussen, C, Madsen, N. T. and Støier, S. (2011). A new system for Sticking Control ('VisStick'). *57th International Congress of Meat Science and Technology*, 7–12 August 2011, Ghent, Belgium.

Brandt, P., Rousing, T., Herskin, M. and Aaslyng, M. D. (2013). Identification of *post-mortem* indicators of welfare of finishing pigs on the day of slaughter. *Livestock Science*, 157, 535–44.

Brandt, P. and Aaslyng, M. D. (2015). Welfare measurements of finishing pigs on the day of slaughter: A review. *Meat Science*, 103, 13–23.

Correa, J. A., Torrey, S., Devillers, N., Laforest, J. P., Gonyou, H. W. and Faucitano, L. (2010). Effects of different moving devices at loading on stress response and meat quality in pigs. *Journal of Animal Science*, 88, 4086–93.

Correa, J. A., Gonyou, H. W., Torrey, S.,Widowski, T., Bergeron, R., Crowe, T. G., Laforest, J. P. and Faucitano, L. (2013). Welfare and carcass and meat quality of pigs being transported for two hours using two vehicle types during two seasons of the year. *Canadian Journal of Animal Science*, 93, 43–55.

Council Regulation (EC) No. 1099/2009 of 24 September 2009 on the protection of animals at the time of killing.

Dalmau, A., Rodriguez, P., Llonch, P. and Velarde, V. (2010). Stunning pigs with different gas mixtures: Aversion in pigs. *Animal Welfare*, 19, 325–33.

EFSA AHAW Panel (EFSA Panel on Animal Health and Welfare) (2004). Opinion on the Scientific Panel on Animal Health and Welfare (AHAW) on a request from the Commission related to welfare aspects of the main systems of stunning and killing the main commercial species of animals, p. 270. doi:10.2903/j.efsa.2004.45.

Gade, P.B. and Christensen, L. (1999). Automatic handling at lairage improves welfare. *Meat International*, 9(4), 17–19.

Gade, P. B. (2004). Pre-slaughter handling. In W. Klinth Jensen (ed.), *Encyclopedia of Meat Sciences*, pp. 1012–20. Oxford: Elsevier.

Grandin, T. (2010). Auditing animal welfare at slaughter plants. *Meat Science*, 86, 56–65.

Gronskyte, R., Clemmensen, L. H., Hviid, M. S. and Kulahci, M. (2015). Pig herd monitoring and undesirable tripping and stepping prevention. *Computers and Electronics in Agriculture*, 119, 51–60.

Gronskyte, R., Clemmensen, L. H., Hviid, M. S. and Kulahci, M. (2016). Monitoring pig movement at the slaughterhouse using optical flow and modified angular Histograms. *Biosystems Engineering*, 141, 19–30.

Hambrecht, E., Eissen, J. J., Nooijen, R. I. J., Ducro, B. J., Smits, C. H. M., den Hartog, L. A. and Verstegen, M. N. (2004). Preslaughter stress and muscle energy largely determine pork quality at two commercial processing plants. *Journal of Animal Science*, 82, 1401–9.

Kristensen, L., Støier, S., Würtz, J. and Hinrichsen, L. (2014). Trends in meat science and technology: The future looks bright, but the journey will be long. *Meat Science*, 98, 322–9.

Larsen, H. D., Blaabjerg, L. O., Lykke, L. and Weihe, S. H. (2015). Video surveillance of CO_2-stunning of finishing pigs in groups. *HSA International Symposium 2015: Recent Advances II*, p.53.

Llonch, P., Dalmau, A., Rodriquez, P., Manteca, X. and Velarde, A. (2012). Aversion to nitrogen and carbon dioxide mixtures for stunning pigs. *Animal Welfare*, 21, 33–9.

Lykke, L., Arnmark, P. and Borggaard, C. (2010). One incident every twentieth year. *Fleischwirtschaft International*, 4, 10–11.

McKinstry, J. L. and Anil, M. H. (2004). The effect of repeat application of electrical stunning on the welfare of pigs. *Meat Science*, 67, 121–8.

Mota-Rojas, D., Becerril, M., Lemus, C., Sánchez, P., González, M., Olmos, S. A., Ramirez, R. and Alonso-Spilsbury, M. (2006). Effects of mid-summer transport duration on pre-and post-slaughter performance and pork quality in Mexico. *Meat Science*, 73, 404–12.

Mota-Rojas, D., Becerril-Herrera, M., Roldan-Santiago, P., Alonso-Spilsbury, M., Flores-Peinado, S., Ramírez-Necoechea, R., Ramírez-Telles, J. A., Mora-Medina, P., Pérez, M., Molina, E., Soní, E. and Trujillo-Ortega, M. E. (2012). Effects of long distance transportation and CO_2 stunning on critical blood values in pigs. *Meat Science*, 90, 893–8.

Nielsen, S. S., Michelsen, A. M., Jensen, H. E., Barington, K., Opstrup, K.V. and Agger, J. F. (2014). The apparent prevalence of skin lesions suspected to be human-inflicted in Danish finishing pigs at slaughter. *Preventive Veterinary Medicine*, 117, 200–6.

Nowak, B., Mueffling, T. V. and Hartung, J. (2007). Effect of different carbon dioxide concentrations and exposure times in stunning of slaughter pigs: Impact on animal welfare and meat quality. *Meat Science*, 75, 290–8.

Parotat, S., Arnold, S., Wenzlawowicz, M. Von and Luecker, E. (2015). Pigs slaughter: Detecting signs of life prior to scalding by means of optical flow. *HSA International Symposium 2015: Recent Advances II*, p.23.

Støier, S., Aaslyng, M. D., Olsen, E. V. and Henckel, P. (2001). The effect of stress during lairage and stunning on muscle metabolism and drip loss in Danish pork. *Meat Science*, 59, 127–31.

Troeger, K., Moje, M. and Schurr, B. (2005). Kontrolle der Entblutung. Voraussetzung für eine tierschutzkonforme Schweineschlachtung. *Fleischwirtschaft*, 85, 107–10.

Troeger, K. and Meiler, D. (2006). Entwicklung eines praxisgerechten Verfahrens zur Kontrolle der Tötung von Slachtschweinen durch Blutentzug. *Mitteilungsblatt der Fleischforschung Kulmbach*, 45 (171), 15–22.

Troeger, K. (2008). Tierschutzgerechtes Schlachten von Schweinen: Defizite und Lösungsansätze. *Tierärztliche Praxis*, 36, S34–S38.

Van de Perre, V., Permentier, L., De Bie, S., Verbeke, G. and Geers, R. (2010). Effect of unloading, lairage, pig handling, stunning and season on pH of pork. *Meat Science*, 86, 931–7.

Velarde, A., Gispert,M., Faucitano, L., Diestre, A. and Manteca, X. (2000a). Survey of the effectiveness of stunning procedures used in Spanish pig abattoirs. *Veterinary Record*, 146, 65–8.

Velarde, A., Gispert, M., Faucitano, L., Manteca, X. and Diestre, A. (2000b). The effect of stunning method on the incidence of PSE meat and haemorrhages in pork carcasses. *Meat Science*, 55, 309–14.

Velarde, A., Gispert,M., Faucitano, L., Alonso, P., Manteca, X. and Diestre, A. (2001). Effects of the stunning procedure and the halothane genotype on meat quality and incidence of haemorrhages in pigs. *Meat Science*, 58, 313–19.

Velarde, A., Fàbrega, E., Blanco-Penedo, I. and Damlau, A. (2015). Animal welfare towards sustainability in pork meat production. *Meat Science*, 109, 13–17.

Warriss, P. D., Brown, S. N., Adams, S. J. M. and Corlett, I. K. (1994). Relationships between subjective and objective assessments of stress at slaughter and meat quality in pigs. *Meat Science*, 38, 329–40.

Welfare Quality® (2009). Welfare Quality® Assessment Protocol for Pigs (Sows and Piglets, Growing and Finishing Pigs). Welfare Quality® Consortium, Lelystad, Netherlands, p. 122.

Index

www.ingramcontent.com/pod-product-compliance
Ingram Content Group UK Ltd.
Pitfield, Milton Keynes, MK11 3LW, UK
UKHW021826100926
8556IPUK00003B/180

9 781786 760968